Environmental Science and Engineering

Environmental Science

Series editors

Rod Allan
Ulrich Förstner
Wim Salomons

For further volumes:
http://www.springer.com/series/3234

External Reviewers

The editors gratefully acknowledge the following reviewers for their valuable contributions to this volume.

Name	**Affiliation**
Luke Bixler	University of Redlands
Pierluigi Coppola	Autentys International
Robert Coulson	Texas A&M University
Frank Davis	University of California, Santa Barbara
Eric Forsman	USDA Forest Service Research
Kirsten Gallo	USDA Forest Service
Pablo F. de Arroyabe	Universidad de Cantabria (Spain)
Miles Hemstrom	USDA Forest Service Research
Kevin James	US National Park Service
Russell Johnson	US Geological Survey
Chris Jordan	National Marine Fisheries Service
Brenda McComb	Oregon State University
Ruben Ortiz	University of Redlands
George Pess	National Marine Fisheries Service
H. Michael Rauscher	USDA Forest Service Research (retired)
R. Neil Sampson	Vision Forestry LLC

An Overview of the Ecosystem Management Decision-Support System

Keith M. Reynolds and Paul F. Hessburg

Abstract By way of introduction, this chapter provides a general overview of the Ecosystem Management Decision Support (EMDS) system, including a brief account of its development history, key factors that have motivated its development, and more central topics such as concepts, principles, and functionality. We conclude the chapter with discussions on applications involving multiple spatial scales, ways in which the technology can support the modern planning process, critical design factors behind the relative success of the system, and experiences drawn from design and use of the system.

Keywords Ecosystem management decision support · Environmental analysis · Planning · Logic model · Decision model · Spatial decision support · EMDS

1 Introduction

In the following section, we give a brief account of the history of decision-support systems (DSS) for the benefit of readers who may not have much formal background in the subject. There are many excellent texts on the subject, so the cursory description here is only to help place EMDS in the broader context. Next, we provide brief accounts of DSS in environmental management and the origins of EMDS.

K. M. Reynolds (✉)
U.S. Department of Agriculture, Forest Service, Pacific Northwest Research Station, Corvallis Forestry Sciences Laboratory, 3200 SW Jefferson Way, Corvallis, OR 97331, USA
e-mail: kreynolds@fs.fed.us

P. F. Hessburg
U.S. Department of Agriculture, Forest Service, Pacific Northwest Research Station, Wenatchee Forestry Sciences Laboratory, 1133 N Western Ave., Wenatchee, WA 98801, USA
e-mail: phessburg@fs.fed.us

K. M. Reynolds et al. (eds.), *Making Transparent Environmental Management Decisions*, Environmental Science and Engineering,
DOI: 10.1007/978-3-642-32000-2_1, © Springer-Verlag Berlin Heidelberg 2014

Chapters "Use of EMDS in Conservation and Management Planning for Watersheds through Measuring Biological Sustainability via a Decision Support System: Experiences with Oregon Coast Coho Salmon" of this volume present a diverse set of specific applications of EMDS in environmental analysis and planning. By way of introduction to these later chapters, this chapter provides a general overview of the system including a brief account of its development history, key factors that have motivated its development, as well as more central topics such as concepts, principles, and functionality. We conclude the chapter with discussions on applications involving multiple spatial scales, ways in which the technology can support modern planning processes, critical design factors behind the relative success of the system, and experiences drawn from design and use of the system.

2 The Origins of Decision-Support Systems

Early pioneering work in the DSS field was carried out at various institutions from the late 1950s to the late 1960s, including theoretical studies on decision making at the Carnegie Mellon Institute, work on the technical aspects of interactive computer systems at the Massachusetts Institute of Technology, and the first DSS applications at the Harvard Business School. Simon (1947, 1960) was instrumental in setting the stage for the evolution of DSS by providing the necessary context for understanding and supporting decision-making processes. Power (2008) provides an excellent historical overview on the origins of DSS, including descriptions of early work at Harvard by Scott Morton (1967, 1971), on computer-aided support for business managers; an historical turning point. Other important milestones in the conceptual development of DSS include the works of Scott Morton (1967), Gorry and Scott Morton (1971), Davis (1974), Keen and Scott Morton (1978), Sprague (1980), Bonczek et al. (1981), and Sprague and Carlson (1982). The work of Bonczek et al. (1981) is particularly significant because these authors articulated for the first time what has become the most enduring definition of a DSS. They identified four essential components that were common to all DSSs:

> 1. A language system (LS) that specifies all messages a specific DSS can accept; 2. A presentation system (PS) for all messages a DSS can emit; 3. A knowledge system (KS) for all knowledge a DSS has; and 4. A problem-processing system (PPS) that is the software engine that tries to recognize and solve problems during the use of a specific DSS (Power 2008).

3 Decision Support in Environmental Management

Environmental management has been a hotbed of DSS development since at least the early 1980s. By 1989, Davis and Clark (1989) were able to catalogue about 100 systems related to environmental management; subsequent reviews of systems suitable for forest management (Mowrer 1997; Schuster et al. 1993) catalogued many more. Somewhat later, Oliver and Twery (2000) and Reynolds et al. (2000) laid theoretical and practical groundwork for applying DSSs to the more formidable goal of ecosystem management.

The majority of what might be called first-generation systems for use in environmental management (1980s) were typically hard-coded, and designed to address relatively fine-focused and well-defined problems such as supporting silvicultural prescriptions (e.g., practices concerned with forest cultivation) for individual species (Rauscher et al. 1990), or pest management for specific pests on specific species (Twery et al. 1993), which partly accounts for the seeming plethora of systems by the mid 1990s. However, especially over the past 15 years, there has been a pronounced trend toward development of far fewer, but more general purpose, multi-functional systems like EMDS. This trend was significantly enabled by rapid advances in computing hardware and software systems engineering. Equally important, natural resource organizations have been called upon to effectively address the complex issues of ecosystem management (Rauscher 1999).

4 Development History

EMDS development began in 1994 as a research and development project of the Pacific Northwest Research Station, a unit of the US Department of Agriculture, Forest Service. Federal and university scientists from various institutions developed the initial design specifications for the system, and early versions were implemented under contract with the Environmental Systems Research Institute (ESRI).[1]

EMDS 1.0 was released in 1997. This version supported spatially explicit, logic-based landscape analysis, and was implemented as an extension to ESRI's ArcView 3.x geographic information system (GIS).

Version 2.0, released in 1999, added a major new component, dubbed the "Hotlink browser," to the logic-processing component by which system users could graphically trace the logical derivation of model conclusions in an intuitive graphic interface. Whereas version 1.0 had essentially been a "black box," version 2.0 made the inner workings of the logic processor transparent to end-users, which significantly increased interest among users.

The ESRI product line began undergoing a major transformation in the late 1990s, culminating in the contemporary ArcGIS, which first appeared around 2002. Production on EMDS 3.0 began in 2000 to keep pace with the new object-based implementation of ESRI products. In addition to the original EMDS code being re-implemented in Microsoft Visual Basic to support the new ArcObjects

[1] The use of trade or firm names in this publication is for reader information and does not imply endorsement by the US Department of Agriculture of any product or service.

framework of ArcGIS, EMDS 3.0 introduced another major component that supported decision modeling as a complement to the logic modeling. At the release of version 3.0 in 2002, EMDS now provided an integrated solution to landscape analysis and planning.

Between the release of version 4.1 in 2009 and the release of version 3.0 in 2002, another major milestone in EMDS development occurred: In 2005, the USDA Forest Service and the University of Redlands (Redlands, CA; www.institute.redlands.edu/emds) signed a memorandum of understanding, under which the university (and the Redlands Institute, in particular) would assume the stewardship of EMDS. More or less contemporaneously, commercial software developers who had been instrumental in delivering the logic and decision components of the system agreed to join the Forest Service and university in a private non-profit development consortium now known as the EMDS Consortium.

Finally, at the release of version 4.2, no new functionality was introduced, but several enhancements were implemented: Version 3.0 code was re-implemented in Microsoft.net, which represents an important intermediate step toward eventually delivering the system as a web service ("EMDS 5.0 and Beyond"). Other major enhancements included implementation of a companion stand-alone edition that runs independently of ArcMap; a new, more intuitive interface built on the workflow concept; and project structures implemented around geodatabases and contemporary database management systems, such as SQL Server and Microsoft Access.

5 Motivations

In recent decades, significant global attention has focused on addressing current and potential future problems with the sustainability of ecosystems, especially since release of the Brundtland Report (WCED 1987). This Report brought international attention to the concept of sustainable development defined through environmental protection, economic growth, and social equity. The subsequent United Nations (UN) Conference on Environment and Development (UNCED 1992 in Rio de Janeiro), and its successor in Johannesburg in 2003, have sharpened the focus and galvanized resolve at many levels of government from international forums like the UN down to local levels. Natural resource management agencies at many levels have adopted principles of ecosystem management, based on their best current understanding of ecosystem dynamics, and are beginning to adapt their management practices accordingly.

The concept of ecosystem management (Overbay 1992) has been with us now for about 30 years. The terms "ecosystem management" (Jensen and Everett 1994), "adaptive management" (Holling 1978; Walters 1986), and "sustainable management" (Maser 1994) are closely connected with each other in the natural resource literature. Ecosystem management has been defined as

> The careful and skillful use of ecological, economic, social, and managerial principles in managing ecosystems to produce, restore, or sustain ecosystem integrity and desired conditions, uses, products, values, and services over the long term (Overbay 1992).

There are many other definitions of ecosystem management in the literature, many of which describe it as a plan or strategy. However, we prefer Overbay's definition (i.e., the use of principles) on the grounds that few if any of the extant descriptions, regardless of how detailed, actually describe a process indicative of planning or strategizing. On the other hand, adaptive management (Holling 1978; Walters 1986) describes a process for implementing ecosystem management, which is why we say that these two terms are closely connected: one describes a set of principles for managing ecosystems, while the other describes a process for the implementing these principles. Subsequently, we shall refer to the two concepts collectively as adaptive ecosystem management (Everett et al. 1994). The Overbay definition also is succinct: it is clear from the definition that the goal behind the application of these principles is ecosystem sustainability in the broad sense of the Brundtland Report (WCED 1987).

By about 20 years ago, the integrated concept of adaptive ecosystem management had been enthusiastically embraced by both the scientific and management communities of the natural resource disciplines. Then-Chief of the Forest Service (US Department of Agriculture), Dale Robertson, declared in 1988 that henceforth the National Forest System was to be managed according to the principles of ecosystem management. Heads of other federal natural resource agencies in the US and elsewhere soon followed suit, and managers set about implementing ecosystem management, often with the eager assistance of scientists who appreciated the experimental nature of active adaptive management (Lee 1999) in particular. Integrated ecological assessments (Christensen et al. 1996) were designed to provide a structured process from formulation of issues to assessment to implementation to adaptive management (Bourgeron et al. 2009). Some 25 years following the Brundtland Report, however, it probably is safe to say that managers and scientists now generally feel a sense of disillusionment with adaptive ecosystem management. There are few good examples in which active adaptive management has been successfully implemented, and progress in learning to manage complex ecosystems more effectively by this approach has been agonizingly slow. Slowing the implementation of ecosystem management is the argument that the general use of the historic range of variability as a reference is not always achievable. Consequently, a shift from ecosystem management to ecosystem stewardship has recently been advocated to sustain the capacity to provide ecosystem services under conditions of uncertainty and change (Chapin et al. 2009a, b). Ecosystem stewardship explicitly includes the acknowledgement of tradeoffs between efficiency and flexibility and between immediate and long-term benefits (Chapin et al. 2009b; Liu et al. 2007). In practice, most of the methodologies used in ecosystem stewardship are shared with ecosystem management. In particular, EMDS is particularly well positioned to assess uncertainty and the trade-offs mentioned above.

You may well be wondering what all of the above has to do with EMDS. In many ways, the swirling maelstrom of principles, methods, and objectives briefly alluded to above in the period from 1985 to 1995 was in fact the context within which the system was conceived (ca 1994). In this context, many good ideas were being advanced, and even a few great ones. Unfortunately, few if any advanced beyond a conceptual model, and the utility of such models is almost always limited by their vagueness, ambiguity, and imprecision (Gustafson et al. 2003). Even as specifications for EMDS were beginning to take shape, it was already becoming clear that successful application of adaptive ecosystem management was turning out to be elusive. Thus, a primary motivation behind EMDS design from the beginning was delivery of a practical decision-support system for adaptive ecosystem management.

6 Concepts and Principles

A few key concepts and principles pertinent to EMDS are discussed in this section as a prelude to subsequent discussion on EMDS structure and functionality.

6.1 Decision-Support System

For the purposes of this volume, we adopt the definition of a decision support system (DSS) from Holsapple (2003, p. 551):

> A computer-based system composed of a language system, presentation system, knowledge system, and problem-processing system whose collective purpose is the support of decision-making activities.

Two key attributes in Holsapple's definition are a problem-processing system and purposeful support of a decision-making process. A decision-making process is a method that guides an individual or group through a series of tasks from problem identification and analysis to design of alternatives and selection of an alternative (Mintzberg et al. 1976).

Systems that generally fulfill the Mintzberg and Holsapple definitions include multi-criteria decision analysis (MCDA) systems that implement the Analytic Hierarchy Process (AHP) and similar MCDA methods. These knowledge-based systems provide a framework for applying procedural or reasoning knowledge to decision problems, and, perhaps somewhat more arguably, to optimization systems. However, while geographic information, spreadsheet, and database systems may be critical components, or even the foundation of a DSS, it stretches the definition of a DSS beyond usefulness to classify these types of applications as DSSs. Numerous simulation systems have been developed to support many aspects of planning (see, for example, Schuster et al. 1993), but most should be considered as potential tools employed in a DSS as opposed to DSSs per se.

6.2 Logic and Inference

By the start of the 20th century, important milestones had been reached in what we now recognize as modern science. Russell (1903) had laid the foundations of modern set theory and logic. Although Popper (1934) is perhaps most commonly associated with the principles of modern hypothesis testing, it was Peirce (1931–1935) in the 1890s who described the essentials of hypothesis testing that are still with us today. Peirce drew upon classical logic (Aristotle, 350 BCE) to develop a theory of inquiry, in parallel with the early development of symbolic logic for which he is much better known, to address contemporary challenges in scientific reasoning. Peirce's theory describes three fundamental modes of reasoning: abductive, deductive, and inductive inference.

Most contemporary scientists get sufficient training in the principles of deductive (a conclusion is derived from premises known to be true) and inductive (a conclusion is inferred from multiple observations) inference, so these terms need no detailed explanation here. However, abductive inference (inference based on robust explanation) is probably less familiar because it represents the path of inference less traveled in modern science. It is certainly much less emphasized. Abductive inference is concerned with the initial stages of scientific inquiry, and could be thought of as hypotheses that explain observed phenomena. Conceptual models are one example of what is meant by abductive inference. At one time or another, most scientists dabble in conceptual models because abductive inference is, as suggested by Peirce, an important part of the scientific process. Abductive inference has powerful heuristic value, often quickly leading to researchable hypotheses that would otherwise resist surfacing via deductive or inductive methods. Nonetheless, abduction has been largely ignored as a formal mode of inference by the scientific community, at least within disciplines concerned with natural resource management because it is usually conducted in an informal manner (e.g., conceptual models), and is therefore not perceived by scientists as rigorously scientific. Modern knowledge-based systems overcome many of the inherent limitations of less formal approaches to abductive inference.

NetWeaver ("NetWeaver") was first developed in 1991 to ease knowledge engineering tasks by giving a graphical user interface to the ICKEE (IConic Knowledge Engineering Environment) inference engine developed at Pennsylvania State University by Bruce J. Miller and Dr. Michael C. Saunders. The first iterations were simply a visual representation of the logic model stored in a LISP-like syntax. NetWeaver quickly evolved into an interactive interface in which the visual environment was also capable of editing the models and saving them in the ICKEE file format. Eventually NetWeaver became "live" in the sense that it could evaluate the models in real time.

A NetWeaver logic model graphically represents a problem to be evaluated as networks of topics, each of which evaluates a proposition. The formal specification of each topic is graphically constructed, and composed of other topics (e.g., premises) related by logic operators such as *and*, *or*, *not*, etc. NetWeaver topics and operators return a continuous-valued "truth value," that expresses the strength of evidence that the operator and its arguments provide to a topic or to another logic operator (Miller and Saunders 2002). The specification of an individual NetWeaver topic supports potentially complex reasoning because both topics and logic operators may be specified as arguments to an operator. Considered in its entirety, the complete logic specification for a problem can be thought of a mental map of the logical dependencies among propositions. In other words, the model amounts to a formal logical argument in the classical sense (Halpern 1989).

Cognitive theory suggests that human beings have two fundamental modes of reasoning: logical and spatial (Stillings et al. 1991). Interesting things happen when logic is implemented graphically.

First, the knowledge of individual subject-matter experts engaged in knowledge engineering often is not fully integrated when dealing with complex problems, at least initially. Rather, this knowledge may exist in a somewhat more loosely organized state, a sort of knowledge soup with chunks of knowledge floating about in it. A common observation of knowledge engineers experienced in graphically designing knowledge bases is that the process of constructing a graphic representation of problem-solving knowledge in a formal logical framework seems to be synergistic, with new insights into the expert's knowledge emerging as the process unfolds.

Second, synergies similar to those observed in organizing the reasoning of individual subject-matter experts can also occur in knowledge engineering projects that require the interaction of multiple disciplines. For example, many different specialists may be involved in evaluating the overall health of a watershed. Use of a formal logic system, with well-defined syntax and semantics, allows specialists to represent their problem solving approach in a common language, which in turn facilitates understanding of how all the various perspectives of the different specialists fit together and accomplish an evaluation.

6.3 *Knowledgebase (or Logic Model)*

In modern parlance, the concept of a knowledgebase is now commonly understood to mean an organized body of information. For example, many software companies offer knowledgebases on their web sites to assist users with various aspects of using the software. However, as originally defined by Walters and Nielsen (1988), the term meant a formal specification for the interpretation of information. In this volume, the term knowledgebase is used in the latter sense, and is synonymous with the term "logic model," which some knowledgebase-system developers prefer as more descriptive. Hereafter, in the remainder of this chapter, we use the term logic model, but we use it interchangeably with knowledgebase.

In terms of the Holsapple (2003) definition of a DSS, logic models represent the knowledge system. The problem-processing system (or engine) is the DSS component that processes the logic model to generate interpretations of the state of a real-world system that is being modeled. Ideally, the construction of a logic model is under the control of the engine, which guarantees that the model is ontologically committed to the engine (Gruber 1995). In other words, the model conforms to the semantics and syntax of the engine.

Logic models can be rule-based, object-based, or some combination of the two. In either case, they implement a reasoning process that supports formal arguments (Halpern 1989). In simplest form, a logical argument can be represented by a conclusion (or proposition) that is to be tested, and a set of premises, each of which contributes evidence for or against the conclusion.

6.4 Decision Model

Environmental assessments provide essential background information about ecosystem states and processes, and are thus a useful starting point for applying adaptive ecosystem management. As a logical follow-up to ecological assessment, managers may wish to identify and set priorities for ecosystem maintenance and restoration activities. Decision models such as the AHP (Saaty 1992) and the Simple Multi-Attribute Rating Technique (SMART; Edwards 1977; Edwards and Newman 1982) provide a bridge from assessment to planning by helping managers to establish rational priorities for management activities ("Criterium DecisionPlus").

Environmental assessments generally deal with a broad array of topics that include biophysical, social, and economic dimensions. Ideally, the same circumspection should carry over into processes used to identify and set priorities for maintenance and restoration activities derived from an assessment. AHP and SMART decision models are discussed together in this chapter because they complement each other in the design of large and complex decision models.

7 EMDS Environment

At version 4.2, EMDS runs on 32- and 64-bit versions of Windows XP, Windows Vista, and Windows 7. In addition, there are two editions of EMDS 4.2. The ArcMap edition, similar to version 3.0, is implemented as an extension to ArcMap, whereas the new stand-alone edition has a custom interface implemented by the Redlands Institute (University of Redlands, Redlands, CA) and runs directly on ArcEngine. The advantage of the ArcEngine edition over the ArcMap edition is that the user interface is greatly simplified (for the GIS-averse), and licenses are an order of magnitude cheaper. On the other hand, the stand-alone edition foregoes immediate access to the full geoprocessing power inherent in ArcMap. Both editions of version 4.2 are compatible with the ArcGIS 9.2, 9.3, and 10.0 platforms.

8 Core EMDS Components

EMDS integrates a decision engine from InfoHarvest, Inc. (www.infoharvest.com) and a logic engine from Rules of Thumb, Inc. (www.rules-of-thumb.com) as core components. The NetWeaver engine ("NetWeaver") performs logic-based evaluation of environmental data, and logically synthesizes evaluations to infer the state of landscape features, such as watersheds (e.g., watershed condition). The decision engine ("Criterium DecisionPlus") prioritizes landscape features with respect to user-defined management objectives (e.g., watershed restoration), using summarized outputs from NetWeaver, as well as additional logistical information considered important to the decision maker(s).

9 Project Structure

Each EMDS project has a well-defined structure that can be summarized as follows:

- A project may contain one or more assessments. An assessment is defined by the user by a set of spatial data layers and a spatial extent selected on those layers. Different assessments are required when the layers participating in the assessments represent different combinations of layers, different extents selected on the same set of layers, or the same combinations of layers but with data from different assessment dates.
- Each assessment may contain one or more analyses. Each analysis has an associated logic model specified by the user. Generally, there is a one-to-one association between assessments and analyses, but multiple analyses can be run in the same assessment to compare structural variants of a logic model.
- Each analysis may contain one or more scenarios. Users are not allowed to alter the data inputs to an analysis after the assessment area has been defined, but data and logic structure can be edited within scenarios to explore "what-if" types of questions.
- Each analysis or scenario may contain one or more decision analyses. Generally, there is a one-to-one association between analyses or scenarios and their associated decision analyses, but multiple decision analyses can be run in the same logic analysis or scenario to compare structural variants of the decision model.

10 EMDS Functionality

The previous section on project structure implies some rather obvious basic functionality (e.g., create assessments, create and run analyses, etc.). However, some functionality falls outside the scope of the structural description, and some of the implied functionality bears further elaboration.

10.1 Analyses and Scenarios

To set up an analysis, the user chooses one or more topics from the logic outline to be included in the analysis. When the run command is executed, the logic processor first traces the logic dependencies from the highest logic level(s) selected down to the data requirements as defined in the logic model. Data requirements from the logic model (names and aliases) are then matched against the names and aliases of attribute fields in the spatial data layers previously associated to the assessment by the user. A name-matching dialogue is then presented to the user to aid them in reviewing, correcting, or filling in matches, as necessary, before an analysis is run.

Once a logic analysis or scenario has been run, any data input or evaluated logic topic can be displayed in the map pane of the system interface. Other functions associated with analyses include:

- Customization of map symbology, which persists within the scope of a project.
- Querying of landscape features to display the evaluated state of the logic in a graphic interface provided by the logic engine.
- A utility to review missing data, and derive priorities for missing data.
- Exporting of map products to various graphic formats.
- Creation of additional scenario(s) based on the current analysis.

10.2 Decision Models

After a logic-based analysis or scenario has been run to evaluate the state of landscape features in the assessment area, this same set of features can then be prioritized for management activities. This is generally based on summarized results generated from the logic-based analysis, as well as any additional data inputs that may be technically or logistically relevant to a decision. Examples of data inputs that may not be considered in the logic evaluation, but that may enter into the decision phase, include factors related to environmental consequences or effects, performance criteria, feasibility or efficacy considerations, and social or economic cost or benefit consideration, among others.

The decision component of EMDS is called the Priority Analyst, which is responsible for configuring a decision model, and displaying its results. The configuration and display tasks are each handled by an associated wizard. Within the configuration wizard, the user is guided through a series of dialogs to:

- Assign a Criterium DecisionPlus (CDP, InfoHarvest, Seattle, WA) model to the decision analysis.
- Map field names between the CDP model and the attribute table generated by the logic engine.
- Optionally, adjust weights on decision criteria.
- Specify options for error handling.

A second display wizard presents the user with a set of tabs in which to review:

- The overall priority score derived for each landscape feature in the assessment.
- The derivation of criterion scores in terms of the contributions made to a criterion by its subcriteria.
- An analysis of model sensitivity or, conversely, model robustness.
- How unit changes in data inputs trade for one another in terms of changing priority scores (tradeoff analysis).

Tables and graphs displayed in the tabs of the display wizard can be exported to a variety of formats. The configured model, including all input records, can be exported to CDP for further analysis. In addition, upon exiting the display wizard, a map of priority scores can be output to the map pane of the system interface.

10.3 Multiple Spatial Scales

Many contemporary decision problems in natural resource management cannot be adequately addressed at a single spatial scale. For example, watershed analysis may require analysis of watersheds and of stream segments within watersheds (Gallo et al. 2005). Similarly, a national forest fuels analysis may require analysis of forest units and regions (Reynolds et al. 2009). Other types of problems may suggest the need for three or more scales of analysis that must be spatially integrated. EMDS projects support multi-scale analyses to the extent that multiple scales of analysis can be accommodated within a single project, but this functionality could be developed further. Currently, it is left to the user to manage integration across scales. However, it is not difficult to conceive of a wizard feature that could assist with scale integration.

10.4 System Role in Adaptive Ecosystem Management and Planning Processes

So far as we are aware, no one has yet fully exercised the capabilities of EMDS with respect to supporting adaptive ecosystem management in particular, or planning processes more generally, so this section is more prospective in nature as opposed to an historical account.

The adaptive management process has been described as a cycle that begins with monitoring and then proceeds through evaluation, planning, and implementation (Holling 1978; Walters 1986). However, any cyclical process requires initialization (i.e., there is always a cycle 0). The process has traditionally been launched by recognition (within the natural resource community) of a significant environmental concern. This leads to identifying key questions that need to be addressed, and an enumeration of data requirements that should be addressed by monitoring. The manner in which data requirements are derived is important. If it is done in an ad hoc or undisciplined way, there is no formal roadmap that traces the dependencies between key questions, intermediate ecosystem states and processes that influence those questions, and a logical trace to needed data. The latter situation is inherently problematic because of the time typically needed to acquire the relevant data. If 5 years have elapsed, half the scientists, specialists, and resource managers initially involved in the process are likely to have moved on, and those that are left may well have lost track of the original rationale, and are left wondering what to do with the data.

Logic models can be a significant aid, in this context, by providing a formal specification that maps the dependencies from key questions, through all the intermediate states, down to data (i.e., a roadmap). Such models serve as a form of institutional memory. A significant point here is that modeling is often an afterthought, when in fact we argue that it should be the first step in the process, in which case the data requirements are derived from the model specification. Furthermore, if the process is implemented in the manner suggested, there is also a clear specification of what to do with the data once they are collected and it's time to evaluate the state of the ecosystem.

In the spatial context of EMDS, the basic products of evaluation are maps of NetWeaver scores, as well as distributions of NetWeaver scores for the set of features that comprise the assessment area. One approach to planning is to employ the Priority Analyst to derive strategic priorities for the management of the landscape features, given the summary results from the logic evaluation and other logistical information important to managers, as discussed earlier. However, there are alternative approaches to planning that can be implemented in EMDS. For example, given the current observed state of the landscape, there may be multiple competing alternative strategies that might be implemented to restore it to a more desirable state. If the consequences of these strategies can be projected into the future (e.g., with simulation systems), and the results attributed back to the GIS layers that fed the original EMDS analysis, one can evaluate the performance of each of the alternatives relative to each other and the original system state (each alternative would be represented by an EMDS assessment). Although one might visually compare the maps associated with each alternative, a more straightforward approach would be to compare the distribution of outcomes across alternatives. In fact, using the distribution data from the assessment of each alternative, standard statistical tests could be brought to bear. With respect to the planning phase, simply comparing the distribution of outcomes of alternatives may not be sufficient. As with more basic EMDS analyses, there may be additional logistical factors that need to be considered, in which case the application of decision models may again be useful. However, the Priority Analyst is no longer the appropriate tool in this case, because it deals with priorities of spatial features, whereas the evaluation of a set of strategic alternatives as described above is no longer a spatial problem. Here, one could revert to the more classical application of CDP models.

EMDS also has a potentially useful role to play in assessing plan performance. To set the stage, let us suppose that an initial landscape evaluation was performed at year 5 (we are continuing the earlier example, and assuming it took 5 years to get to the first evaluation), and, following a planning exercise, a strategy for landscape improvement was selected and implemented. It is now 5 years later (year 10), the monitoring data have been updated to reflect the current landscape condition, and managers want to assess how well the plan is performing. This situation differs in only minor details from the discussion about comparing the expected performance of alternative management strategies. Whereas prior analysis compared expected outcomes of the alternatives, now analysis is comparing two realizations of the evaluated system state over time. Thus, here again, standard statistical tests could be applied to draw inferences about the significance of changes in the distribution of outcomes. Continuing the earlier discussions about the impact of time, 10 years have elapsed from start to present, there has been almost a complete turnover of staff, and the argument for methods that instill institutional memory into the process become more compelling.

11 Critical Factors in the Success of EMDS

The following five factors have been critical in the success of the EMDS system, and provide some additional insights into the functionality of the system.

11.1 Generality

EMDS is a framework within which developers can design customized applications that implement logic and decision models to address many different kinds of questions related to natural resource management, and at whatever spatial scale(s) may be relevant to associated questions. Because of its implementation as a general framework, EMDS has been used in numerous natural resource applications around the world since 1997 (for a fairly comprehensive compilation of published accounts see http://en.wikipedia.org/wiki/Ecosystem_Management_Decision_Support). The design of EMDS as a general solution framework accounts for much of the success of the system since its introduction in 1997.

11.2 Transparency

Rational and repeatable processes are a foundation of decision support, but are not sufficient to achieve maximum effectiveness. EMDS design also has placed a premium on transparency, and we argue that this has been another key ingredient in the relative success of the system as a tool for decision support. Transparency has two important dimensions. First, models should be fully self-documenting in terms of revealing data limitations, underlying assumptions, and particulars of development, with this information accessible to users via the system interface. Second, logic and decision engines should be able to reveal the derivation of answers, via an intuitive user interface, so results can be transmitted to stakeholders who may have limited technical expertise in the subject area. As an example, EMDS solutions have been used to address important national issues in the US precisely because senior agency officials were able to effectively communicate results to Congress or federal oversight agencies.

11.3 Simplification

Logic and decision models in an EMDS application complement one another. A logic model focuses on the question, "What is the state of the system?" A decision model focuses on the question, "Given the state of the system, what can be done about it?" Logistical issues of significance to managers are not pertinent to the first question, but they are important to the second. One consequence of separating the overall modeling problem in EMDS into two complementary models has been that each model is rendered conceptually simpler. The logic model only evaluates the status of a system according to its attributes, whereas the decision model primarily considers attributes of special interest to resource managers. In addition, a logic model can be used as a preprocessor to a decision model, which leads to better handling of the abstractions and complexities of modern natural resource management decisions.

11.4 Abstraction and Complexity

Logic-based approaches to environmental analyses have proven useful in EMDS for accommodating the levels of abstraction and complexity commonly encountered in contemporary decision-support contexts. Abstraction is an issue even in problems that are predominantly biophysical in nature. For example, watershed analysis, wildfire potential, and landscape integrity are all examples of decision-support topics that are both inherently abstract and complex, but essentially biophysical in scope. The application of logic to such problems has been successful in EMDS because the only limitation to use of logic is that one must be able to reason about solutions in terms of chains of conclusions and underlying premises. The case for logic-based solutions becomes still more compelling when the complexity of the problem increases with the need for integrated analyses spanning biophysical, social, and economic dimensions.

12 Experiences in Design and Use

At version 4.2, EMDS is a fully operational decision-support system for environmental analysis and planning, but, by design, it remains a work in progress. The original vision for the system was that it ultimately should provide complete support for the full adaptive management process, but complete specification of such a system, following concepts of waterfall design, was considered impractical, and the development team opted instead to implement an incremental and evolutionary approach to design, following the principle of "build a little and test a little." Based on logic-based processing, versions 1 and 2 implemented decision support for monitoring and evaluation. However, as these early versions were put into application by users, we soon observed that users were attempting to simultaneously evaluate environmental conditions and derive priorities for management (e.g., planning). Such approaches created considerable confusion about how to model the problem. Only after some reflection on these use cases did it become apparent that providing for a separate but complementary decision component might have the potential to simplify integration of evaluation and planning. Subsequent experiences with version 3, which added a decision engine to support planning, confirmed the utility of the solution.

More recently in 2009, EMDS analyses to support budget allocation for forest-fuels management for the US Department of Interior led to recognition of another opportunity by which to improve and extend the functionality of the system. Between 2007 and 2009, solutions had been designed by Department scientists, technical specialists, and mid-level managers to set priorities for level of effort with respect to forest fuels treatment. The set of models was primarily oriented toward biophysical considerations, but included some additional socioeconomic factors as well. Nevertheless, senior managers were primarily interested in using the modeling results to allocate budgets to agencies and regions of the Department. Conflating these two different purposes could easily lead to undesirable outcomes because the existing solutions did not explicitly account for economic and institutional factors needed to support budget allocation decisions. Discussions around this dilemma lead to the realization that a manifold was needed, that is, an additional layer of decision models that consumed results from the existing decision process, to process the multiple purposes for which solutions were being developed.

13 Conclusion

One of the advantages of an application development platform like EMDS is that it enables users to decompose complex, multi-dimensional analysis and decision-making processes into manageable components that conceptually frame and transparently present planning results. Decision-support systems like EMDS require users to formally represent how they logically structure a problem analysis, which allows them to more effectively communicate the rationale of decisions made on the basis of their analyses. Once they are built, logic models can be readily adapted in the future, based on new information, and what has been learned from applied activities. Hence, EMDS applications are highly useful adaptive management tools, as well.

EMDS is an application development framework within which users can design logic and decision models to address many different kinds of questions related to landscape evaluation, at whatever spatial scale(s) may be relevant to their questions (Reynolds et al. 2003). EMDS was intentionally designed for the purpose of integrating environmental analysis and planning processes. Models developed in EMDS can provide decision support for multi-level and multi-dimensional landscape analyses through the use of logic and decision engines integrated with the ArcGIS1 9.2, 9.3, and 10.0 geographic information systems (GIS, Environmental Systems Research Institute, Redlands, CA).

The NetWeaver logic engine evaluates landscape data against a logic model designed by the user in the NetWeaver Developer system (Miller and Saunders 2002), to derive logic-based interpretations of complex ecosystem conditions. A decision engine evaluates outcomes from the logic model and other feasibility, valuation, and efficacy criteria related to management actions, against a decision model for prioritizing landscape treatments in light of these criteria, built with its development system, CDP. CDP models implement the Analytic Hierarchy Process (AHP; Saaty 1992), the Simple Multi-Attribute Rating Technique (SMART; Kamenetzky 1982), or a combination of the two processes.

The logic and decision models in EMDS are complementary; the logic model focuses on the question, "What are the states of the system?", and the decision model focuses on the question, "Given the states of the system, what might be done about it?" Logistical issues are not pertinent to the first question, but they are vital to the second. One consequence of separating the overall modeling problem into two complementary models is that each model is made simpler: the logic model considers the status of the topics under evaluation; the decision model takes the status of the system and then places it in social and operational contexts that can further inform decision-making. The decisions rendered need only be partially based on system conditions; they can also be based on social context and human values. Decision-makers set priorities for treatment or remedial action using the AHP and/or SMART utilities in CDP. After priorities have been fully derived by the decision model, decision-makers can then review the results and observe the relative contributions of ecological states and social contexts to alternative decisions. This may be accomplished either through sensitivity analysis or by developing various management or treatment scenarios that alternatively weight priorities as they might contribute to a decision. Users adjust the weightings when evaluating alternative scenarios and their trade-offs.

The process of landscape evaluation can be simply rendered or include representative complexity. This feature is critical to ecosystem evaluations because system structure and functionality may exist at several scales that are influential to species large and small, their habitats, and other ecological patterns and processes. A failure to consider the contributions of potential management treatments on conditions, patterns, or processes at these scales can have a large, distributed impact. Thus, having a platform like EMDS allows the user to develop and structure knowledge about the levels and dimensions of landscapes, and how various components interact, enabling them to substantially increase the quality, complexity, and transparency of their thinking. It also enables them to catch themselves in non-critical or erroneous thinking because the logic representing how they think about their systems is formally specified, and can be reviewed and adapted with learning. When outputs of the logic or decision models are revealed that are inconsistent with expectations, the user can then delve into the structure of the logic to determine where the error(s) occurred. This is often an iterative process and it results in improved reasoning and model representation.

References

Aristotle, prior analytics (trans: Jenkinson AJ). MIT Archives, 350 BCE, Cambridge. http://classics.mit.edu/Aristotle/prior.html. Accessed Aug 2009

Bonczek RH, Holsapple CW, Whinston AB (1981) Foundations of decision support systems. Academic Press, New York

Bourgeron PS, Humphries HC, Riboli-Sasco L (2009) Regional analysis of social-ecological systems. Nat Sci Soc 17:185–193

Chapin FS III, Kofinas GP, Folke C (eds) (2009a) Principles of ecosystem stewardship: resilience-based natural resource management in a changing world. Springer, New York

Chapin FS III, Carpenter SR, Kofinas GP et al (2009b) Ecosystem stewardship: sustainability strategies for a rapidly changing planet. Trends Ecol Evol 25:241–249

Christensen NL, Bartuska A, Brown JH, Carpenter S, D'Antonio C, Francis R, Franklin JF, MacMahon JA, Noss RF, Parsons DJ, Peterson CH, Turner MG, Moodmansee RG (1996) The report of the ecological society of America committee on the scientific basis for ecosystem management. Ecol Appl 6(1996):665–691

Davis G (1974) Management information systems: conceptual foundations, structure, and development. McGraw-Hill, New York

Davis JR, Clark JL (1989) A selective bibliography of expert systems in natural resource management. AI Appl 3:1–18

Edwards W (1977) How to use multi-attribute utility measurement for social decision making. IEEE Trans Syst Man Cybern 7(1977):326–340

Edwards W, Newman JR (1982) Multi-attribute evaluation. Sage, Beverly Hills

Everett R, Hessburg P, Jensen ME, Bourgeron PS (1994) Old forests and dynamic landscapes. J For 92:22–25

Gallo K, Lanigan SH, Eldred P, Gordon SN, Moyer C (2005) Preliminary assessment of the condition of watersheds under the Northwest Forest Plan. General technical report PNW-GTR-647. U.S. Department of Agriculture, Forest Service, Pacific Northwest Research Station, Portland

Gorry A, Scott Morton MS (1971) A framework for information systems. Sloan Manage Rev 13:56–79

Gruber TR (1995) Toward principles for the design of ontologies used for knowledge sharing. Int J Hum Comput Stud 43:907–928

Gustafson E, Nestler J, Gross L, Reynolds K, Yaussy D, Maxwell T, Dale V (2003) Evolving approaches and technologies to enhance the role of ecological modeling in decision-making. In: Dale V (ed) Advances in ecological modeling. Springer, New York

Halpern DE (1989) Thought and knowledge: an introduction to critical thinking. Lawrence Erlbaum Associates, Hillsdale

Holling CS (ed) (1978) Adaptive environmental assessment and management. Wiley, Chichester

Holsapple CW (2003) Decision support systems. In: Bidgoli H (ed) Encyclopedia of information systems, vol I. Academic Press, New York, pp 551–565

Jensen ME, Everett R (1994) An overview of ecosystem management principles. In: Jensen ME, Bourgeron PS (ed.) Eastside forest ecosystem health assessment, vol. 11, Ecosystem management: principles and applications. General technical report PNW-GTR-318. US Department of Agriculture, Forest Service, Pacific Northwest Research Station, Portland, 376 p

Kamenetzky R (1982) The relationship between the analytical hierarchy process and the additive value function. Decis Sci 13:702–716

Keen PGW, Scott Morton MS (1978) Decision support systems: an organizational perspective. Addison-Wesley, Reading

Lee KN (1999) Appraising adaptive management. Conserv Ecol 3:3

Liu J, Dietz T, Carpenter SR et al (2007) Complexity of coupled human and natural systems. Science 317:1513–1516

Maser C (ed) (1994) Sustainable forestry: philosophy, science, and economics. St Lucie Press, Delray Beach

Miller BJ, Saunders MC (2002) The NetWeaver reference manual. Pennsylvania State University, College Park

Mintzberg H, Raisinghani D, Theoret A (1976) The structure of unstructured decision processes. Admin Sci Quart 21:246–275

Mowrer HT (1997) Technical compiler, Decision support systems for ecosystem management: an evaluation of existing systems, General technical report RM-GTR-296. US Department of Agriculture, Forest Service, Rocky Mountain Forest and Range Experiment Station, Fort Collins

Oliver CD, Twery M (2000) Decision support systems: models and analyses. In: Johnson NC, Malk AJ, Sexton WT, Szaro RC (eds) Ecological stewardship: a common reference for ecosystem management. Elsevier Science Ltd., Oxford, pp 661–686

Overbay JC (1992) Ecosystem management. In: Taking an ecological approach to management. WO-WSA-3. US Department of Agriculture, Forest Service, Washington, pp 3–15

Peirce CS (1931–1935) Collected papers of Charles Sanders Peirce, vols. 1–6. In: Hartshorne C, Weiss P (eds) Harvard University Press, Cambridge

Popper K (1934) Logik der Forschung. Springer, Vienna

Power DJ (2008) Decision support systems: a historical overview. In: Burstein F, Holsapple CW (eds) Handbook on decision support systems, vol. 1. Springer, Berlin, pp 121–140

Rauscher H (1999) Ecosystem management decision support for federal forests in the United States: a review. Ecol Manage 114:173–197

Rauscher HM, Benzie JW, Alm AM (1990) A red pine forest management advisory system: knowledge model and implementation. AI Appl 4:27–43

Reynolds KM, Rodriguez S, Bevans K (2003) EMDS 3.0 user guide. Environmental Systems Research Institute, Redlands

Reynolds K, Bjork J, Hershey RR, Schmoldt D, Payne J, King S, DeCola L, Twery M, Cunningham P (2000) Decision support for ecosystem management. In: Johnson NC, Malk AJ, Sexton WT, Szaro RC (eds) Ecological stewardship: a common reference for ecosystem management. Elsevier, Oxford, pp 687–721

Reynolds KM, Hessburg PF, Keane RE, Menakis JP (2009) National fuel-treatment budgeting in US federal agencies: capturing opportunities for transparent decision-making. Ecol Manage 258:2373–2381

Russell BAW (1903) Principles of mathematics. Cambridge University Press, Cambridge

Saaty TL (1992) Multicriteria decision making: the analytical hierarchy process. RWS Publications, Pittsburg
Scott Morton MS (1967) Computer-driven visual display devices: their impact on the management decision-making process. PhD Dissertation, Harvard Business School, Cambridge
Scott Morton MS (1971) Management decision systems: computer-based support for decision making, PhD Dissertation. Graduate School of Business Administration, Harvard University, Cambridge
Schuster EG, Leefers LA, Thompson JE (1993) A guide to computer-based analytical tools for implementing National Forest plans, general technical report INT-296. U.S. Department of Agriculture, Forest Service, Intermountain Research Station, Ogden
Simon HA (1947) Administrative behavior. MacMillan, New York
Simon HA (1960) The new science of management decision. Harper and Row, New York
Sprague RH Jr (1980) A framework for the development of decision support systems. Manage Inform Syst Quart 4:1–26
Sprague RH Jr, Carlson ED (1982) Building effective decision support systems. Prentice-Hall, Englewood Cliffs
Stillings NA, Feinstein MH, Garfield JL et al (1991) Cognitive science: an introduction. MIT Press, Cambridge
Twery MJ, Elmes GA, Schaub LP, Foster MA, Saunders MC (1993) GypsES: a decision support system for gypsy moth management. In: Liebhold AM, Barrett HR (eds) Spatial analysis and forest pest management, general technical report NE-175. U.S. Department of Agriculture, Forest Service, Northeastern Forest Experiment Station, Radnor, pp 56–64
Walters C (1986) Adaptive management of renewable resources. Macmillan, New York
Walters JR, Nielsen NR (1988) Crafting knowledge-based systems. Wiley, New York
WCED (World Commission on Environment and Development) (1987) Our common future. Oxford University Press, Oxford

NetWeaver

Michael C. Saunders and Bruce J. Miller

Abstract NetWeaver is one of the foundational technologies of the Ecosystem Management Decision Support (EMDS) system. NetWeaver's graphical interface, real time evaluations, fuzzy-logic-based measures of uncertainty, and overall ease of use led to it being chosen as a major component of EMDS. By way of background on this stand-alone knowledge engineering tool, and in order to provide enhanced perspective on EMDS and its inner workings, we first present a description of the origins of NetWeaver, and how and why it was developed.

Keywords NetWeaver · Fuzzy logic · Decision support

1 Background

A computer-based model of human expertise is called a knowledgebase. NetWeaver is a knowledgebase development system that provides a graphical environment in which to construct and evaluate knowledgebases (Saunders and Miller 1997) built with dependency networks (see definition below). Before the advent of NetWeaver, we built knowledgebases by drawing dependency networks on whiteboards or flipcharts. These drawings were then handed off to a knowledge engineer to code. Before C++ was available, the dependency network drawings were hard coded as if-then statements in the C language.

M. C. Saunders (✉)
Department of Entomology, Pennsylvania State University, 501 Agricultural Science and Industries Building, University Park, PA 16802, USA
e-mail: mcs5@psu.edu

B. J. Miller
Rules of Thumb Inc., 11817 Cedar Mill Rd., North East, PA 16428, USA
e-mail: bjmiller@gmail.com

K. M. Reynolds et al. (eds.), *Making Transparent Environmental Management Decisions*, Environmental Science and Engineering,
DOI: 10.1007/978-3-642-32000-2_2, © Springer-Verlag Berlin Heidelberg 2014

The next stage of development was to separate the coding of knowledgebases from their code implementations. This meant creating a LISP-like scripting language to represent the knowledgebases and the creation of an inference engine to read and implement the new knowledgebases. In computer science, an inference engine is an application that acts to interpret a knowledgebase. The scripting language implementation had two main advantages: (1) a non-programmer could easily code a knowledgebase, and (2) a knowledgebase could be edited independently from the inference engine.

This early implementation led to large increases in our ability to produce knowledgebases. However, limitations of this approach became evident when a project came our way that involved 100s of dependency networks drawn by a subject matter expert (SME). Hand coding the scripts revealed two flaws: they were tedious to edit (but not nearly as tedious as hard-coding!), and it was difficult to verify that they accurately depicted the hand-drawn dependency networks.

Facing an overwhelming amount of work, a graphical dependency network editor was conceived. Enter NetWeaver. Conceived and written over winter break of 1991, it originally only displayed edited dependency networks, saving the knowledgebases in the same scripting system. But now graphical editing was possible: click a button to add a node, click a button to move a node. It was truly a what-you-see-is-what-you-get editor, so the flow of changes was easy to handle, and verification was as simple as comparing the hand-drawn dependency network to the one on the screen.

Once the dependency network was coded correctly, the next issue was ensuring that a given dependency network represented its intended logic correctly. The chosen solution was to integrate the inference engine into NetWeaver so that the user could test the graphical model with data. Now, knowledgebases could be designed and verified live. No more disjointed design, coding, and testing loops, and a SME could independently build knowledgebases and get real-time verification.

Over the years, NetWeaver has evolved and its capabilities have expanded. To deal with missing data and with qualitative concepts, we developed evaluation algorithms based upon fuzzy math (Saunders et al. 2005) that worked well (see below for more details).

In the early years of the 21st century, NetWeaver underwent a thorough rewrite to NetWeaver2. NetWeaver2's main improvements were:

- Internationalization—for both development and deployment.
- Runtime applications—deployable self-contained knowledgebase applications.
- Documentation—robust internal and exported documentation at all levels of the knowledgebase, and with all the capabilities of a modern word processor.
- Binary file format—to make the software smaller, faster, more extensible.
- Knowledgebase security—password protection for various aspects of the knowledgebase.
- Automated documentation—exporting the knowledgebase to HTML documents, complete with all linkages and embedded documentation.

(see http://rules-of-thumb.com/development_plans for a more complete review)

2 NetWeaver Concepts

Knowledgebase systems come in a variety of forms, but the dominant types currently in use are rule-based systems (Parsaye and Chignell 1988). Knowledge representation in NetWeaver, in contrast, is based on object-oriented fuzzy-logic networks (dependency networks). These types of dependency networks offer several significant advantages over the more traditional rule-based representation.

Compared to rule-based knowledgebases, NetWeaver knowledgebases are easier to build, test, and maintain because their underlying object-based representation makes them modular. The modularity of NetWeaver knowledgebases, in turn, allows the designer to incrementally evolve complex knowledgebases from simpler ones. Modularity also allows interactive knowledgebase debugging at any and all stages of development, which expedites the process.

Finally, fuzzy logic provides a formal and complete calculus for knowledge representation that is less arbitrary than the "confidence factor approach" (Negoita 1985), used in rule-based systems, and more parsimonious than bivalent rules.

Although the term "fuzzy logic" has a distinctly esoteric ring to it, the concept is actually quite simple. Fuzzy logic provides a metric for expressing the degree to which an observation on some variable belongs to a set that represents that variable. Alternatively, one might say that fuzzy logic is concerned with "aboutness." To make the concept clear, consider the following example.

Everyone has some concept of what it means to be an adult. For legal purposes, an adult, in western cultures is often defined to be a person who is 21 years old or older. A rule-based system dealing with legal issues can easily accommodate this bivalent definition: if a person is 20 years, 11 months, and 30 days old, they are not an adult, but if the person is at least one day older than 21 years, they are an adult. This characterization of adultness is sufficient if the concept of adult is limited to a simplistic legal one. However, if by adultness we instead are really interested in expressing something more complex such as an individual's emotional maturity, then the simple bivalent rule for determining adultness is no longer adequate. Most people would agree that a five-year-old has no, or at best minimal, adult qualities. In a 13-year-old, however, we might begin to see at least some early signs of adult characteristics. Some 18-year-olds demonstrate many adult qualities (they act very "grown up"). Conversely, most people can think of at least a few 25-year-olds they have met in their life that could not be called particularly emotionally mature. Thus, as a first step toward improving the characterization of adultness, one might construct a simple fuzzy membership curve that translates age into degree of membership in the set "adult."

We indicated that fuzzy logic allowed a more parsimonious knowledge representation than that which is possible with rule-based systems. The reason is simple. A single fuzzy membership function is sufficient to express the full spectrum of adultness. In contrast, rule-based systems are inherently bivalent, meaning that a rule is either true or false. To more precisely characterize adultness in a rule-based system, one would need to define, say, five age categories

Table 1 The NetWeaver logical node types and their function

OR	An OR node is true when any one of its antecedents is true. It is false when all of its antecedents are false. Functionally, it passes the value of its most true antecedent
AND	An AND node is true when all of its antecedents are true. It is false when any one of its antecedents is false. Functionally, it performs a weighted average of the values of its antecedents unless one of the antecedents is fully false. Compare this with the next definition of UNION
UNION	A UNION is true when all of its antecedents are true. It is false when all of its antecedents are false. As a practical distinction between AND and UNION nodes, antecedents to AND function like limiting factors, whereas antecedents to UNION function like compensating factors
NOT	A NOT node simply inverts the value of its antecedent
SOR	A SOR node (sequential OR) is a special class of node designed to select between alternative decision scenarios where there is a definite hierarchy of quality level associated with each possible data gathering method. In other words, the SOR node is a data route selector; it provides a method for selecting the best choice of paths within the scope of the currently given data. For example, the preferred path may involve decision making on the basis of acid neutralizing capacity (ANC), but if ANC is missing, then the decision can be based on an alternate parameter such as conductivity or pH. Connections to the antecedents of a SOR node are represented with dotted lines to indicate their relative position in the hierarchy
XOR	A XOR node (exclusive OR) is true when one and only one of its antecedents is true

corresponding to different levels of adultness, and each category would require a rule. Moreover, if our rule base also dealt with intelligence, and this attribute similarly had five categories (ranging from brilliant to ignorant, for example), then to jointly consider both adultness and intelligence in our rule base could require as many as 25 additional rules. In contrast, in a fuzzy-logic-based representation of this more complex situation, we only need one more fuzzy curve and perhaps a new network object to jointly evaluate the two fuzzy curves. So, in our example, two fuzzy curves have an expressive power that is equal to or better than 35 (10 + 25) rules. To summarize, the number of rules needed to adequately represent possible outcomes explodes approximately combinatorially, whereas the number of fuzzy curves and related objects needed to describe the same problem in an object-oriented fuzzy logic representation increases approximately linearly.

NetWeaver uses OR, AND, UNION, NOT, and XOR and SOR logic nodes to define the logical dependency of a network on antecedent networks, and on data links (Table 1). A data link is a type of dependency network object in NetWeaver. A data link is where data are interpreted by an argument (i.e., simple data links), or where data are used in a mathematical expression, the result of which is interpreted by an argument (i.e., calculated data links). If no data links antecedent to a network use fuzzy arguments, then the operation of these nodes conforms quite closely to their usage in conventional logic. The only real difference in this context between these nodes, as used in NetWeaver, and in standard logic is that true = 1, false = −1 in NetWeaver.

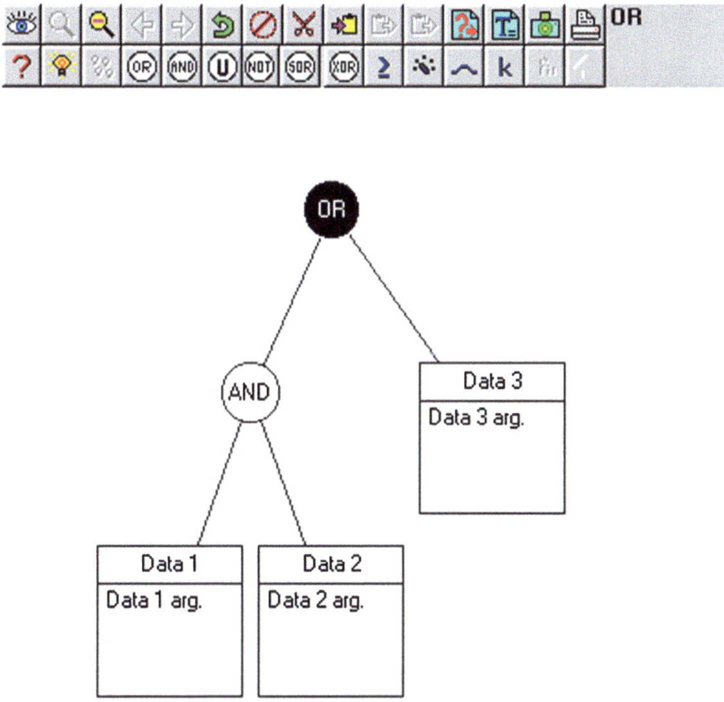

Fig. 1 A dependency network as represented in NetWeaver. In this dependency network, there are three data links represented by the squares at the bottom of the figure. Each of the data items is evaluated relative to the degree to which it satisfies its arguments. The network can be read as a rule as follows: "IF Data 1 satisfies the argument Data 1 arg. AND Data 2 satisfies the argument Data 2 arg. OR Data 3 satisfies the argument Data 3 arg. THEN the assertion is true." The degree to which the assertion is true is a function of the degree(s) to which the individual data satisfy their arguments and the types and arrangements of the logical nodes used within the network

NetWeaver allows simple or calculated data links to take fuzzy arguments to determine a data value's membership in a fuzzy set. In order for fuzzy set membership to be propagated through a knowledgebase, the definitions of the conventional logical operators OR, AND, NOT, and XOR have been extended to handle measures of fuzzy-set membership (Table 1). The SOR node object is unique to Net-Weaver, and we describe its operation in a later example.

As previously mentioned, models in NetWeaver are based on dependency networks which are graphical depictions of rules (Fig. 1). At the bottom of a dependency network are data links (e.g., Data 1, Data 2), which are used to hold, fetch, or modify raw data. There are two types of data links; simple and calculated. Simple data links fetch and hold data from various sources (databases, GIS map layers, direct data input, environmental variables, and other sources). Calculated data links modify data (e.g., calculate an ecological index or a mathematical relation from raw data) through networks of calculation nodes chosen from a toolbox of arithmetic,

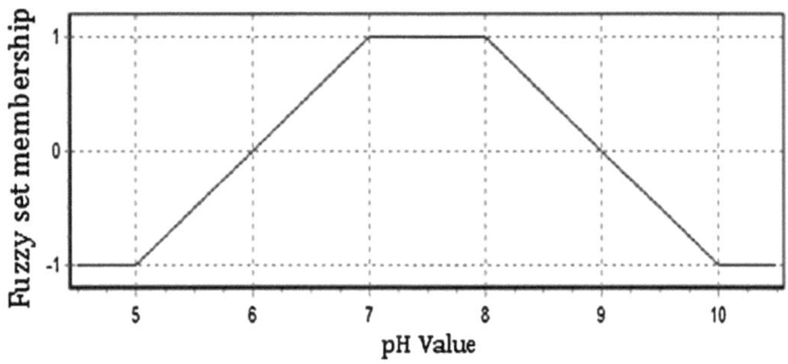

Fig. 2 A fuzzy argument used for interpreting the pH value of a stream. The fuzzy membership is shown on the Y axis with −1 indicating no fuzzy set membership (i.e., False) and 1 indicating complete membership in the fuzzy set (i.e., True). For this example, pH values between 7 and 8 fully satisfy the argument and indicate that the pH is indicative of a healthy stream. pH values less than 5 and greater than 10 are unacceptable pH values for a healthy stream

trigonometric, selection, summation, and other tools. Both types of data link are visually represented as a square object in a dependency network.

To provide a "trueness" level that can be used in a dependency network, the data within a data link are compared to an "argument." Arguments can be reference conditions, ecological thresholds, ecological index set points, or other types of indicator measures (e.g., single values or ranges of pH values, ranges of water-temperature values). NetWeaver provides two types of arguments, the standard argument and the fuzzy argument. The standard argument compares data values against an argument to return a TRUE or FALSE value (or undetermined when data are absent). An example of a standard argument is presence (TRUE) or absence (FALSE) of a particular species. The fuzzy argument compares the data values against a fuzzy set membership function that returns a level of trueness based on the degree of membership in the fuzzy set. In NetWeaver, fuzzy set membership is measured on a scale of −1 (no membership in the fuzzy set TRUE, which is equivalent to 100 % FALSE), to 0 (UNDETERMINED in the case of no data, or if there are data provided, it represents 50 % membership in the fuzzy set TRUE), to 1 (complete membership in the fuzzy set, which is equivalent to 100 % or completely TRUE). There are four break points provided to define a fuzzy argument within a data link, each of which can be defined to be TRUE, UNDETERMINED, or FALSE. An example of a fuzzy argument is the range of pH that is ideal to support aquatic organisms (Fig. 2).

3 Why Use NetWeaver Knowledgebases

Knowledge-based reasoning is a general modeling methodology in which phenomena are described in terms of abstract entities and their logical relations to one another (Holsapple and Whinston 1996). There are two basic reasons for using knowledge-based reasoning:

- The entities or relations involved in the problem to be solved are inherently abstract, so that mathematical models of the problem are difficult or even impossible to formulate.
- A mathematical solution is possible in principle, but current knowledge is too imprecise to formulate an accurate mathematical model.

Both cases are common. The first case naturally arises when the nature of the problem involves relatively abstract entities. These problems may simply be easier to solve with logic. The second case arises very frequently, particularly when dealing with ecosystems, because there are an almost unlimited number of relations of potential interest. Agencies, academia, and others have developed numerous mathematical models to describe some of the important relations of interest to ecosystem management, but many relations have not been studied in sufficient detail to provide generally applicable mathematical models. However, there is often a wealth of human experience in these same institutions that can be drawn upon to develop useful, more qualitative, knowledge-based models to guide decision making.

Another valuable aspect of knowledge-based reasoning that makes it ideal for use in environmental assessment is that such systems can provide clear reasoning with incomplete information. The NetWeaver engine provides partial evaluations of system states and processes based on available information, and provides useful information about the influence of missing data, which can be used to improve the logical completeness of an assessment.

In its most basic form, a NetWeaver knowledgebase is a collection of dependency networks. It can also include such things as supporting documentation and hyperlinks. A NetWeaver knowledgebase represents relations among concerns, system states and processes, and data requirements. Uses of dependency networks include:

- Evaluation of the truth value of assertions about system states and processes, given existing data.
- Identification of data requirements for an analysis.
- Ranking of missing data in order of relative importance to the analysis.

One of the virtues of a dependency-network representation is that a single knowledgebase may incorporate a very wide variety of topics. This is particularly valuable in the context of ecological assessments in which topics of interest might include, for example, many different topics and subtopics related to terrestrial vegetation and wildlife habitat conditions, native fish population status, available recreation opportunities, water and air quality conditions, visual and aesthetic concerns, and commercial concerns or opportunities. The number of topics and the interrelations that can be represented in a knowledgebase is only limited by the state of knowledge held by SMEs, and by a computer's dynamic memory.

4 Evolving Knowledgebases

NetWeaver is a rigorous, object-oriented, knowledgebase development system. One of the more practical implications of object-oriented knowledgebases is that it is very easy to start with simple knowledge representations and gradually to evolve them into large, complex systems because they are extremely modular. A basic modeling principle that has motivated development and application of object-oriented technology, in general, is well captured by Gall (1978):

> A complex system that works is invariably found to have evolved from a simple system that worked… A complex system designed from scratch never works and cannot be patched up to make it work. You have to start over, beginning with a working simple system.

NetWeaver is designed to build a knowledgebase in an incremental and evolutionary fashion. The best possible advice we can give to the novice user is to avoid designing large, complex systems at the outset, and instead, start with a simple knowledgebase, built from a small number of dependency networks, and gradually evolve this simple representation into a more complex representation of the problem.

As another practical matter, we have found that it is almost always best to start at the top with the highest-level (primary) dependency networks that apply to the problem domain, and develop the structure downward. To get started, create and document at least a few of the primary dependency networks. It is not necessary to identify an exhaustive list of primary networks at the outset. Because of the modular structure of NetWeaver knowledgebases, new dependency networks can easily be added later in development without upsetting overall knowledgebase structure.

For each primary network in the knowledgebase, create antecedents. Unless it is an unusually simple knowledgebase, there will usually be at least one or two levels of antecedents before a chain of dependencies terminates in a data link.

In a completed knowledgebase, you will normally want to be sure that each chain of dependencies terminates in a data link. However, while a knowledgebase is under development, it is always possible to evaluate a network object, regardless of how complete the network structure is.

As a knowledgebase structure evolves, you will probably find occasion to use existing antecedents (both dependency networks and data links). Multiple occurrences of dependency networks and data links in a single knowledgebase are not a problem because NetWeaver objects are reusable. In fact, the presence of a dependency network in two or more other networks within the same knowledgebase is an important mechanism by which networks are interrelated through shared antecedents.

5 Applications

NetWeaver has been used to develop a broad array of knowledge-based models, many of which are detailed in this book and in Wikipedia (http://en.wikipedia.org/wiki/Ecosystem_Management_Decision_Support). For purposes of illustration, we refer to a recent effort that sought to characterize watershed conditions within the Delaware Water Gap National Recreation Area (DEWA) and Upper Delaware Scenic and Recreational River (UPDE) (see Mahan et al. 2011 for a complete description of this model).

The focus of this modeling effort was on natural resources at these two park units; however, our assessment was conducted at the watershed scale in each park. The assessment was developed to assist superintendents and natural resource managers with: (1) strategic planning, (2) general management planning, (3) park reporting on land health goals, and (4) overall natural resource management and conservation.

As a positive consequence of employing a NetWeaver modeling approach for this systematic natural condition assessment, we delivered not only the required final report complete with tables and maps, but also an operational system that park managers can revisit as new data become available. Periodic re-running of the delivered NetWeaver model will allow managers to generate new reports to assess the new conditions and to identify any trends.

All data used in this assessment were compiled from relevant reports, scientific literature, and data files, and were initially managed using an online Wiki tool. We used these data and information available in the scientific literature to develop thresholds for our overall natural resource assessment model. We also used a variety of GIS-based analytical models to synthesize landscape data and develop indices of landscape condition across the two parks. These landscape condition indices included input data on impervious surfaces, forest fragmentation, and land use within watersheds at both parks.

The overall assessment examined a variety of aquatic and terrestrial ecosystem components and their interactions. Components included: chemical and physical elements of water quality, biologic elements of water quality, and forest landscape condition elements. For the water quality chemical and physical elements, we used a water quality index to assess water quality in each watershed. For the water quality biologic elements we used the Ephemeroptera, Plecoptera, Tricoptera (EPT) and the Hilsenhoff indices. Finally, for the landscape forest condition elements of our model, we used percentage area in forest, and percentage area with impervious surfaces measures. In addition, the DEWA/UPDE model contains natural resource elements that were not included in the overall assessment per se, but may be useful for management of park resources. For instance, our model included information on the number of stream crossings, dams, road miles, and rare species per watershed, and in each park.

In most cases, data did not come from single sources. For many elements, the data sources were varied. For example, pH values for any given watershed could

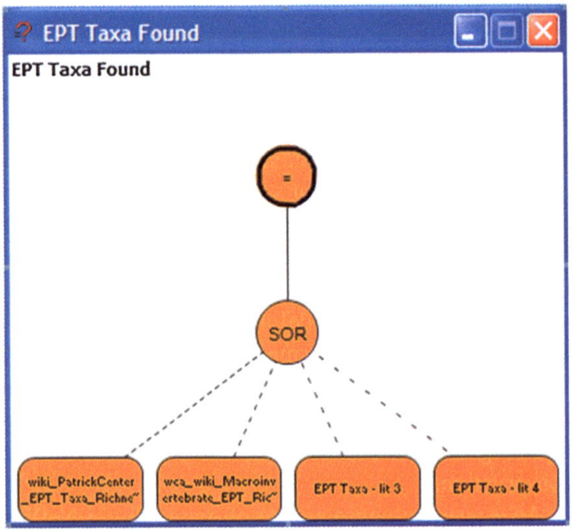

Fig. 3 Use of a sequential OR (SOR) node for the preferential selection of data from multiple data sources

have come from a handful of published reports, National Park Service (NPS) curated databases, or from gauge-station data of the US Geological Survey. NetWeaver facilitated the prioritization of these disparate data sources for a watershed based on the available data for a given watershed. The ability to aggregate data sources greatly enhanced the coverage of the analyses.

In some cases, multiple analyses were performed using competing analytical measures to observe differences in results, such as when using EPT or Hilsenhoff index when evaluating water quality, with respect to available aquatic insect taxa (US EPA 2002). Where data were sufficient, the results of the competing methods could be compared. Where data were lacking for one or the other method, the results from the method with sufficient data were used to represent conditions, according to NPS preference.

The logical nodes that are available for use within NetWeaver are shown in Table 1. In addition to purely logical operators such as AND, OR, and NOT, there are other operators that can be useful when dealing with multiple sources of data that vary in terms of desirability, accuracy, quality, and relevance. In the DEWA/UPDE model, the sequential OR (SOR) node was used frequently. This node (Fig. 3, EPT taxa found) provides the modeler with the ability to specify the order in which the model will use data, and in so doing, consistently apply the best available data to the model. In this example, the data for extant EPT taxa within a given watershed could be found in at least four reports. Some of these reports (and the data within) were preferred over others, often based on the originating agency of the report, report recency, and other factors. In Fig. 3, we show how a SOR node can be used to ensure that the most preferred data source is used. The SOR node always ensures that the left-most source of extant data is used. If there are no data present, the SOR node seeks its value from the next data source to the right. In the case of the DEWA/UPDE watershed model, any EPT value from any of the four sources was evaluated the same. However, cases arise whereby a modeler may wish to interpret less reliable data in a more conservative manner than would be applied to more reliable data. This can be easily implemented by attaching a data link or dependency network to the SOR node that is designed to evaluate these less reliable data sources.

6 Conclusions

NetWeaver software provides graphic tools for constructing executable dependency networks that permit both forward- and backward-chained reasoning. Because the inference engine is integrated, networks can be evaluated in real-time with nodes changing color to indicate their changing "trueness" levels. This ability to peer into the logical workings of a knowledge network greatly optimizes the knowledge engineering process by:

- Providing the ability to run and evaluate freshly elicited knowledge in the presence of the domain expert(s).
- Enabling the knowledge engineer to trace the logic structure from data to conclusions.
- Allowing the knowledge engineer to quickly identify and edit errors and inconsistencies in the logic.
- Providing consistent analyses across a landscape.
- Providing intelligent prioritization of data and results.
- Allowing competing analytical methods to be employed simultaneously.

References

EPA US (2002) Methods for measuring the acute toxicityo of effluents and receiving waters to freshwater and marine organisms. Document no. EPA-821-R-02-012, Washington

Gall J (1978) Systemantics: how systems work and especially how they fail. Pocket books, New York, 158 pp

Holsapple C , Whinston A (1996) Decision support systems: a knowledgebased approach. West Publishing Co., Eagan, 713 pp

Mahan CG, Miller BJ, Saunders MC, Young JA (2011) Assessment of natural resources and watershed conditions for Delaware Water Gap National Recreation Area and Upper Delaware Scenic and Recreational River. U.S.D.I National Park Service, Natural Resource Report NPS/NER/NRR 2011/429–NPS 620/108557; 647/108557, July 2011, 90 pp

Negoita CV (1985) Expert systems and fuzzy systems. Benjamin Cummings Pub. Co., San francisco, 190 pp

Parsaye K, Chignell M (1988) Expert systems for experts. Wiley, New York, 462 pp

Saunders M C, Miller BJ (1997) A graphical tool for knowledge engineers designing natural resource management software: NetWeaver™. Proc ACSM/ASPRS Ann Conv Res Technol 4:380–389

Saunders MC, Sullivan TJ, Nash BL, Tonnessen KA, Miller BJ (2005) A knowledgebased approach for classifying lake water chemistry. Knowl Based Syst 18(1): 47–54

Criterium DecisionPlus

Philip J. Murphy

Abstract Multi-criteria decision analysis (MCDA) supports decision-making when multiple conflicting criteria must be considered. Whether a decision involves choosing a single alternative (e.g., where to site a water treatment plant) or the prioritization of many alternatives (e.g., on what subset of stream reaches should restoration be focused), MCDA provides a methodology to structure and analyze the decision process. An MCDA engine has been integrated in the EMDS since 2002. As part of the EMDS workflow, MCDA models created using the Criterium DecisionPlus (CDP) authoring tool can be run against map features in a study area, to generate analysis graphs, tables and maps showing the decision model outputs. This chapter introduces MCDA and discusses considerations both when designing MCDA models using CDP, and running them in the EMDS.

Keywords CDP · MCDA · Evaluation · Prioritization · Decisions · AHP · SMART

1 Multi-Criteria Decision Analysis and EMDS

In the first section of this chapter, I provide a brief introduction to multi-criteria decision analysis (MCDA). In the second section, I describe two popular approaches to MCDA and how the MCDA-modeling tool, CDP, implements them, and what the decision maker can learn from them. In the subsequent section, I describe how CDP models are used in EMDS and some considerations for their design. In the final section, I point the reader to later chapters of this book, where examples of the use of MCDA in EMDS are discussed.

P. J. Murphy (✉)
InfoHarvest Incorporated, 25155, Seattle, WA 98165, USA
e-mail: philip.murphy@infoharvest.com

K. M. Reynolds et al. (eds.), *Making Transparent Environmental Management Decisions*, Environmental Science and Engineering,
DOI: 10.1007/978-3-642-32000-2_3,

1.1 Multi-Criteria Decision Analysis

Multi-criteria decision analysis is concerned with decisions amongst alternatives when decision makers have multiple, often competing criteria, that they require the alternatives to address.

One of the earliest descriptions of multi-criteria decision analysis (MCDA) was provided in 1771 in a letter by Ben Franklin (Franklin 1956). A friend had asked him how to decide whether a course of action should be pursued or not. Franklin described a structured decision process he himself used. He would write pros and cons in two columns, estimate their respective weights and eliminate pros and cons that seemed equal, so that he could see "where the balance lies."

Modern MCDA generalizes and formalizes this approach (Keeny and Raiffa 1976), providing an axiomatic underpinning for its assumptions and expressing its methods as either mathematical algorithms or decision rules (Malczewski 1999). Given that decision making is a practical and ubiquitous task of bureaucracies and enterprises everywhere, various approaches that more or less encompass Raiffa and Keeney's academic MCDA framework have been popularized—for example, Even Swaps (Hammond et al. 1998), Smart Choices (Hammond et al. 1999), Value Focused Thinking (Keeney 1992), Decision by Advantages (Suhr 1999) and more loosely, The Balanced Score Card (Kaplan and Norton 1996).

1.2 MCDA and Spatial Decision Support Systems

MCDA techniques are very useful for spatial decision support, whether choosing one location amongst a group of alternative sites or simultaneously prioritizing many geographic features—e.g., lakes, rivers, roads—in an area. Geographical Information Systems (GIS) appeared in the late 1960s (Coppock and Rhind 1991) and digitally implemented the map overlay approach to spatial decision making pioneered by Ian McHarg (Longley et al. 2001). Map layers representing unique attributes co-existing in a common space were developed, one for each decision criterion. A weight was associated with each criterion layer, and the value of each layer at a given point is multiplied by that criterion's weight. Finally, those weighted values are summed to produce a new layer, whose values can be interpreted as decision scores, and the new layer becomes a decision surface (ESRI 1995; Malczewski 1999). Decision surfaces can be used to select the most suitable areas to locate an alternative, or prioritize a suite of management actions. MCDA is highly useful for formalizing any GIS process that recognizes the centrality of competing objectives, and the need to weigh their relative importance in spatial decision making.

Over the last decade, MCDA has been applied to hundreds of environmental decisions (Huang et al. 2011). It is well suited to ecological management where multiple criteria associated with environmental, socioeconomic, or technical factors have to be considered. The Ecosystem Management Decision Support (EMDS) system provides both a well-defined decision workflow ("An Overview of the Ecosystem Management Decision-Support System") and a spatial decision-support development environment that enables analysts and land managers to create and seamlessly apply MCDA models to mapped features of any system.

1.3 Two Core Uses of MCDA in EMDS

The popular book "Blink" (Gladwell 2005) notwithstanding, decision making is a social *process*. A decision making process is formally described as a workflow of activities that may lead to an expected outcome. There are many possible workflows for decision making, but most are variations on a core workflow comprised of the following sequenced steps: *Issue Articulation* ≫ *Decision Process Design* ≫ *Condition Assessment* ≫ *Alternative Design* ≫ *Choice* (SDS Consortium 2008).

1.3.1 MCDA for Completing the Condition Assessment Step

In "An Overview of the Ecosystem Management Decision-Support System" and "NetWeaver", the use of knowledge or logic modeling in EMDS to generate estimates of the strength of evidence in support of high functioning processes and primary components of an ecosystem was described. For the purpose of decision making, a synthesis of the current conditions of the ecosystem is needed to implement the *condition assessment* step of the workflow. Then, if a change to the ecosystem can be forecast to modify the underlying input data of the current analysis—captured in EMDS as scenarios (see "An Overview of the Ecosystem Management Decision-Support System")—the same evaluation model can be re-applied to the changed state of the system. This makes it possible to compare the current and alternative scenario spatial evaluation outputs to ascertain, feature by feature, across the study area, where the proposed change would improve or degrade the current ecosystem.

EMDS helps users articulate reasoning about landscape features—e.g., stands of trees, watersheds, river reaches—to (typically) arrive at estimates of the strength of evidence, or truth level, for assertions about the condition of a process or subsystem of the ecology. To create single or multiple estimators, the user must still (a) interpret each primary hypothesis and the strength of evidence supporting it, in terms of a quantitative metric, and (b) weigh and combine the interpreted metrics into a single number for each feature. In EMDS, these synthesis steps can be implemented either using the logic modeling of NetWeaver ("NetWeaver") or through explicit multi-criteria decision analysis, in which the features are the alternatives, the primary hypotheses become criteria, and their strength of evidence serves as ratings of the feature against those criteria. The latter approach is most appropriate when the relative weighting of ecosystem components does not emerge from the modeling of underlying processes of the ecosystem, but are imposed as informed judgments of the relative importance of the components in a particular decision context—the 'evaluation models' of the Steinitz planning workflow (Steinitz 2012).

1.3.2 MCDA for Supporting the Choice Step

More typically, in the current EMDS system, MCDA is employed to partially implement the *choice* step of decision workflows. Rather than evaluating individual actions that change the state of the system, decision makers instead prioritize subcomponents of the ecosystem for restoration work, conservation, or preservation. To do this, they combine the estimates of the current condition of subcomponents of ecosystem with independent estimates of the feasibility of conducting restoration work or applying preservation ordinances, and the likely efficacy of any work or ordinance on those subcomponents. The relative weights decision makers apply to these competing needs, feasibilities, and efficacies, when looking for the highest priority subcomponents, involve expert judgments that reflect constraints and aspirations that do not emerge from the ecosystem modeling directly, making MCDA a suitable approach to support this prioritization step.

1.4 The History of MCDA in EMDS

In 2002, multi-criteria decision analysis was incorporated into version 3 of EMDS by integrating the Priority Analyst—a computation engine and visualization component that executes and displays MCDA models authored in Criterium DecisionPlus. Criterium DecisionPlus (CDP) is a Microsoft Windows-based tool first released in 1993 by Sygenex Corporation, then acquired in 1995, and currently supported by InfoHarvest Incorporated of Seattle, WA.

At the time, CDP was selected over other available Microsoft Windows™-based MCDA programs because it offered utility functions for rating alternatives against criteria (see below), pairwise preference elicitation, and uncertainty analysis. Most importantly, it had an associated MCDA engine ready that could be integrated into other Window-based programs.

The Priority Analyst becomes available to EMDS users after they have executed a NetWeaver logic model ("NetWeaver") in the analysis step of the EMDS workflow ("An Overview of the Ecosystem Management Decision-Support System"). The outputs from applying the MCDA model to map features in the study area are tables and maps showing the decision scores of those map features.

2 Multi-Criteria Decision Analysis and CDP

This section describes the implementation of MCDA in the software application Criterium DecisionPlus, and decision and analysis outputs common to MCDA applications.

2.1 MCDA Decision Hierarchies in CDP

2.1.1 A Generic Multi-Criteria Decision Model

Figure 1 represents a generic multi-criteria decision model captured in Criterium DecisionPlus. How an alternative meets the overall goal of a decision can be estimated by how well each alternative rates against a set of weighted criteria. Criteria can be successively broken down into sub-criteria, as needed, until a level of granularity is reached at which the decision makers feel comfortable rating alternatives directly against these lowest criteria, based on available information and experience. Those criteria in the models against which alternatives are directly rated are called *lowest criteria*. The alternatives to be considered in making this decision are shown at right, with links connecting them to the lowest criteria.

In general, sub-criteria require more detailed knowledge of the decision problem. In CDP, a decision hierarchy is graphically represented, which renders decisions more readily comprehensible to decision makers and stakeholders. Graphical representation also reinforces discipline in problem structuring, and supports processes that breaks decisions into manageable components.

Alternatives are rated against the lowest criteria (the sub-criteria in Fig. 1 above). Assigning weights to the various criteria and sub-criteria completes the decision model. At that point, mathematical calculation can be performed to determine the preference of the alternatives using algorithms specific to the rating methods. These are described in the next section.

The design tasks that must be completed by an analyst creating an MCDA model are: (1) create the structure of the decision hierarchy, (2) decide how weights will be elicited, (3) decide how preferences will be aggregated, and (4) define the metrics for each lowest criterion against which the alternatives will be weighted.

2.1.2 Designing the Structure of an MCDA Decision Hierarchy

Almost any decision making formulation can be represented using a decision hierarchy, including Goal—Objective—Criteria, Strengths—Weaknesses—Opportunities—Threats (SWOT), and threat-based risk evaluation.

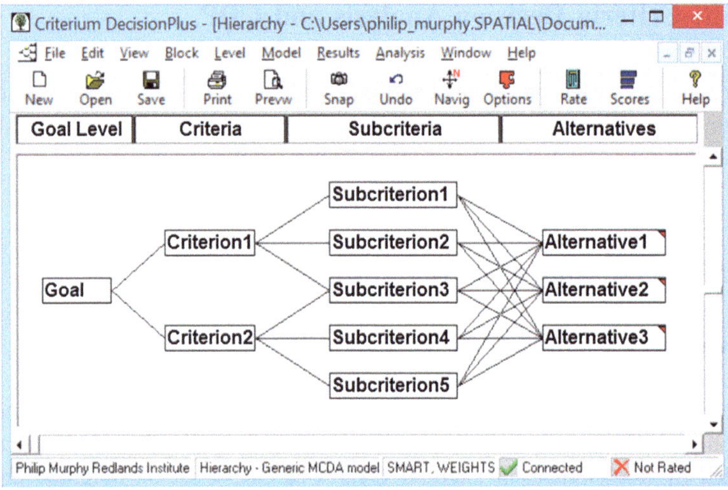

Fig. 1 An example of a typical decision hierarchy that is laid out (arbitrarily) from *left* to *right*. Notice the overall goal node on the *left*, and then primary criteria that are further refined by sub-criteria. The rightmost sub-criteria, the lowest criteria, are those against which alternatives are directly evaluated

The Goal—Objectives—Criteria approach is one that many have found useful. In this formulation, the model designer (or analyst) uses the brainstorming capability within CDP to ask a decision-making group to articulate the goal of their decision opportunity—e.g. prioritize a forest area for action (Reynolds and Peets 2001), or select the best programmatic alternative in a NEPA process (GSNM 2010). The designer then asks what objective(s) does the group hope to achieve with the decision, and which criteria can be used to evaluate how well alternatives achieve objectives? In the best brainstorming process, concrete metrics (quantitative or qualitative) are identified for those criteria. This last step of developing metrics can be left until the later step at which alternatives are rated, but such omission often contributes to vagueness when defining criteria, which must be resolved later in the ratings step.

There are at least two key points about structuring a decision hierarchy that are important to those designing them. First, a sub-criterion may contribute to two or more criteria, and the result is not double counting provided that the criteria represent independent contributions to the value of an alternative within the decision model. For instance, in a model of the value that infrastructure components contribute to the Nation, hydro-electric power generated may contribute to both energy independence and clean air, two independent criteria. Second, some criteria may have more levels of sub-criteria below them than others. Consequently, not all lowest criteria reside together at the lowest level of a hierarchy.

Figure 2 shows a decision hierarchy for prioritizing watersheds in the Chewaucan River Basin in Oregon. This was part of a multi-scale analysis that

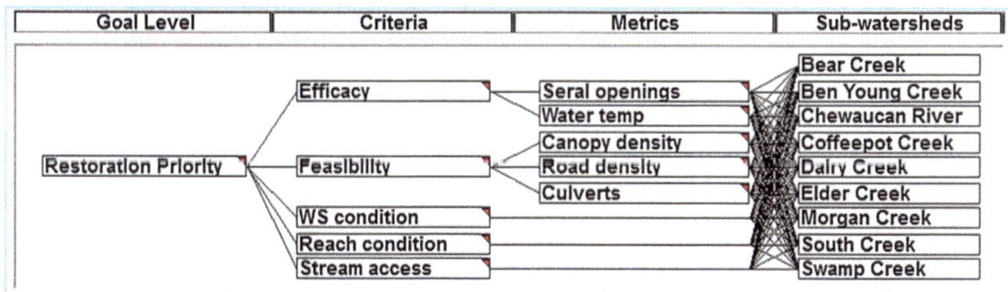

Fig. 2 A decision hierarchy for prioritizing watersheds in the Chewaucan Basin for riverine habitat restoration (Reynolds and Peets 2001)

prioritized both watersheds and stream reaches for protection and restoration (Reynolds and Peets 2001). An MCDA model was developed for each scale. The watershed prioritization model in Fig. 2 featured top-level criteria of efficacy, feasibility, water shed condition, reach condition and stream access. From the figure, it is apparent that some of these criteria are lowest criteria, some are not. The decision score for each watershed was interpreted as its priority for restoration actions. This MCDA model will be used as an example throughout this chapter.

2.2 *Two Standard Methods for Multi-Criteria Decision Analysis*

CDP implements two decision-making methodologies commonly used by corporate and governmental decision-making bodies: the Analytic Hierarchy Process (AHP, Saaty 1992a) and an implementation of Multi-attribute Utility Theory called Simple Multi-Attribute Rating Technique (SMART, Von Winterfelt and Edwards 1986). Both of these methodologies use decision hierarchies with weighted, linear aggregation; the key difference lies in the *alternatives rating techniques employed*—the ways in which the ratings of alternatives on the lowest criteria scales are converted into priority values (Fig. 3). In CDP, the choice of how criteria weights will be elicited—the hierarchy preference technique—is independent of the rating technique. In this chapter, I refer to the AHP and SMART rating techniques; however, in reality, these are two complete decision making methodologies with significant epistemological differences between them.

2.2.1 The Analytic Hierarchy Process

The Analytic Hierarchy Process (AHP) was developed in the 1970s by Dr. Thomas L. Saaty at the University of Pennsylvania, Wharton School of Business, and is a method of organizing information and judgments to select a preferred alternative (Saaty 1992b; Dyer and Forman 1991). To make any decision, decision criteria are

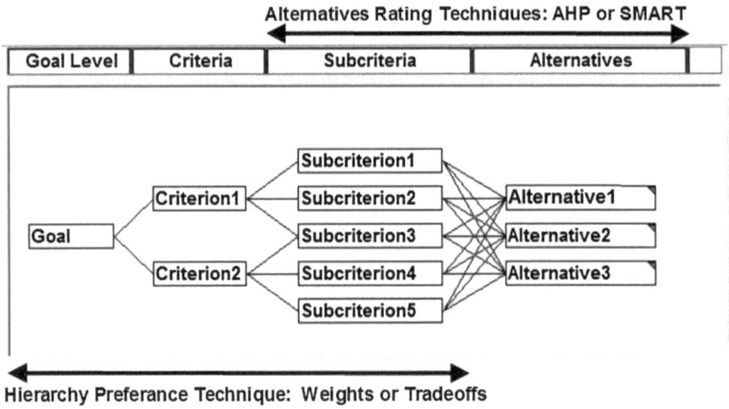

Fig. 3 Hierarchy Preference Techniques are used to set the preferences of all criteria in a decision hierarchy; Alternative Ratings Techniques are used to rate alternatives against the lowest criteria (attributes) in the decision hierarchy. They can be chosen independently of each other

grouped and then prioritized. In AHP, the most important criteria are grouped at the highest level, closest to the single overall goal of the model, with sub-criteria grouped below, which further define their parent criteria.

Graphically in the AHP, a decision problem is constructed as a layered network diagram starting with the goal, then showing at the next lower level, the first level of criteria to be considered, followed by further lower layers of sub-criteria as warranted, and finally the alternatives. Such a diagram represents a decision hierarchy of elements (goal, criteria and alternatives) that are to be considered in selecting the preferred solution(s). It may be displayed vertically or horizontally as in Fig. 1.

In the classic "distributed" AHP mode (see next section), the user rates alternatives either using the pairwise comparison method or directly on a user defined scale, both of which calculate the priority value of an alternative against a criterion as a relative value, with the priority values for all alternatives against a criterion summing to 1 for that criterion.

2.2.2 Simple Multi-Attribute Rating Technique

The Simple Multi-attribute Rating Technique (SMART) is another technique implemented in CDP. It originates from the work done in Multi-Attribute Utility Theory (MAUT) by Edwards in 1977 (Von Winterfelt and Edwards 1986).

When using SMART in decision making, the decision problem is broken down into objectives with attributes, and single-attribute evaluations of the alternatives are constructed by means of value measurements. As for the AHP, a value tree structure is created to assist in defining the problem. The lowest criteria are referred to as attributes. Values are determined for each alternative against each attribute. Finally, weighted linear aggregation of the model provides results facilitating comparison of the alternatives.

SMART provides a straightforward way of employing MAUT techniques. In it, a direct rating procedure is applied to assess alternatives against single attribute values, and the resulting priorities are aggregated using weighted sums to calculate the preference for each alternative. The rating procedure uses nonlinear monotonic functions in assigning priority values to the attributes.

Many other general methods have been developed for operationalizing the MCDA process (Triantaphyllou 2000, Belton and Stewart 2002). Each has advantages and disadvantages in terms of theory, ease of use, fidelity to the decision makers' context, and susceptibility to misuse. The two methods that CDP implemented are widely used (Saaty and Forman 1992; Haung et al. 2011), are most directly connected to the underlying theories of decision making, and have the simplest underlying mathematics, which is important for end-user verification of a systematic and logically derived decision-making process.

2.3 Hierarchy Preference Techniques

No matter how complex the hierarchy structure, the basic algorithm in multi-criteria decision analysis is to multiply the priority value of an alternative against each lowest criterion by the relative importance of that criterion. The resulting products are summed across all lowest criteria to provide an overall decision score for that alternative. The decision score measures how well any alternative satisfies the overall decision model.

For hierarchical models with intermediate layers of criteria, the user assigns preferences to all sub-criteria that contribute to a parent criterion (the one nearest the goal node). These preferences are converted, according to the decision rules of the specific preference technique, to priority weights for each subcriterion. The priority weights of that set of contributing criteria add up to one. This process is repeated for all parent criteria with sub-criteria. The priority weight of the parent criterion is then applied to the priority weights of its subcriteria through multiplication, onwards from the goal (which has nominal weight of one), down through all sub-criteria sets until the lowest criteria are reached. The priority weight of a lowest criterion is then the sum of the weights obtained by repeated multiplication of the weights along all pathways that stretch from the goal node at the top to that lowest criterion (see Fig. 4a). For a concise algebraic description, see Golden et al. (1989). One important consequence of this approach, in which the priority value of the parent criterion is "distributed" (Forman 1993) amongst its sub-criteria, is that, when the derived priorities of all the lowest criteria are summed, the result is 1. In other words, the priority of the goal is fully distributed amongst the lowest criteria of the decision hierarchy, regardless of how many levels of intervening criteria are in the model.

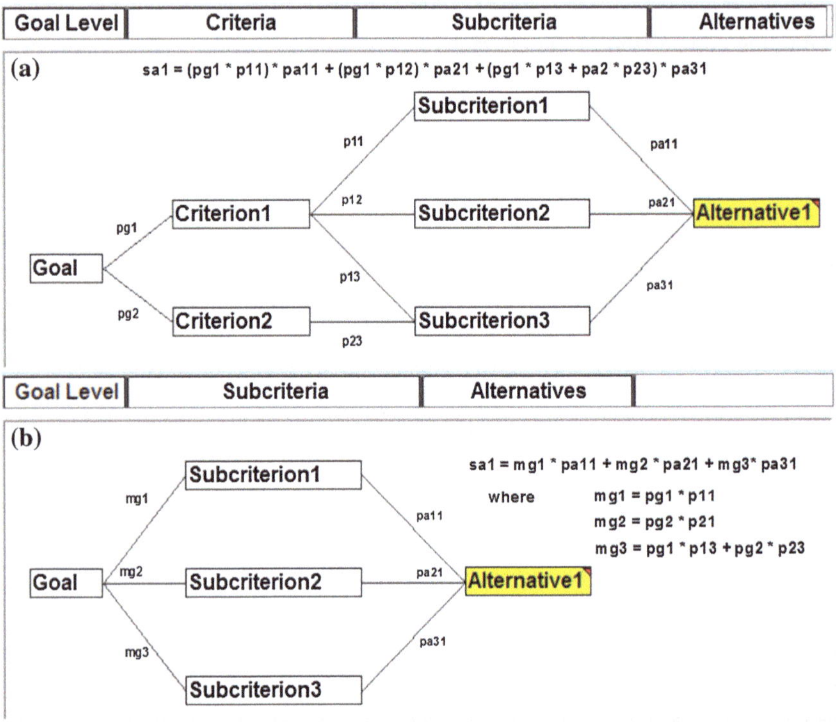

Fig. 4 **a** The decision score sa1 for Alternative1 in a multilevel hierarchy is obtained by traversing all paths from the goal to the alternative, calculating the products of the priority weights along the paths, then multiplying that product by the priority of the alternatives (pa11, pa21, pa31) and finally summing those products over all the pathways. **b** This is equivalent to a two level model where the model weights of the lowest criteria (mg1, mg2 and mg3) are calculated from the original priority weights by summing their products over corresponding pathways

The derived priority weights of the lowest criteria are referred to as model weights—as any complex decision hierarchy that uses linear aggregation can be replaced by a two-level—goal and lowest criteria—model with those model weights without changing the decision score of the alternatives (see Fig. 4b).

2.3.1 Hierarchy Preference Elicitation Techniques

CDP supports two decision analysis techniques to establish the relative importance of the lowest criteria or attributes.

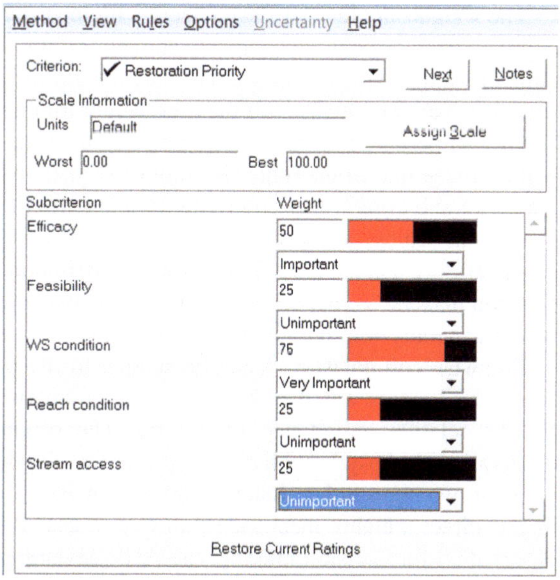

Fig. 5 Graphical user interface for eliciting direct weights for the objectives contributing to the goal of "Restoration Priority" in the watersheds prioritization model in the Chewaucan Basin study. The figure shows three weights input mechanisms—numeric, verbal and graphical

2.3.2 Hierarchy Preference Technique: Weights

In the weights preference technique, the user assigns the relative importance of each sub-criterion with respect to the other sub-criteria of a given parent criterion as a weight. Typically, a user would start at the Goal, and ask the decision makers to judge the relative importance of each of the criteria directly beneath the Goal, and then repeat this process for every criterion with sub-criteria in the model. Once complete, CDP multiplies these preferences down the structure of the hierarchy, so that each lowest criterion is assigned a relative importance—a model weight—with respect to all other lowest criteria. There are a number of methods available to elicit the weights of criteria.

Direct weights. In the direct weights elicitation method, decision analysts enter weights, one per sub-criterion of a parent criterion, directly on a scale of their design. The weights input mechanism may use any combination of numeric, verbal or graphical scales (Fig. 5). Numeric scales are fully characterized by minimum and maximum values and the orientation of the scale—whether higher values correspond with higher preference, or not. Verbal scales are comprised of an ordered set of text items, with each item being interpreted as having a numeric value, equally spaced from the minimum to the maximum of the numeric scale, or the reverse, depending on the desired orientation of the numeric scale. When using a graphical scale, the user sees a colored slider for each sub-criterion, where again the left of the slider corresponds to the minimum of the numeric scale, and the right to the maximum, or the reverse, depending on the desired orientation.

The weights are linearly normalized, with the priority weight, which will be the relative weight used in the weighted sum algorithm, of each sub-criterion being its numeric value divided by the sum of the numeric weights of all the sub-criteria. As such, these are ratio scales—if the user weight for one sub-criterion is twice that of another sub-criterion in the same rating set, then the priority of that sub-criterion will be twice that of the other. By dint of their derivation, the priority weights of the set of sub-criteria that contribute to a parent criterion add to one.

Custom scales can be created for both the numeric and verbal mechanisms. Designers are encouraged to choose an elicitation mechanism that suits their decision makers. Verbal scales should use language that the organization's experts would generally use, but keeping in mind the ratio nature of the scales.

Pairwise comparison. A key technique in the original form of the Analytic Hierarchy Process was the use of pairwise comparisons to elicit expert judgment for the relative importance of criteria. The origin of this approach was the observation that psychologically people are more adept at comparing the relative value of two items rather than assigning them each a value on an abstract scale (as in the direct weights mechanism above) (Saaty 1992b). To operationalize this observation for decision making, the AHP developed both a nine-point scale for pairwise comparison, and a mechanism for extracting a set of direct weights from the $n\ (n-1)$ unique pairwise estimates that result when there are n sub-criteria in a rating set. In addition, the theory developed an inconsistency index based on the individual pairwise comparisons that indicates raters' consistency throughout a set of pairwise comparisons (Saaty 1992b).

The nine-point scale ranges from 1 to 9. Comparing sub-criterion c1–c2, a value of 1 expresses the judgment that c1 and c2 are equally preferable. A value of 9 indicates that c1 is nine times more preferable than c2. If c1 is twice as preferable to c2, and c2 is three times as preferable as c3, then consistency in the method requires that c1 be six times more preferable than c3. Because the decision makers compare the criteria in pairs, their judgments may not be consistent—when comparing c1 against c3, they may directly judge c1 as five times more preferable to c3. In AHP, the inconsistency index is calculated from the $n\ (n-1)$ pairwise comparison matrix. The indexed value is then divided by the average inconsistency of randomly generated matrices of the same size, to arrive at an inconsistency ratio. When the inconsistency ratio is more than 10 %, it is considered advisable to reexamine the pairwise judgments to decrease inconsistent judgments before proceeding with the decision (see Golden [1998] for a concise description of the underlying mathematics, and Saaty (2007) for a summary of the case for the psychological underpinnings of pairwise comparisons).

When eliciting judgments from decision makers, a formal pairwise verbal scale is often used that corresponds to the 1–9 numeric values (Fig. 6). CDP then assigns the corresponding numbers and executes the principal eigen-value calculation that generates the n normalized priority values.

For the designer, when using pairwise comparisons, there are no explicit design choices to make. The pairwise comparisons method assumes that the sub-criteria

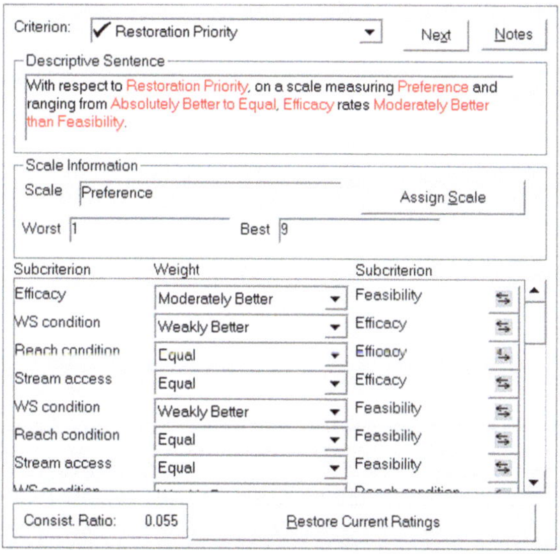

Fig. 6 Pairwise comparison of the criteria that contribute to the goal of "Restoration Priority" in the watersheds prioritization model in the Chewaucan Basin study

being evaluated will not have some sub-criteria that are orders of magnitude more important than others. If that seems likely, it would suggest that some sub-criteria are misplaced in the hierarchy.

2.3.3 Hierarchy Preference Technique: Direct Tradeoffs

Another way to establish the relative importance of the lowest criteria is by directly setting their relative tradeoffs.

In any difficult decision, some decision objectives always conflict. For example, consider the common desire to have the highest quality at the lowest price. The essence of the multi-criteria decision is the tradeoff between such competing objectives, and this is most directly measured by the numerical tradeoffs between the lowest criteria, which provide metrics for those objectives. For instance, a reasonable tradeoff between cost and schedule for a construction proposal might be $100,000 for 1 month. This means the user is willing to pay $100,000 to avoid a 1-month delay in a particular project. The user would proceed with pairwise tradeoff judgments, pair by pair, until all lowest criteria have been traded against each other. At that point, CDP can calculate the relative importance of all lowest criteria in the model, and the weights of the hierarchy that would support those tradeoffs. This direct tradeoffs technique, available in CDP since version 3.0 (1995), is similar to the way priorities of alternatives determine criteria weights in the distributive mode in the Analytic Network Process (Saaty 2006).

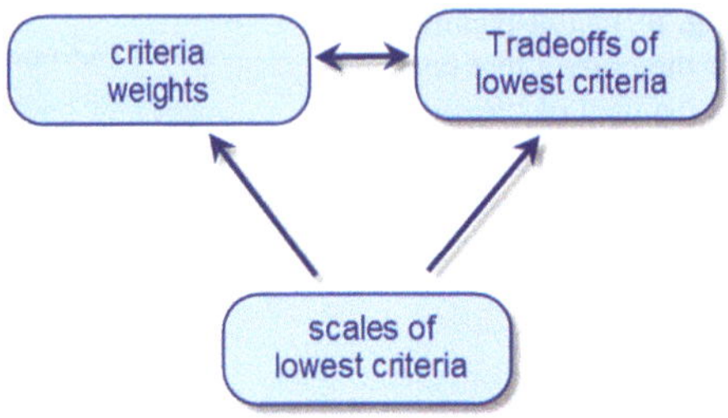

Fig. 7 For decision hierarchies where the goal and criteria form a pure tree structure, and the scales of the lowest criteria are known, then the weights and tradeoffs of lowest criteria are deterministically interrelated—full knowledge of one determines the other uniquely

2.3.4 Interchangeability Between Hierarchy Preference Techniques

Mathematically, if both the scales of the lowest criteria and the relative importance of the lowest criteria are known, the resulting tradeoffs of lowest criteria can be calculated (Fig. 7). Conversely, if the scales and the tradeoffs are known, it is possible in CDP, for hierarchies that are true tree structures (no criterion is a sub-criterion of more than one parent criterion) to calculate the relative importance of the lowest criteria (by uniquely determining the priority weights at every level of the hierarchy). Provided the scales of the lowest criteria are unchanged, the decision maker can switch between weights and tradeoffs preference techniques; for true tree structures, changing from one method to the other will not change the relative importance of the lowest criteria in the decision model.

2.3.5 Comparing Hierarchy Preference Elicitation Techniques

Numerical tradeoffs have the great advantage that their values can be compared to company or industry norms, and (with care) to published research. If adjustments to the tradeoffs entered by the decision makers need to be made, it is simpler to do so using the direct tradeoffs than the weights elicitation technique. In the latter case, the user must perform mental gymnastics to figure out what weight changes will generate the required changes in derived tradeoffs.

On the other hand, asking stakeholders to provide a direct tradeoff value can be intimidating. They may not understand what is meant by a tradeoff, or they may not be familiar enough with the decision space to hazard a good estimate, or both. Furthermore, if many of the lowest criteria are qualitative in nature, a tradeoff such as 1 unit of Reputation for 3 units of Quality may be meaningless or difficult to assess.

For hundreds of years, people have made decisions that involve directly weighing the importance of one criterion against another, and deciding on the alternative that performs best against the most important criteria (Franklin 1956; Keeny and Raifa 1976). The weights elicitation technique may thus provide a more intuitive approach than direct tradeoffs when many of the lowest criteria are qualitative in nature, or decision-makers are uncomfortable providing tradeoff judgments. Preferences established using weighting techniques can be explored using the tradeoff analysis discussed below. As decision makers come to understand tradeoff analysis, they are often more willing to switch to the direct tradeoff technique for establishing hierarchy preference.

2.4 Alternative Rating Techniques

Two methods for determining the priority values of alternatives against a lowest criterion (or attribute) are described below.

In 1993, CDP was the first commercial package to offer both SMART and AHP methodologies for rating alternatives. The author was able to verify the work of Kamenetzky (1982), and implement the relationship between weights, scales and tradeoffs so that a user can, within limits, change between SMART and AHP ratings methods on the fly. While the calculated decision scores for each alternative are very different under the two methods, we at InfoHarvest have found that, except under limited conditions, the rankings of alternatives were unchanged by such a switch.

2.4.1 Normalizing Ratings in MCDA Models

In order to aggregate the contributions of criteria with differing scales, the decision model must provide a method for handling differing ratings scales on an equal footing. For example, if a decision in the healthcare field involved criteria that used dollars and human lives as scales in the same model, the model needs to handle both very different scales consistently so that meaningful weights or tradeoffs can be established. In both AHP and SMART, the alternatives rating technique itself handles this through normalization, in which all scales are converted to a common internal scale that takes a value between 0 and 1.

2.4.2 Normalizing Alternatives in AHP

In AHP, when using a direct rating method to obtain the comparable cross criteria ratings or priorities, the user-provided ratings values under a lowest criterion are divided by the sum of the ratings of all the other alternatives against that criterion. This guarantees that no matter what the values or units of measurement of the original user ratings are, all priorities will fall between 0 and 1, the priorities of all alternatives sum to 1. A normalization method with these properties is called a *relative* normalization—the priority value of an alternative against a lowest criterion depends on the ratings of all the alternatives under that criterion. In the case where pairwise comparisons of ratings are used to prioritize alternatives under a lowest criterion, the derived direct weights already show this relative normalization. The relative normalization of ratings in AHP, coupled with the distributed nature of the weights of the lowest criteria, guarantee that the sum of the decision scores of all the alternatives themselves sum to 1—the entire preference of the model is *distributed* over the alternatives (Saaty 2007; Forman 1993). In CDP, the AHP alternatives rating technique utilizes this relative normalization technique.

This is a simple and effective method that allows synthesis across criteria whose scale units are completely different. The distributed nature of the preference scores means that the decision score of any one alternative in the model depends on the array of alternatives that are being considered. For instance, in an AHP model that utilizes a relative normalization, if a model is created for an initial set of alternatives, adding a new alternative later can change the relative scores of each alternative in the original set.

2.4.3 Normalizing Alternatives in SMART

SMART doesn't use a relative method for scaling user defined rating units to a priority scale from 0 to 1. Instead, decision analysts define their own method for doing this using a value function. A value function uses mathematical functions to transform ratings on the user's input scale to priorities on the common model scale. Much research has been done in determining the most appropriate value functions for a given decision problem (Von Winterfelt 1986). CDP provides three value functions for transforming user ratings to priorities for each attribute: a linear function, an exponential function, and a piecewise linear function. These three functions provide sufficiently broad choices for most decision making, although they are all monotonic, to the chagrin of fuzzy logic champions.

2.4.4 Rank Reversal of Alternatives

MCDA models whose alternatives are rated using the relative normalization technique described above can exhibit a controversial behavior called *rank reversal*. Basically, in such a decision model, if a new alternative is introduced, it affects the normalized decision scores of all alternatives. So even though the alternative introduced may itself not be high scoring, it may cause a shift in ranking of the previously highest scoring alternatives. Over the decades, this behavior has offended both decision theorists and engineers. A good overview of rank reversal is provided by Tversky (1969), and its occurrence in other MCDA approaches by Triantaphyllou (2000) and Wang and Triantaphyllou (2008). In more limited cases, the same phenomenon arises in SMART (Salo and Raimo 1997), where an alternative is introduced that requires a new minimum or maximum of an attribute scale, the priorities of all alternatives for that attribute will change, possibly leading to rank reversal.

In 1993, Forman introduced the *ideal mode* for the AHP, in which the relative normalization of alternatives described above was relaxed in favor of scales with appropriate units. In this mode, the MCDA decision model utilizes relative AHP weights for criteria, but linear SMART value functions for ratings. Foreman allowed that the original AHP with relative normalization of alternatives described above, often referred to as the *distributive* mode (of prioritization of lowest criteria to alternatives) may also be appropriate to some decision contexts. In particular, he suggests (Forman and Gass 2001) that the ideal mode should be used when there is no scarcity of alternatives, and the distributed model be used when there is. This is discussed further in the section on considerations when developing an MCDA model for use in EMDS.

2.4.5 The Decision Hierarchy Structure for SMART and AHP in CDP

The structure used to model decision problems in SMART is called a value tree, or objective hierarchy. The difference between a SMART value tree and an AHP hierarchy is that the value tree is a true tree structure, restricting each sub-criterion to be connected to only one higher level criterion. In AHP, a sub-criterion can connect to more than one higher level criterion. However, if the user changes the rating technique from AHP to SMART, the user is not forced to restructure the hierarchy as a true tree structure. The difference occurs at the attribute level with the normalization of alternatives under different attributes.

However, if the Direct Tradeoffs preference elicitation approach is used to set hierarchy preferences, the hierarchy must again be a true tree structure. Otherwise the paired tradeoff comparisons cannot be used to calculate the unique set of criteria priority weights that would reproduce those same values of the tradeoffs between lowest criteria.

2.5 *Results and Analysis*

When an MCDA model is complete—hierarchy designed, lowest criteria scales established, weights or tradeoffs elicited, alternatives rated against lowest criteria—then the decision scores for each alternative can be calculated. When sorted by diminishing decision scores, the alternatives are said to be prioritized. Figure 8 shows that in the Chewaucan River Basin study, the Bear Creek watershed received the highest priority, over three times that of the South Creek watershed, which received the lowest priority. For decision opportunities that require a single alternative to be chosen, the alternative with the highest decision score is usually the recommended alternative.

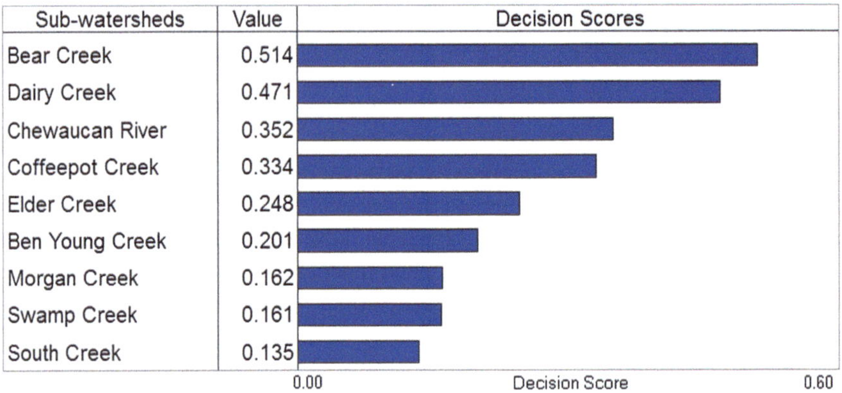

Fig. 8 Bar chart of decisions scores of the watersheds for the watersheds prioritization model in the Chewaucan Basin study

The "A" in MCDA is for analysis, and that is where MCDA tools provide their core worth—helping the user validate, as far as possible, the decision scores calculated for each alternative. Those scores depend on the weights, ratings, and structure of the decision model.

2.5.1 Analysis of Contributions by Criteria

The most straight-forward analysis is simply the decomposition of the aggregated preference values at any node in the model into the contributions by the criteria from individual levels in the models. In this analysis, the user selects any node in the decision hierarchy—goal, criterion, sub-criterion—and any level in a decision hierarchy, and observes how the aggregated preferences of the alternatives for that node are made up of the contributions from each criterion in the selected level.

For the Goal node, the aggregated preferences are the decision scores themselves. Figure 9 displays the contributions to the decision scores from the criteria in the Criteria level for the Upper Chewaucan River prioritization example. From the graph, it is evident that two watersheds with the highest priority are differentiated from the others due to the contribution they receive from the criterion for watershed condition (WS Condition). In the study, the strategic approach was taken that watersheds in good condition, as evaluated in a NetWeaver model, were given extra priority. The question for the decision makers is—do the rankings make sense? Is there a pair of alternatives well known to the decision makers whose relative rank is the opposite of what was expected? When perceived anomalies are noticed, drilling down through the levels can reveal whether weights, user ratings, or normalization are in need of correction, or perhaps that the initial feeling that the alternatives' rankings were reversed is dispelled.

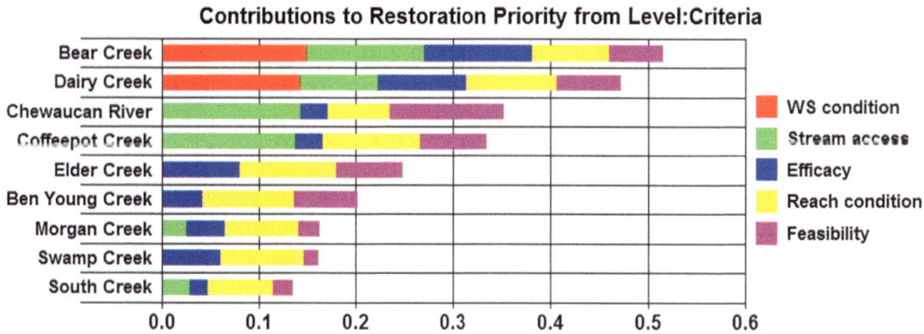

Fig. 9 Stacked *horizontal bar* chart show the decomposition of the decision score for each alternatives into the contributions from all criteria in the Criteria level for the watersheds prioritization model in the Chewaucan Basin study

Reference Attribute: Seral openings

Tradeoff	Scale Units	Worst	Best	Relative Weight	Name
1.00	% seral openings	30.00	15.00	100%	Seral openings
0.18	Degrees Fahrenheit	75.00	64.00	400%	Water temp
-6.37	% canopy density	50.00	100.00	52%	Canopy density
0.06	Rd miles per sq mile	4.70	1.60	318%	Road density
0.26	Number of culverts	5.00	0.00	129%	Culverts
-0.01	NetWeaver Metric	-1.00	1.00	1500%	WS condition
-0.03	NetWeaver Metric	-1.00	1.00	500%	Reach condition
-0.03	NetWeaver Metric	-1.00	1.00	500%	Stream access

Fig. 10 The table shows the tradeoffs of the ratings scales of all the lowest criteria compared to a reference lowest criterion in the watersheds prioritization model in the Chewaucan Basin study. For instance, in this MCDA model, a one percent increase in seral openings along the creek, is weighted to be equivalent to a 6.37 % decrease in canopy density

2.5.2 Tradeoffs of Lowest Criteria

As discussed above, where the SMART alternative ratings technique is used, once the scales for the lowest criteria are defined, value functions are set, and all weights captured, the tradeoffs per alternative can be calculated and displayed in a "tradeoffs of lowest criteria" analysis table. Figure 10 provides an illustration of one such table. It shows how many units of each lowest criterion give rise to the same increase in decision score for the top ranked alternative. Since the magnitude of the increase effects the answers, the increase in the decision score considered is defined as that which a one unit increase in a reference lowest criterion would provide. In the figure, the lowest criterion of Seral openings was selected as the reference criterion, and the table shows that, to cause the same decrease in decision score that an increase of 1 % in Seral openings would produce, a decrease of 6.37 % in canopy density would be required. These values can reveal unforeseen tradeoffs that were established during weights elicitation—e.g., a 1 week delay in schedule being traded off for $2 million in project costs. If the context of the decision provides decision makers with some expectation of the tradeoffs for some of the lowest criteria, (based on, for example, organization norms or industry best practices), tradeoffs can be identified as unacceptable or fanciful. In the previous example, if the time frame for delivery is 5 years, and the projects being considered cost around $10 million, it would appear that the decision makers are putting an unusually high importance on schedule compared to cost. The insights in this table can reveal unacceptable judgments of the importance of criteria that are otherwise difficult to identify by examining only the overall decision scores and rankings.

2.5.3 Sensitivity to Weights

There may be many weights to be elicited, depending on the complexity of a decision hierarchy. Sensitivity analysis can provide some insight as to which weights decision scores are most sensitive, and, more usefully, to which weights the decision scores are relatively insensitive. Weights are the numerical embodiment of elicited judgments, and as such may represent the greatest source of uncertainty in an MCDA model. Knowing which weights decision scores are insensitive to means uncertainty in those weights may be ignored, and decision makers can focus deliberations and validation on those to which the decision scores are sensitive.

CDP provides a sensitivity analysis method based on varying a single weight at a time, and then estimating how large a change—the critical value—would have to occur in that weight before the current top ranking alternative is overtaken by another. To place the sensitivity analyses for all the weights on the same scale, regardless of the user-defined weights scale (direct or pairwise) used for elicitation, CDP calculates that critical value in terms of changes in the locally normalized values of the weights—the critical priority values. All the weights are then ordered by increasing critical priority values, to provide decision makers with comparable estimates of the overall sensitivity of the entire model to these "one-at-a time" changes in weights, and to draw their attention to those weights to which the decision scores are most sensitive.

3 Conducting MCDA in EMDS

3.1 The Priority Analyst

In EMDS, the MCDA methodology is introduced in the Priority Analyst step of the EMDS workflow ("An Overview of the Ecosystem Management Decision-Support System" and "EMDS 5.0 and Beyond"). When the user has completed an analysis using logic modeling ("NetWeaver"), they can create an MCDA prioritization of the map features by creating a Priority Analysis. A priority analysis uses the topics and calculated data links (from the logic model employed in the analysis step) as inputs to the lowest criteria for an MCDA model, with each map feature in the study area treated as an alternative.

The EMDS analyst uses CDP to create the full decision hierarchy, but with only a default, or a few example features (recommended!) included as alternatives. A typical CDP model in EMDS combines topics from NetWeaver logic models that estimate the truth level of an assertion about a condition of map features, along with original attributes of the map features that can be useful to characterizing feasibility and effectiveness of actions that might be taken to improve conditions of the feature. Such attributes can be included and managed in the NetWeaver logic model as *calculated data links* (see Sect. 2.2 of "An Overview of the Ecosystem Management Decision-Support System"). The analyst maps the lowest criteria in the decision hierarchy to selected topics and data links either as an explicit mapping table when they invoke the Priority Analyst in EMDS and load the CDP model, or by including their names in the external identifier (XID) attribute of the lowest criteria in CDP. In EMDS, a complete list of topic or data link mapping to lowest criteria is called the CDP Mapping Table.

In such an MCDA model, the weights reflect expert opinion as to the relative contributions of criteria related to conditions, effectiveness and feasibility criteria when prioritizing features for restorative action or management emphasis. The resulting decision scores for individual features can be sorted to provide an estimate of the relative priority of all map features.

From the EMDS workflow manager ("An Overview of the Ecosystem Management Decision-Support System"), the analyst launches the Priority Analyst—an MCDA graphical user component—that uses the embedded CDP-engine to load the CDP model and mapping table, and calculates the decision scores for all the map features considered in the analysis step. The Priority Analyst provides the analysts with the key MCDA analysis functionality discussed above—contributions by criteria, tradeoffs and sensitivity to weights—for the many (possibly 1000s) of map features. The decision scores of the features can then be used to generate a map layer that is added to EMDS's map table of contents. The default symbology of that map layer helps the analyst and stakeholders see which features have the highest priority, and where they are located. A "Hot Links" function makes it possible for the user to click on any feature in the map layer and see the contributions of top level criteria to the decision score for that feature.

3.2 *Special Considerations When Developing an MCDA Model in EMDS*

3.2.1 Normalizing Ratings of Alternatives

In Forman and Gass (2001), they argue that the "ideal mode" for normalizing ratings of alternatives is appropriate when there is no scarcity associated with the selection of the alternatives—the selection opportunity is "open". If there is an absolute limit to a resource being distributed amongst the alternatives—and the example they provide is a government assigning its entire gold reserve to all segments of the population—they argue that the classic or "distributed" AHP normalization should be used for the alternatives, as the selection opportunity is "closed."

If the purpose of the EMDS project is to select a single alternative—e.g. siting a single water treatment plant—no scarcity is involved. However, in a more typical ecological restoration or preservation model, a subset, often 10–20 %, of all map features is to be selected for action—whether restoration, preservation, mitigation or removal. While limitations on the amount of resources available (funding, people, materials) will constrain the number of features in the subset, the decision process is not "closed"—there is no expectation that all map features considered will have some portion of a finite resource allocated to them. Consequently the SMART normalization for alternatives is generally suitable for CDP models that will be used to prioritize map features for action in EMDS.

If the analyst wishes to execute a closed analysis, and use the decision scores to proportionately allocate resources to all alternatives, then, as well as imposing classic AHP normalization on the alternatives in the CDP model, the analyst must also ensure that the lower end of the scales for each lowest criterion be set at the value at which an alternative under that criterion presents no value to the decision makers, and not just at the minimum value of all the alternatives on that criterion. This ensures that the decision scores are calculated on a ratio scale, a prerequisite if scores are to be used directly for allocation purposes.

On a practical level, any normalization of alternatives that requires the ratings of all possible alternatives to generate a priority value for one, imposes computational challenges when the number of alternatives climbs into the millions, and representational challenges if the decision scores for all alternatives is always to sum to 1.

3.2.2 Hierarchy Preference Techniques

Regardless of the alternative rating technique chosen, any of the weights elicitation methods may be used. When decision analysts from the Forest Service use MCDA in EMDS (see "Evaluating Wildfire Hazard and Risk for Fire Management Applications" and "Landscape Evaluation and Restoration Planning"), they tend to use pairwise comparison to establish preferences. Two reasons the author prefers using either direct weights (with a swing weights approach) or direct tradeoffs, is that it is easier for decision makers to understand the relationship between the weights elicited and the underlying variations in the ratings scales that give meaning to those weights (see Malczewski 2000).

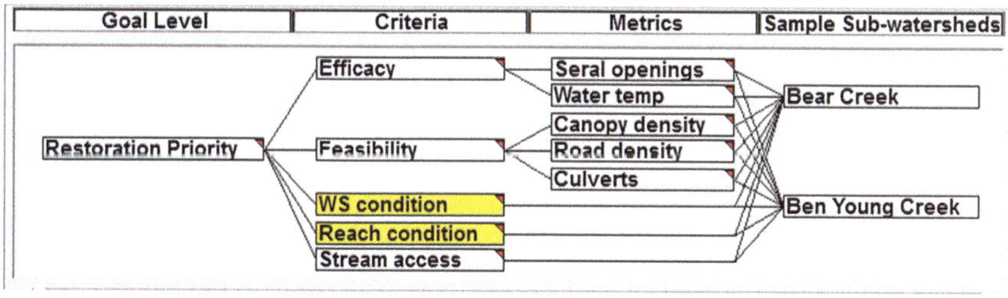

Fig. 11 The lowest criteria highlighted in yellow in this MCDA model are mapped to the topics of Watershed and (Stream) Reach condition that are topics from the logic model that assess the state of watersheds and, in aggregate, stream reaches in the Chewaucan Basin. The other lowest criteria take their values from attributes of the sub-watersheds

3.2.3 Scales for Rating Alternatives

In many EMDS analyses, key topics that are outputs of the NetWeaver models are estimates of the truth value (see "NetWeaver") of assertions about the current state of the system, recorded on a continuous numeric scale that ranges from −1 (100 % False) to 1 (100 % True). Such topics are often used in CDP prioritization models to represent the current state of the system at different scales, for example the lowest criteria of stream and reach conditions in the CDP model to prioritize watersheds in the Chewaucan Basin example (see Fig. 11).

There is no guarantee that such products of logic modeling provide classic ratio scales. The analyst may want to consider developing value functions to transform such truth value scales to scales that more closely approximate ratio scales. How successfully this can be accomplished is worth further research.

For lowest criteria that are based on attributes of the map features, and passed through the NetWeaver model as data links, NetWeaver provides a broad set of mathematical functions to transform the raw attribute values to values that better match the ratings scales the analyst designed for the CDP model. Since CDP has no built in tools for data preparation, this EMDS functionality that passes such calculated data on to an MCDA prioritization model is very valuable, as it completes any needed data transformations before the data are used in the CDP model.

3.2.4 Handling Data and Mapping Errors

When the analyst maps a lowest criterion to a truth value of a condition or data link from an analysis, the corresponding data values for some features may be missing, or lie outside the rating scale range defined for the lowest criteria, or may be of the wrong type altogether—for example, a numeric value is encountered where the scale of the CDP lowest criterion specified a text input. The Priority Analyst provides various options as to how the CDP engine should handle the above situations, with options that range from stopping processing and raising error dialogs, to employing default handling—e.g., substituting the average value of the ratings scale—and allowing the entire feature set to be processed and then later providing flags and statistics on data errors encountered. Given that there may be thousands of map features to be prioritized, the analyst needs to decide on the appropriate error handling strategy for each prioritization project.

While CDP has the capability to propagate uncertainty in ratings of alternatives into uncertainty in corresponding decision scores, that functionality is not yet available in the Priority Analyst, and the interplay with strength of evidence for the topic outputs from NetWeaver is a subject of ongoing research.

4 Examples of MCDA in this Book

Four of the nine chapters in Part II of this volume include some form of decision modeling in the applications presented. Keane et al. ("Evaluating Wildfire Hazard and Risk for Fire Management Applications") describe the use of decision models in the context of prioritizing federal forest lands for fuel-treatment planning. Hessburg et al. ("Landscape Evaluation and Restoration Planning") describe the use of decision models in several applications designed for landscape evaluation and restoration planning. In the latter two chapters, the Priority Analyst component of EMDS was used to support priority setting of landscape elements. Two other chapters present interesting variations on the use of decision models in the EMDS context. In an application for siting industrial parks in the Cantabria region of Spain (Puente, "Planning for Urban Growth and Sustainable Industrial Development"), weights derived from the AHP were directly built into the NetWeaver logic models rather than using the two-step process of EMDS v3 and later ("An Overview of the Ecosystem Management Decision-Support System"). Finally, Bourgeron et al. ("The Integrated Restoration and Protection Strategy of USDA Forest Service Region 1: A Road Map to Improved Planning") describe a three-tiered decision model for prioritizing sub-watersheds for integrated restoration and protection planning in the Northern Region of the USDA Forest Service. Here, the AHP and SMART formulae used to compute priority scores for the sub-watersheds were programmed directly into a Microsoft Excel spreadsheet by user preference, and were intentionally designed to replicate some of the very basic functionality of the Priority Analyst.

Finally as described in "EMDS 5.0 and Beyond", in EMDS 5.0, the CDP engine will be upgraded to support services and distributed processing, and the outputs of the CDP Engine can be used as the inputs for other engines. In addition there are plans to introduce several improvements to the CDP engine to better support portfolio analysis—decision support for choosing a subset of alternatives that provide the maximum benefit for a given cost in resources. These improvements would include (1) identifying a sub branch of the hierarchy as cost(s), and another as benefit(s), so that the results could be ordered by cost-benefit ratios;

(2) sensitivity analysis that focuses on the stability of the portfolio itself under changes in weights, and (3) a contributions analysis that is better integrated with the spatial map of features that are alternatives.

References

Belton V, Stewart T (2002) Multiple criterion decision analysis: an integrated approach. Kluwer Academic Publishers, Dordrecht

Coppock J, Rhind D (1991) The History of GIS. In: Maguire D, Goodchild M, Rhind D (eds) Geographical information systems. Longman Scientific and Technical, Harlow, pp 21–43

Dyer RF, Forman EH (1991) An analytic approach to marketing decisions, Prentice Hal, Englewood Cliffs

Edwards W (1977) How to use multiattribute utility theory for social decision making. IEEE Trans Syst Man Cybern 7:326–340

ESRI Understanding GIS: The Arc/INFO Method. GeoInformational International, Cambridge 1995

Forman E (1993) Ideal and distributed synthesis modes for the analytic hierarchy process presented at the International Federation of Operations Research, Lisbon Portugal, July 1993

Forman E, Gass S (2001) The analytic hierarchy process—an exposition. Oper Res Informs 49(4):469–486

Franklin B (1956) Mr. Franklin: A selection from his personal letters. Contributors: Whitfield J. Bell Jr., editor, Franklin, author, Leonard W. Labaree, editor. Yale University Press, New Haven

Gladwell M (2005) The power of thinking without thinking. Little, Brown and Company, New York

Golden BL, Harker PT, Wasil EA (1989) The analytic hierarchy process—applications and studies. Springer, New York

GSNM (2010) Summary of multi-criteria decision support process used in developing the giant sequoia national monument draft EIS, Giant sequoia national monument, Draft environmental impact statement Appendix-J, USDA Forest Service 2010

Hammond JS, Keeny RL, Raiffa H (1998) Even swaps: a rational method for making tradeoffs. Harvard Bus Rev 76(2):137–138, 143–148, 150

Hammond JS, Keeny RL, Raiffa H (1999) Smart choices: a practical guide to making better decisions. Harvard Business School Press, Boston

Huang IB, Keisler J, Linkov I (2011) Multi-criteria decision analysis in environmental sciences: ten years of applications and trends. Sci Total Environ 409(19):3578–3594 (1 Sept 2011)

Kaplan R, Norton D (1996) The balanced scorecard. Harvard Business School press, Boston

Kamenetzky R (1982) The relationship between the analytic hierarchy process and the additive value function. Decis Sci 13:702–716

Keeney R, Raiffa H (1976) Decisions with multiple objectives: preferences and value tradeoffs. Wiley, New York

Keeney RL (1992) Value-focused thinking: a path to creative decisionmaking. Harvard University Press, Cambridge

Longley PA, Goodchild MF, Maguire DJ, Rhind DW (2001) Geographic information systems and science. Chichester, England, pp 313–314

Malczewski J (1999) GIS and multicriteria decision analysis. New York, Wiley, pp 198–204

Malczewski J (2000) On the use of weighted linear combination method in GIS: common and best practice approaches. Trans GIS 4:5–22

Reynolds K, Peets S (2001) Integrated assessment and priorities for protection and restoration of watersheds. In: Proceedings of the IUFRO 4.11 conference on forest biometry, modeling and information science, Greenwich, 26–29 June 2001

Saaty TL (1992a) Multi-criteria decision making—the analytic hierarchy process. RWS Publications, Pittsburg

Saaty TL (1992b) Decision making for leaders. RWS Publications, Pittsburg

Saaty TL (2006) Theory and applications of the analytic network process. RWS Publications, Pittsburgh

Saaty T (2007) The analytic hierarchy and analytic network measurement processes: applications to decisions under risk. Eur J Pure Appl Math [Online] 1:122–196 (12 Sept 2007)

Saaty T, Forman E (1992) The hierarchon: a dictionary of hierarchies. RWS Publications, Pittsburgh

Salo AA, Hamalainen RP (1997) On the measurement of preferences in the analytic hierarchy process. J Multi-Criteria Decis Anal 6:309–319

SDS Consortium, Spatial Decision Support Knowledge Portal. http://www.spatial.redlands.edu/sds. Accessed May 2008

Steinitz C (2012) A framework for geodesign. ESRI Press, 380 New York Street, Redlands, pp 60–63

Suhr J (1999) The choosing by advantages decision making system. Quorum Books, Westport

Triantaphyllou E (2000) Multi-criteria decision making: a comparative study. Kluwer Academic Publishers (now Springer), Dordrecht, The Netherlands, p 213

Tversky A (1969) Intransitivity of preferences. Psychol Rev 76:31–48

Von Winterfelt D, Edwards W (1986) Decision analysis and behavioral research. Cambridge University Press, Cambridge

Wang X, Triantaphyllou E (2008) Ranking irregularities when evaluating alternatives by using some ELECTRE methods. Omega 36:45–63

Part II
EMDS Applications

Use of EMDS in Conservation and Management Planning for Watersheds

Sean N. Gordon

Abstract Some of the earliest applications of the Ecosystem Management Decision Support (EMDS) system were in the field of watershed assessment and planning. This chapter reviews nine cases in which EMDS has been applied to watershed management and three additional cases of the application of similar multi-criteria evaluation approaches that were at least partially inspired by EMDS. These cases are compared using the following major themes: participation, objectives, definition of watershed condition, temporal and spatial aspects, indicators and evaluation criteria, and model structuring. The cases show the model complexity and variability that is possible when trying to operationalize an ambiguous concept, such as watershed condition. EMDS has been quite successful in helping organizations structure the concept of watershed condition, but all efforts faced significant challenges with obtaining the desired data and setting evaluation criteria. Some possible areas for future software development are suggested, but many difficulties in building good models are more conceptual in nature, and thus, would likely be best addressed though improved information sharing among practitioners.

Keywords Conservation · Planning · Watershed analysis · Habitat · Fish populations · Population viability

1 Introduction

The importance of water for human uses and ecological processes has led to a variety of methods for assessment and planning based on watersheds (aka catchments). Historically, the singular focus of calculating water yields for human uses

S. N. Gordon (✉)
Institute for Sustainable Solutions, Portland State University, P.O. Box 751, SUST, Portland, OR 97207-0751, USA
e-mail: sean.gordon@pdx.edu

K. M. Reynolds et al. (eds.), *Making Transparent Environmental Management Decisions*, Environmental Science and Engineering,
DOI: 10.1007/978-3-642-32000-2_4,

made such assessments tractable to statistical methods; however, increasing interest in water's role in supporting natural ecosystems has made assessment more complex. For example, evaluating habitat conditions for fish populations has proven challenging. A wide variety of factors (e.g., availability of large wood, pools, riparian and upslope vegetation) at several scales of observation are thought to influence habitat, but the data needed to correlate these indicators with fish populations are generally unavailable over large landscapes. Furthermore, the species of concern in the Northwest, where most of the applications of EMDS discussed in this chapter have taken place, are primarily anadromous, introducing ocean habitat conditions and migration barriers as confounding variables. As a result, few studies have attempted such correlations at landscape scales (but see Kaufmann and Hughes 2006; Pess et al. 2002).

Land managers and regulatory agencies need ways to evaluate current aquatic habitat conditions and the likely future effects of proposed commodity extraction and restoration activities on water quality and fish populations. In the USA, such evaluations are required by federal laws, particularly the Clean Water (CWA 1977) and Endangered Species Acts (ESA 1973). A variety of methods have been developed to address this need. The US Environmental Protection Agency (EPA) and associated state agencies (e.g., Oregon Department of Environmental Quality) have developed and applied water quality regulatory standards under the CWA (Hayslip et al. 2004; Hubler et al. 2009). Similarly, the national regulatory agency for fisheries has developed standards for evaluating habitat attributes (NOAA Fisheries 1996).

This chapter reviews nine cases in which the EMDS software has been applied to watershed management, and an additional three cases that have used conceptually similar approaches. The objectives are to compare these cases along a few key dimensions, as well as to summarize lessons learned about the use of such models and suggest areas for possible software development.

2 Motivation for Using EMDS in this Context

Use of EMDS for watershed assessment was precipitated by two weaknesses in previous approaches. First, for regulatory clarity, assessors used absolute thresholds to evaluate each stream or watershed metric (e.g., a watershed temperature of 17 °C would pass, but 18° would fail), even though biological effects are graduated along gradients. Second, they did not go beyond the independent evaluation of each parameter, so it was unclear how several parameters might be integrated to determine overall condition or status. EMDS provided a solution framework to both of these problems with its use of gradient-based evaluation criteria and integration of multiple measures, using a variety of logic functions. Additionally, EMDS can easily combine quantitative information and more qualitative expert knowledge, when explicit measures or relationships are not available.

3 EMDS Applications in this Problem Domain

3.1 Northwest Forest Plan

Some of the earliest EMDS applications were developed for watershed assessment and planning on federal forest lands in the Pacific Northwest of the United States. In particular, the initial development period of EMDS coincided with development of the Northwest Forest Plan (NWFP), one of the earliest and largest ($\sim$10 million hectares) applications of ecosystem management principles in the US. It consists of a coordinated set of guidelines for federal lands in the continental northwestern United States (USDA and USDI 1994). The area covered by the NWFP was principally determined by the range of the northern spotted owl (*Strix occidentalis caurina*), which was listed as a threatened species in 1990 under the national Endangered Species Act (USFWS 1990). The planning process also developed an aquatic conservation strategy to protect aquatic ecosystems in anticipation of possible listings of a number of fish species.

Early on, EMDS was proposed to meet the need for watershed assessment and planning for the NWFP (Reynolds et al. 1996). However, the involvement of numerous federal agencies in the NWFP, each with a different concept of watershed monitoring, led to a drawn-out process of finalizing the aquatic monitoring strategy (Reeves et al. 2003). Despite this delay, the watershed monitoring program was established in 2000 and an initial assessment using EMDS was completed in 2001. The first major report on NWFP monitoring (Gallo et al. 2005) addressed the first 10 years of the plan (1994–2003). To cover the large, diverse area, the watershed monitoring team worked with local experts and created six subregional EMDS assessment models (Gordon and Gallo 2011). NWFP monitoring reports are planned for every five years, and a second iteration of the modeling and assessment process using EMDS was recently completed (Lanigan et al. 2012).

3.2 National Forest Planning

National forests managed by the US Department of Agriculture Forest Service (USFS) make up the largest portion of the NWFP area, and development of EMDS in the Pacific Northwest research branch of the USFS (Reynolds et al. 2002), along with adoption for the NWFP, has led to a number of applications developed at the national forest scale. One of the first was an assessment of the Chewaucan watershed in south central Oregon USA (Reynolds and Peets 2001). The assessment demonstrated how multiple scales of data (subwatersheds and stream reaches) could be integrated, and users of EMDS will recognize it as the tutorial case.

Under federal law, national forests are required to create a comprehensive management plan and update this plan every 15 years (NFMA 1976). The actual rules for defining these plans have changed numerous times in the past decade due to attempted updates and court challenges (http://www.fs.usda.gov/main/planningrule/history), but all versions have included language on the protection of species. A few national forests in the Pacific Northwest are engaged in plan revision and have been using EMDS to help assess and plan for "aquatic ecological sustainability." This approach was piloted on the Okanogan-Wenatchee National Forest, which is documented in Reiss et al. (2010).

3.3 California North Coast Watershed Assessment Program

Government agencies at the state level also share responsibility for the protection of water quality and conservation of aquatic species. In 2001, the state of California initiated the North Coast Watershed Assessment Program to evaluate the status of coastal watersheds that historically supported anadromous fish. Through connections to the Northwest Forest Plan process, the California program decided to use EMDS to assess overall watershed and in-stream conditions for fish. They produced models for instream habitat and sediment delivery, a key component of watershed condition, as determined by the EPA (Dai et al. 2004; Walker et al. 2007).

3.4 Other Northwest Agencies and Nonprofits

In Oregon, an EMDS-based watershed assessment process has been applied to at least two other major efforts for which publications are available. First, a broad partnership between federal and state agencies and nonprofits used EMDS to develop restoration priorities for the Sandy River, a major Columbia River tributary, near Portland, Oregon (SRBWG 2006, 2007). Second, the Portland-based nonprofit Wild Salmon Center undertook an evaluation of 5,500 km^2 of watersheds in northwestern Oregon to evaluate and influence the implementation of a "Salmon Anchor Habitats" strategy being implemented by the state's forestry department (Miewald et al. 2008).

3.5 Environmental Protection Agency

The Pacific Northwest Research Station (USDA Forest Service) and EPA cooperatively developed one of the earliest EMDS applications for watershed assessment and monitoring of ecological states and processes (Reynolds et al. 2000).

Much of this application was concerned with assessment of stream characteristics to provide decision support for managing nonpoint source (NPS) pollutants under the EPA's Total Maximum Daily Load (TMDL) program. TMDLs specify the maximum daily amount of pollutants allowed meeting state water quality standards, and the TMDL program allocates pollution control responsibilities among pollution sources in a watershed. NPS TMDL parameters considered critical to aquatic ecosystems include streamflow, stream temperature, nutrients, stream sediment, and in-channel stream habitat. Given the objectives of the EPA water quality assessment program, the primary logic topics included in design were watershed processes, watershed patterns, general effects of human influence, and specific effects of human influences on aquatic species.

3.6 Other Regions

While most applications of EMDS to watersheds have taken place in the Pacific Northwest, there has also been an extensive application to planning in Region 1 of the USDA Forest Service (Montana, North Dakota, and northern Idaho), and similar multi-criteria techniques have been adopted by two national efforts. The USFS Region 1 application was initiated as part of the forest planning process to evaluate ecosystem sustainability across a number of national forests and adjacent lands, and then to help prioritize restoration activities. Aquatic integrity was one of five submodels built for this effort (the others being terrestrial integrity, fire danger, social opportunity, and economic integrity) (Jensen et al. 2009).

One Oregon and two national watershed assessment initiatives have also been implemented using similar multi-criteria techniques, which appear to have drawn from the EMDS applications, although they do not use the software specifically. I have included these cases because of their national significance and similarity in methods. In 2007, the nonprofit, Trout Unlimited, launched its Conservation Success Index, which is a multi-criteria approach to evaluating the status of a variety of coldwater fish species to facilitate protection, restoration, reintroduction, and monitoring efforts. The index includes 20 indicators grouped into four categories. To date, assessments of numerous species have been carried out in five broad regions of the USA (California Salmon, Sierras, Intermountain West, Desert Southwest, Midwest, and East) (Williams et al. 2007; Trout Unlimited 2007). The second national effort is the USDA Forest Service National Watershed Condition Classification (WCC, USDA Forest Service 2010), which was begun in 2006 in response to a federal oversight agency (Office of Management and Budget, OMB) criticism that the Forest Service lacked a nationally consistent method for assessing and prioritizing watershed restoration actions even though it was a major strategic emphasis (US OMB 2006). Each national forest is now required to assess each of their watersheds using a standard set of 12 indicators, which are combined into an overall index score. EMDS was used for a pilot national assessment, but the

FS established a national database system to store and calculate the first full assessment in 2010. Finally, a recent feasibility study for the reintroduction of threatened bull trout (*Salvelinus confluentus*) into the Clackamas River, Oregon used a multi-criteria framework built on earlier experiences with EMDS (Dunham et al. 2011).

4 Experiences Using EMDS for Watershed Assessment

This section reviews the positive and negative aspects of the application of EMDS in the cases mentioned above. It is organized according to a number of major issues involved in preparing and applying such models to structured problem solving. To more compactly reference the cases, bracketed case numbers from Table 1 are used in the following discussion.

4.1 Participation

Problem framing depends on who is involved and the processes used to gather input. Relatively little has been written about this aspect of the cases, beyond providing a brief mention of general types of participants or occasionally a more specific list. The summary of model-building efforts in Table 1 shows that involvement has largely been limited to technical experts in the various aspects of watershed functioning. Participation has ranged from a team of two (a fish biologist and a modeling specialist [case 2]) to a case involving 12 organizations, including nonprofits and local/state/federal government agencies [8]. In fact, many of the cases involve multiple organizations [1, 7, 8, 9, 10], demonstrating that the EMDS framework has been simple and flexible enough to span such organizational boundaries. Even when used within one organization, it has served to integrate multiple disciplinary perspectives [4, 5, 6] and bridge multiple hierarchical levels. As reported in Gordon and Gallo (2011), the concepts involved in multi-criteria modeling did not appear difficult for the experts to grasp, and the modeling process provided a much-needed framework for expressing and combining their knowledge. The EMDS framework was also easy to understand and helped increase the transparency of the assessment process. As one user put it,

> It's a great tool for that [watershed assessment] and everyone sees right there on paper but this is how you did it, whereas some of the other models that I'm aware of, you go into a black box approach and you don't really know what happened with the data. It just spits out a number in the end. (Gordon 2006).

However, Gordon and Gallo (2011) also identified some core challenges related to broader involvement in model-building:

Table 1 Summary of key factors in 12 applied cases of EMDS and related multicriteria methods to watershed assessment and planning

#	Application	References	Participation	Objectives	Definition of watershed condition	Temporal extent	Spatial extent
1	Northwest forest plan watershed condition	Gallo et al. (2005) Gordon and Gallo (2011) Lanigan et al. (2012)	Aquatic experts from federal and state agencies	Monitor status and trends under a regional forest policy	Supports a high diversity and abundance of aquatic and riparian species, particularly salmonids	Historical condition at start of plan (1994) then current conditions at 5-year reporting interval (2005-ongoing)	NWFP area (98,000 km^2)
2	Nestucca Basin, OR	Reynolds and Reeves (2003)	USFS research fish biologist and EMDS specialist	Evaluate salmon habitat suitability	suitable biophysical condition for salmon habitat	Current condition only	Nestucca Basin (5th field hydrologic unit, 664 km^2)
3	Upper chewaucan watershed (Oregon)	Reynolds and Peets (2001)	USFS watershed assessment lead and EMDS specialist	Identify priority restoration areas in a national forest watershed	Assessment: Salmonid habitat suitability Prioritization: Protect the best, then restore the rest	Current condition only	Upper Chewaucan watershed (5th field HUC, 694 km^2)
4	USFS region 6 National forest planning	Reiss et al. (2010)	Team of 4 agency aquatics specialists + consultations with ~20 agency resource specialists and scientists	Support national forest planning requirement to assess contributions to the sustainability of ecosystems and species	Focal species populations in sub-basins are sustainable	Current and historical reference condition	2 pilot 4th-field watersheds (2,800 and 5,500 km^2)
5	USFS region 1 National forest planning	Jensen et al. (2009)	USFS staff specialists representing various disciplines, including landscape ecology, fish biology, and hydrology	Support national forest planning requirement to assess contributions to the sustainability of ecosystems and species	Evaluating ecosystem sustainability (desired conditions) and prioritizing areas for integrated landscape restoration (objectives)	Current condition only	3 Forest Service planning zones (58,000 km^2)
6	USEPA TMDL Analysis	Reynolds et al. (2000)	EPA scientists	Evaluate watershed condition under EPA framework of total maximum daily loads (TMDLs)	Watershed patterns and processes relative to unmanaged conditions and organizational standards	Current condition only	4th-field watershed

(continued)

Table 1 (continued)

#	Application	References	Participation	Objectives	Definition of watershed condition	Temporal extent	Spatial extent
7	California North Coast watershed assessment program	Bleier et al. (2003) Dai et al. (2004) Walker et al. (2007)	5 state agencies and EPA Scientific peer reviews Public input solicited on draft assessment manual	Assist prioritization of both in-stream and upland area watershed restoration and recovery efforts	The overall condition of the stream reach is suitable for maintaining healthy populations of native coho and Chinook salmon and steelhead trout	Current condition only	50,000 km^2, approximately 26,000 km^2 of which are private lands
8	Oregon Sandy River Basin Anchor habitats project	Sandy River Basin Working Group (2006, 2007)	12 organizations, including nonprofits and local/state/federal government agencies	Support maintenance and restoration of salmon and steelhead populations and the habitat that supports them	Best habitats for 4 specific fish species	Current condition only	Sandy River basin (1,316 km^2)
9	NW Oregon salmon anchor habitats	Miewald, van Dyk and Reeves (2008)	Nonprofit-organized, 4-person core science team (ODF, ODFW, USFS, TNC) Consultations with ~10 additional experts	Assess status of watersheds for supporting salmon species and relationships between conditions and management protections	watersheds in "good" condition in areas of high intrinsic potential for multiple species	Current condition only	Oregon North Coast (5,500 km^2)
10	Reintroduction decision framework for Bull Trout, Clackamas River, OR	Dunham et al. (2011)	Clackamas River Bull Trout Working Group (local, state, federal, tribal, utility)	A simple and transparent system for assessing the feasibility of bull trout reintroduction	Combination of current/historical presence, good current habitat conditions, and low future threats	Historical presence Current conditions Future threats	Upper Clackamas River (area not specified)
11	Trout unlimited conservation Success index	Williams et al. (2007)	Trout Unlimited staff and external expert reviewers	Characterization and synthesis of native salmonid status and prioritization of conservation areas Communicate complex assessment data and conservation opportunities to the public	Good distribution, population, and habitat conditions, and security from future threats	Mostly current conditions, but including % historic habitat occupied and future security indicators	Individual species ranges (areas not specified)
12	USDA forest service national watershed condition classification	USDA FS (2011)	USFS Watershed Condition Advisory Team (specification) USFS national forest staff (implementation)	Report on watershed conditions and trends at the national level for all national forests	Assess the state of the physical and biological characteristics and processes within a watershed that affect the hydrologic and soil functions supporting aquatic ecosystems	Current condition only	National forests (45–67,000 km^2)

Table 1 (continued)

#	Analysis units	Types of indicators	Assessment model structure	Prioritization model structure	Criteria source(s)	Weighting	Sensitivity/uncertainty	Tools used
1	1379 Subwatersheds (40–120 km^2) 193 stream reaches	- watershed-level: vegetation, roads, landslide risk - stream reach level: physical and biological factors	Upslope and Riparian Inchannel	NA	Expert knowledge elicitation workshops	Some provinces Roads versus vegetation Riparian versus upslope	Gallo et al. (2005) reported sensitivity of results to individual indicators	Netweaver
2	45 sub watersheds	Upslope: mature vegetation, road crossings, land use Riparian: NFBveg, FBveg Inchannel (low gradient): wood, pools, water temperature, substrate, off-channel habitat	Riparian Upslope Inchannel (low gradient)	NA	Expert knowledge and interpretation of the literature	Capacity to read in structure, but weights not mentioned in text	None	Netweaver
3	6 subwatersheds 104 stream reaches	- watershed-level: upland vegetation cover, road density, fish access, spawning fines, water temperature, reach condition - stream reach level: riparian vegetation (composition, seral stage), bank stability, W-D, large wood, pools Prioritization: efficacy, feasibility, access, watershed and reach conditions	Upland condition Habitat accessibility Stream condition	Protection watershed condition reach condition access efficacy feasibility Restoration watershed condition reach condition feasibility vegetation similarity	Previously published assessment	None described	None	Netweaver DecisionPlus (AHP and SMART)

(continued)

Table 1 (continued)

#	Analysis units	Types of indicators	Assessment model structure	Prioritization model structure	Criteria source(s)	Weighting	Sensitivity/ uncertainty	Tools used
4	13/13/28 subwatersheds (6th field), 3 subbasins (4th field)	- HUC6 habitat: road density (riparian, upland), riparian vegetation, fire risk - HUC6 focal species population: distribution, status, connectivity, nonnatives - HUC4 condition: distribution, patch sizes, HUC6 condition - HUC4 connectivity: population connectivity, habitat connectivity	Condition Distribution Patch HUC6 population HUC6 habitat Connectivity	Used in subsequent prioritization process (ACS key watersheds)	Expert knowledge and interpretation of the literature	Roads along low-gradient streams Road density by sensitive soils	Potential influence of each indicator (−1/+1 boundary analysis)	Netweaver
5	Subwatersheds (6th field, 40–120 km^2)	Watershed condition: water quality, roads, vegetation, stream habitat Fish species status: 7 species Threats: fire, flow, grazing, mining, population, recreation, roads	Watershed condition, Fish species status Threats	value (incl. watershed condition) risk feasibility	Relative to dataset (25/75th percentiles)	Prioritization model weighted for 3 different possible objectives	3 different prioritization objectives	Netweaver CriteriumDecisionPlus

(continued)

Table 1 (continued)

#	Analysis units	Types of indicators	Assessment model structure	Prioritization model structure	Criteria source(s)	Weighting	Sensitivity/ uncertainty	Tools used
6	Subwatersheds (6th field, 40–120 km^2)	processes: streamflow, erosion, fire patterns: upland, valley bottom, channel human influences: roads, dams, diversions, channelization, groundwater extraction, mines, grazing, recreation habitat: baseflow, substrate, temperature, cover	watershed processes watershed patterns human influence aquatic species habitat	NA	(1) reference conditions from unmanaged watersheds (2) standards set by resource management agencies (3) standards set by regulatory agencies	None	None	Netweaver
7	stream reaches CA planning watersheds (10–30 k acres)	Reach condition: water temperature, riparian vegetation, inchannel habitat (pools, substrate), stream flow Upland condition (sediment): roads, land use, land cover, and slope instability	Salmonid habitat Sediment production	Assessment results used to help identify limiting factors in broader prioritization process	Salmonid habitat: Science assessment team Sediment: 10/90th percentiles of dataset	None described	Data quality control procedures	Netweaver

(continued)

Table 1 (continued)

#	Analysis units	Types of indicators	Assessment model structure	Prioritization model structure	Criteria source(s)	Weighting	Sensitivity/ uncertainty	Tools used
8	Stream reaches (then summarized to 5/6th field HUCs)	Fish/redd counts Expert rating EDT habitat score	Fish/redd counts Expert rating EDT habitat score	Assessment model used to determine high priority restoration areas	high performing reaches (relative to dataset)	None	None	Netweaver
9	Subwatersheds (6th field, 40–120 km^2)	Upslope: native vegetation, hydrologic maturity, landslide risk Floodplain: shade, wood recruitment, native vegetation, channel containment Stream: dissolved oxygen, temperature, flow, connectivity, pool and wood frequency, fines, shade Fish: juvenile and spawning surveys Intrinsic potential: flow, valley width, gradient	Upslope Floodplain Stream Fish Intrinsic potential	Assessment results used in broader qualitative process to help identify priority watersheds for protection	Watershed condition: mainly previous modeling efforts + expert knowledge Fish: relative to area average Intrinsic potential: IP model	Yes, by indicator group (Table A1, p. 37)	None	Netweaver

(continued)

Table 1 (continued)

#	Analysis units	Types of indicators	Assessment model structure	Prioritization model structure	Criteria source(s)	Weighting	Sensitivity/ uncertainty	Tools used
10	Analysis area as a whole	Recipient habitat Historical presence Contemporary presence Habitat suitability (temperature) Threats Recolonization potential Donor population Suitability Threats to donor	Feasibility Recipient habitat Donor population	NA	Expert knowledge and interpretation of the literature	All equal	averaged the absolute values of scores for each component to derive a second index of uncertainty. In this case, scores closer to zero would indicate less certainty whereas scores closer to 1 would indicate greater certainty	Excel?

(continued)

Table 1 (continued)

#	Analysis units	Types of indicators	Assessment model structure	Prioritization model structure	Criteria source(s)	Weighting	Sensitivity/ uncertainty	Tools used
11	Subwatersheds (6th field, 40–120 km^2) Also rangewide and subbasin indicators	Range-wide condition: % historic habitat occupied by subbasin, subwatershed, stream order, lakes Population integrity: density, extent, purity, disease vulnerability, life history diversity Habitat integrity: land stewardship, connectivity, physical (roads), water quality, flow Future security: land conversion, resource extraction, climate change, introduced species	Range-wide condition Population integrity Habitat integrity Future security	Uses different combinations of indices to prioritize 4 management actions: - Protection: high population and habitat integrity, low future security - Restoration: moderate to high for population and habitat integrity, high future security - Reintroduction: low population, high habitat integrity and future security - Monitoring: all qualifying for 3 actions above	Expert knowledge and interpretation of the literature	All equal		Google Maps Google Earth

(continued)

Table 1 (continued)

#	Analysis units	Types of indicators	Assessment model structure	Prioritization model structure	Criteria source(s)	Weighting	Sensitivity/ uncertainty	Tools used
12	Subwatersheds (6th field, 40–120 km^2)	Aquatic physical: 303d listed, flow, habitat fragmentation, large wood Aquatic biological: life form presence, native versus exotic species, riparian vegetation condition Terrestrial physical: road density and maintenance, proximity to water, soil productivity, soil erosion Terrestrial biological: fire condition class, loss of forest cover, invasive species, insects and disease, ozone	Aquatic Physical Aquatic Biological Terrestrial Physical Terrestrial Biological	Suggests using assessment results used in broader qualitative process to help identify priority watersheds for protection	Expert knowledge and interpretation of the literature	All equal		Excel Oracle database

> The downside of flexibility and incorporation of local knowledge is the generation of multiple, possibly inconsistent, models. In hindsight, one of the decisions with the most profound impacts turned out to be the geographic scale at which we engaged the experts. We expected some model differences between physiographic provinces due to ecological differences, but holding multiple knowledge elicitation workshops also introduced an unknown amount of variation based on the mental models of the different experts attending each session.

The USFS anticipated this potential problem in implementing their national watershed condition assessment over their 175 management units, and they chose to implement a strict, standardized structure to help achieve consistency. The cost of this "controlled" approach is that a generic model structure may not fit local conditions well. Gordon and Gallo (2011) hypothesized two other approaches: (1) an informational approach, which would provide participants with detailed information on the other models and encourage them to use similar structures as appropriate; and, (2) a post-processing approach, in which, after the workshops, the project team or a small group of regional experts (preferably who are able to attend all the workshops) would review all the models and propose changes to harmonize them, perhaps followed by an additional validation workshop. Given high interest in assessment compatibility, both horizontally and vertically in the organizational sphere, further documentation of this aspect and research in this area is needed.

4.2 Objectives

EMDS has been used to meet the objectives of a number of different watershed-related organizational programs (see Table 1, Objective column). One fundamental difference in the scope of applications is whether they involve only assessment or both assessment and prioritization. Four of the cases focus exclusively on assessment [1, 2, 6, 10]. Five cases occupy a "middle ground", where assessment results are either used directly for prioritization [8], or they are used in a broader, qualitative prioritization process [4, 7, 9, 12]. For example, the USFS National Watershed Condition Framework [12] suggests prioritizing the protection of watersheds with high condition scores that also fall into certain land use categories, such as designated wilderness or municipal water supplies.

Only three cases include formal prioritization models implemented using the multi-criteria approach [3, 5, 11]. The details of these cases are discussed further in the **Model structure** section below. Clearly, EMDS has been seen as a useful tool for both types of objectives; however, it is interesting that only three of the 12 cases have used the multi-criteria methods for prioritization as well as assessment. For the first three cases mentioned immediately above, prioritization appears to simply be outside their scope. The five "middle ground" cases, however, clearly state prioritization as a key objective, so why was the multi-criteria method not used? This question is not answered directly by the texts, but reading between the lines,

I hypothesize two reasons: (1) the factors were deemed too complex or idiosyncratic for formal analysis, and/or (2) it was desired to maintain flexibility for local decision makers. A third possibility is technical: the integration of different software programs for assessment (" Netweaver", Chap. 2) and prioritization (" Criterium Decision Plus", Chap. 3) creates a steeper learning curve for users and adds costs for obtaining versions of both programs needed for model development.

4.3 Definition of Watershed Condition

Despite the shared focus on watershed condition, there are considerable differences among the cases in how this concept is operationalized (Table 1, Definition of watershed condition). In a case study of the NWFP, Gordon and Gallo (2011) identified framing the initial approach to watershed assessment as one of the most challenging aspects of the modeling effort. Multiple federal agencies were involved, each of which had previously developed their own approach.

Notable in the objectives across cases is the degree of emphasis on maintaining native fish species; many of the applications were developed in the Pacific Northwest, USA, where a number of salmonid populations have been listed under the federal Endangered Species Act (ESA). Because the ESA mandates a number of protection measures, these species are important foci for management and policymaking. A major framing decision for planning is whether to focus on fish populations, habitat, or both.

The importance of maintaining fish species is emphasized in all the cases, yet some include abundance metrics in their models, while others do not. The NWFP assessment [1] does not include any abundance indicators because the USFS, as a land management agency, is primarily responsible for habitat rather than population management, the province of regulatory agencies, and habitat is what is most directly affected by management. Additionally, no consistent population data are available over this large area, and the emphasis of the program is on consistent, repeatable trend metrics. The Chewaucan, EPA, and CA North Coast assessments [3, 6, 7] also do not include population data, although it is not stated whether this is for lack of data, or an intentional focus on habitat. In contrast, the USFS aquatic sustainability [4], Oregon anchor habitats [9], and USFS Region 1 [5] models all include fish population metrics. The more limited spatial scale of the first two makes this more tractable. Also, in the case of the USFS aquatic sustainability effort, the governing law requires assessment of the viability of species.

A second broad consideration in the definition of watershed condition is the relative emphasis on ecosystem processes versus ecosystem states. Emphasizing processes acknowledges the dynamic nature of aquatic ecosystems, an area where aquatic assessments have generally lagged behind their terrestrial counterparts (Reeves et al. 1995). A number of the cases frame the problem in terms of processes, but at the same time they still rely on state-based indicators. As Reiss et al. (2010, p. 25) explain,

> The habitat condition component of the HUC [hydrologic unit code] 6 AEC [aquatic ecological condition] model was designed to assess ecological processes, rather than evaluate the habitat needs of any particular species. Upslope and riparian road and vegetation attributes were selected as surrogates for the processes that affect stream habitat[,] and to act as indicators of anthropogenic influence on ecological processes.

A process-based approach also acknowledges that even in "pristine" environments, not all watersheds can be expected to be in a state supporting good fish habitat all of the time. Natural disturbances will change conditions over time, such as landslides or debris flows that may disrupt habitat in the short-term, but provide necessary inputs to in-stream habitats over the longer term. The common solution to these dynamics is to focus on the distribution of watershed scores in an area rather than the score of each individual watershed. These actual distributions are then compared to a reference distribution, for example the start of the NWFP [1] or pre-European settlement [4].

EMDS was flexible enough to accommodate this range of definitions and had the benefit of making definitions of an ambiguous concept more explicit. Users in this domain should be aware, however, of the complexity of this task, as revealed by the attempt of Lanigan et al. (2012) to provide a clear definition:

> The condition of a watershed is defined as "good" if the state of these attributes support a high diversity and abundance of aquatic and riparian species. Many of the physical indicators are chosen for their relevance to native or desired fish species because of these species' roles in driving management policies (including the NWFP itself) and the availability of research related to their habitat needs. However, we attempt to assess indicators relative to the natural potential of the site to provide biotic habitat. A watershed that naturally does not support fish populations (because of elevation or other natural conditions) but has little vegetation disturbance, few roads, good pools, and wood should be evaluated positively. If this watershed loses significant vegetation, even from natural causes (e.g., fire), then the condition rating will go down (it is below its potential).

4.4 Temporal Aspects

Another aspect of problem framing is the temporal extent of interest. Most of the cases focus on the present–current conditions and priorities based on these conditions. A few cases, however, use historical data in different ways. The NWFP assessment [1] is tasked with evaluating changes since the inception of the plan, and so has created a historical baseline assessment of conditions in 1994 as a comparison point for trend. The ability of EMDS to provide a consistent and repeatable metric was a major reason it was chosen. However, it does not currently have utilities that support calculating such score changes over time. In a similar vein, the USFS Region 6 national forest planning model [4] uses a simulated historical condition at watershed scale for a comparison baseline. At the subwatershed level, it includes a few individual indicators that are based on comparisons with historical conditions (focal species status and distribution). The Trout Unlimited Conservation Success Index (CSI) [11] also includes an indicator of fish distributions relative to historical conditions. The NW Oregon Salmon Anchor Habitats assessment [9] discusses potential differences in priorities over time (although the report only provides a single set of priorities):

> First, any conservation strategy must include a short-term plan to protect the watershed processes that maintain current core areas or anchor habitats (Reeves et al. 1995). Second, for longer term conservation it is useful to consider the concept of the "intrinsic potential" (IP) of a stream to provide high quality habitat.

Some models also look to the future. The Trout Unlimited CSI [11] includes a whole branch of the model related to "future threats." These future threats have been drawn from a variety of sources, such as land development and climate change models. In summary, looking at multiple time periods is an important aspect of most natural resource problems. Many cases have figured out ways to integrate temporal aspects into EMDS analyses, but the inclusion of temporal tools would be a useful area for future EMDS development.

4.5 Spatial Aspects

Another fundamental consideration in the development of EMDS models are spatial aspects, including spatial extent, analysis units, and spatial relationships. The spatial extents of the cases range from the 98,000 km^2 of the NWFP [1] to the ~600 km^2 of two single watershed studies [2, 3]. Broader-scale applications appear to have been typically defined by administrative boundaries, such as USFS planning units [1, 5, 12] and a state planning region [7]. Mid-scale assessments seem more tied to particular ecological or hydrologic units, such as one or more major river basins [8, 9, 10] or species ranges [11]. Studies in the smaller size range were generally intended as pilot efforts that could be scaled up to larger regions [2, 3, 4, 6]. The USFS assessments limited their evaluation coverage to federally managed lands, although their new national effort [12] includes qualitative ratings of other lands in shared ownership watersheds. Other assessments included all ownerships but faced the associated challenge of variable data coverage and data quality evenness.

While spatial extents of the studies varied widely, there has been considerably more congruence on the analytical/modeling units used: all cases used either stream reaches (segments) and/or subwatersheds (12-digit hydrologic unit, HUC12). The convergence to the HUC12 is likely due to it being the smallest available hydrologic unit in the watershed boundary component (WBD, Seaber et al. 1987; USDA NRCS 2012) of the national hydrologic dataset (NHD, USGS 2012).

The US government has sponsored development of the NHD to facilitate consistency in hydrologic modeling across agencies and jurisdictions. The NHD provides considerable potential for the compatibility of different watershed assessments. However, it is also subject to continuing updates by state data stewards, so the polygons used for one assessment at one point in time may not be congruent with those produced from a different time period. These changes present a challenge for sampling designs intended for trend analysis, such as in the NWFP [1]. Furthermore, there appears to be little attention to date on clear versioning of the WBD; there have been no version labels (e.g., version 1.4) applied and update dates are rather deep in the metadata.

The spatial unit of "reaches" appeared to be used in four of the twelve studies [1, 3, 7, 8]. Stream reaches have been defined in many different ways, which are exemplified in these cases. Two [1, 7] based reaches on stratified random point samples, with measurements extending a variable length along the channel based on stream width (e.g., 20 bankfull stream widths), but usually limited by fixed minimum and maximum lengths (150–500 m). The Chewaucan case [3] did not specify what the reach represented, but it can be assumed to be based on the USFS Level II survey protocol, which defines reaches by physical characteristics, such as valley width, channel gradient, and sinuosity (USDA FS R6 2009). These reaches tend to be longer (generally 1200 m minimum), and whole streams within the ownership are typically sampled, resulting in a segmented line feature. The Sandy River case [8] also created a segmented stream network for the entire study area, based principally on geomorphic considerations used in one of their submodels (EDT, although details are not provided in the main report). Within this master structure they integrated diverse datasets with original reach lengths varying from <200 m to >10 km. Interestingly, none of the studies mentioned the NHD and its standardized reach segmentation. This lack of standardization presents a problem for the reuse and integration of reach-level datasets.

Many of the studies acknowledge the importance of integrating multiple scales of analysis. All of the studies employing reach-level metrics, except [8], also aggregate these up to HUC12-level summaries using reach length-weighted averaging.

The USFS Region 6 analysis [4] summarized from the HUC12 to the HUC8 (subbasin) level. Their rationale was that subbasins are more consistent with species recovery plans and the size of their administrative ranger districts. Similar to the reach-to-subwatershed integration, the HUC12 assessments were combined with a number of analyses done at the broader HUC8 level, including fish distributions, habitat patch sizes, and population and habitat connectivity. Trout Unlimited's CSI [11] is unique in that it takes data from broader scales (populations, subbasins), and applies these (common) ratings as part of their individual watershed scores. Aggregating data and linking to other scales is a common function in watershed assessment (as well as other natural resource assessments), and therefore a task which EMDS may wish to consider providing specific functions to support in the future. This capability might be accomplished by a tool which allows users to identify different scales of analysis and then enables the selection of their outputs as inputs to other scales. There also appears to have been little investigation or reporting on the accuracy of such aggregation procedures. It would be helpful to create some measure of the statistical uncertainty based on the sampling design. However, such a function broaches the much larger topic of the integration of statistical and fuzzy uncertainties, for which more user education would be most helpful but unfortunately is beyond what this author can provide.

A final spatial aspect that deserves consideration is the use of indicators involving spatial relationships. All of the cases include indicators based on spatial location, and the most common are those assessing vegetation and road conditions in proximity to streams (e.g., riparian vegetation). What is largely absent in these models are indicators involving spatial relationships between individual reaches or watersheds. These second order spatial relationships are important given the connected nature of the stream network between reaches and watersheds (Naiman et al. 1992). Only two indicators along these lines were found in the cases: 1) the USFS region 6 model [4] includes a genetic connectivity metric, which measures degree of connectivity between the HUC6 and other populations within the HUC4, and 2) the Trout Unlimited CSI [11] includes a watershed connectivity indicator, which is based on a count of fish passage barriers downstream from the subwatershed. Such network or neighborhood relationships are common in other natural resource topics as well, and so may be another consideration for future EMDS development.

4.6 Data and Indicators

At the base of a multi-criteria approach, such as EMDS, is the choice of indicators to include in the model. Efforts often begin by mapping out all types of data thought to influence watershed condition, and then paring these down to those with available data. A fundamental modeling decision is the extent to which one leaves in indicators with missing data. One of the strengths of EMDS-Netweaver is that it handles missing data by assigning a neutral value for the indicator (e.g., 0 on the standardized -1 to $+1$ scale). However, if a number of indicators with missing data are left in, they push the model results to an undetermined, neutral value. The California Coastal Assessment [7] sediment model is an example in which there are more indicators without than with data. More discussion of this issue would be helpful. However, later versions of the EMDS-NetWeaver component include an option to turn off indicators with no data (i.e., do not count them rather than assign a zero value).

One strategy for incorporating key indicators that lack empirical data is through the reliance on expert knowledge. Two cases [4, 12], in particular, incorporate a large number of indictors based on expert assessment. Experts may be able to provide accurate assessments for some attributes, and EMDS can at least capture and preserve these results. However, the EMDS framework does not provide a mechanism for capturing the rationale or uncertainty associated with an individual expert's knowledge or opinion. As a result, the challenge of achieving consistency between experts and assessments over time becomes even more difficult. Modelers should create processes and data structures necessary to capture such expert judgment metadata. The EMDS Netweaver and CDP components already have some capacity to store such metadata; possible improvements include further consistency between components and their abilities to export these data in structured formats.

While the use of empirical data for indicators is more transparent and repeatable, it is still subject to a number of potential problems. For one, preparing the raw data for model input is likely to be the most time consuming piece of the assessment. EMDS models can be built relatively rapidly, but the compilation of empirical datasets can be quite time consuming, accounting for as much as 80 % of the time and money used to develop a model. Even when working with existing data, the necessary transformations and concatenations to format data for EMDS inputs can take time. This formatting and transformation is also a large potential source of errors. Special care should be taken with indicator metrics. The units involved in environmental assessment are often complex (e.g., is the measure "miles of riparian road per square mile riparian area" or "miles of riparian road per mile of stream"?). Some metrics can simply be poor choices. For example, the USFS national watershed assessment [12] uses the following riparian road indicator: "percent of road/trail length is located within 300 feet of a stream." Using this measure, a watershed with 100 m of road, all of it along a stream, will be evaluated worse than a watershed with 10 km of roads, where only half are located next to streams. In general, density metrics are preferable (e.g., per stream km or per km^2).

A final issue related to data and indicators relates back to the earlier discussion of process- versus indicator-based models. Conceptually, the preference of watershed experts seems to be for process-based thinking; however, most of the models to date have been structured more around available data using an indicator-based approach. A central reason behind this inconsistency is the apparent redundancy that occurs in process-based structures. A process-based model looking at mass wasting, sedimentation, wood delivery, and shading would likely include potentially quite similar vegetation indicators under each of these processes. This apparent redundancy is unsettling for those trained in statistics, where the problem of multicollinearity requires the elimination of highly correlated independent variables. I argue that indicators should be re-used if they are proxies for different processes. Multicollinearity is not an issue because in an EMDS knowledgebase one is not statistically estimating the influence (i.e., coefficients) of explanatory factors on a dependent or response variable; rather, one is assigning weights that are used to calculate a dependent index (no actual values of "watershed condition" are independently measurable). This is not an issue of EMDS as a software system, but one that needs to be addressed through scientific communication from the EMDS community.

4.7 Evaluation Criteria

Under the multi-criteria approach, each indicator chosen must be evaluated or normalized to a common scale to facilitate their combination. The choice of evaluation criteria is often one of the most difficult and subjective aspects of model building. Three basic approaches are found in the watershed cases. First, expert opinion-based indicators are typically expressed directly using a standardized (e.g., −1 to +1) continuous scale, or through the application of a categorical ruleset. For example, in the USFS R6 "distribution" indicator [4] asks for an expert judgment of the percent of potential habitat occupied, with 0–100 % judgment transformed in a continuous linear fashion to the −1 to +1 scale. Their "population status" indicator asks for categorical judgments, which have standardized values (e.g., Strong population = +1, Depressed population = +0.5, and so on).

A second approach to setting criteria uses an external standard. Specific thresholds are set based on measured or estimated levels of impact, e.g., the −1 threshold is >2.4 road mi·mi^{-2} of watershed, and the +1 threshold is 0 mi·mi^{-2}. Sometimes management or regulatory standards already exist and can be applied. For example, the EPA study [6] makes use of numerous existing water quality regulatory standards. Often, however, no such standards exist: rarely is a definitive, specific impact study available to set these standards, and so the modelers must rely on some combination of expert judgment and synthesis of the literature. As discussed in Gordon and Gallo (2011), tracking this interpretive process is a major challenge, but one that is crucial for the replicability and validity of the model.

Another source of external criteria is called reference conditions, which are usually derived from the range of values found in undisturbed watersheds or historical estimates of site condition. Several cases [4, 6, 11] use this approach for multiple indicators. Given the natural variability of watershed attributes and reasonable management expectations, reference distributions still require some interpretation to set evaluation criteria. For example, Trout Unlimited's CSI indicator "percent of historic[al] watersheds occupied by populations" uses thresholds of 20 %/80 % to define the −1/+1 thresholds. Grey literature reports for some cases [4, 11] provide details on how reference conditions were chosen and established. Such details are not provided for the EPA model [6], likely due to the space constraints of the journal format, which points to the challenge in communicating models through traditional scientific publications.

The third major approach to setting criteria found in these cases is an internal approach, which sets levels relative to the range of values encountered in the analysis area. For example, the USFS R6 analysis [4] used the 25th/75th percentile values to set the −1/+1 thresholds for riparian vegetation, while California North Coast assessment [7] used the 10th/90th percentiles for attributes in its sediment model. The Oregon North Coast assessment [9] used the average spawning fish count for the region as their +1 threshold and zero spawners as the −1 threshold. Walker et al. (2007) refer to it as "a more empirically-based approach to defining reference curves"; however, "empirical" should not be confused with "objective" here, as the thresholds were subjectively derived. Reiss et al. (2010, p. 31) note the potentially hazardous assumption that is inherent in this approach, "By selecting these percentile values, we assumed that 25 % of all of the HUC6s on each forest were in good condition and 25 % were in poor condition with the rest distributed in between."

4.8 Model Structure

After individual indicators are evaluated, they are then combined into an overall model structure, which includes the hierarchical arrangement, aggregation functions, and weights (e.g., see Fig. 1 and Table 1). Model hierarchies are presented in all the cases reviewed here, but the rationale behind them is typically not documented. However, the underlying logic is generally evident in the structures. Most of the assessment model indicators are grouped by watershed process location (upslope, riparian, in-channel) and then indicator domain (roads, vegetation, etc); exceptions are the EPA model [6], which takes a more explicitly process-based approach, and the Sandy River case [8], whose overall structure is determined by the three different major data sources (see Table 1, Assessment Model Structure column).

As mentioned above, three of the cases include explicit prioritization models, which integrate additional indicators to the assessment model results. The Upper Chewaucan case [3] had separate models for protection and restoration goals, both of which included factors relating to feasibility and efficacy of possible actions. The USFS Region 1 prioritization model [5] similarly included feasibility metrics, but also included a section on risk: risk of severe fires (from vegetation attributes) and risk of sedimentation (based on road density).

Environmental context switches also play an important part in some of the watershed model hierarchies. Such switches select different evaluation pathways in the model depending on the value of an environmental attribute. One of the most common examples is the use of stream gradient. In reach-level models, some cases [1, 2] only evaluate certain indicators on low gradient streams ($\leq$3 %), and in some watershed models [1, 4] roads along low gradient streams are evaluated differently from high gradient (>3 %) streams. Scores for roads in areas deemed geologically sensitive also receive from 50 to 100 % higher weighting in some models [1, 4]. Vegetation attributes naturally vary by a number of environmental factors, and elevation and precipitation are used as context switches in a few models [1, 4]. The USFS national model [12] simply allows a few indicators to be designated "not applicable": forest cover, rangeland vegetation, large woody debris, mass wasting. Given the strong spatial patterning in many landscape attributes, it is surprising that more models do not include such context factors. Explanations may include the limited scale of some cases (a relatively homogenous area), or it is an acceptable simplifying factor based on the overall distribution within a watershed. In support of the scale argument, the NWFP [1], as the largest application case, actually developed separate models for six subregions. This complete division allowed considerable variability between regions, however, and it became difficult to discern which differences were truly due to ecological factors versus involvement of different panels of experts. The team is now planning to consolidate the models into one using context switches based on explicit ecological factors.

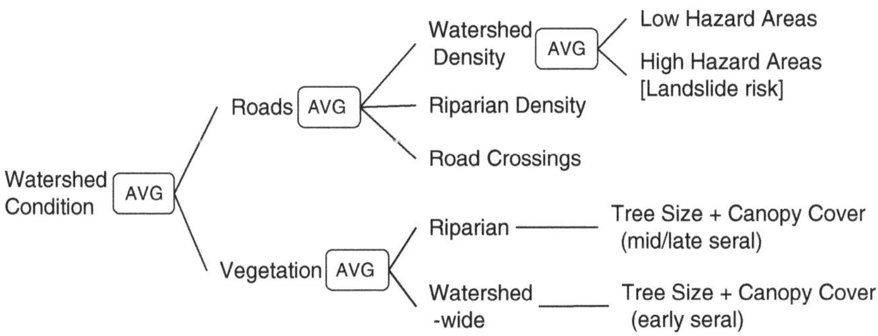

Fig. 1 Example model structure for assessing watershed condition ('AVG' represents use of an averaging aggregation function)

EMDS-Netweaver provides a wide variety of mathematical and logic operators for combining the standardized indicator scores. The most prevalent ones in the watershed models are the UNION and AND operators, which correspond to a partially-compensatory (i.e., average) or non-compensatory (minimum, limiting factor) functions, respectively. As users of Netweaver know, the AND function is rather ingenious and more complex than a simple minimum, but can be difficult to explain. However, NetWeaver also includes a true minimum math operator. The choice between these operators typically has significant consequences for model results. An early draft of the NWFP model [1] used primary ANDs, but the results were judged by experts to be unrealistically poor. The California North Coast assessment [7] found that using non-compensatory operators seemed more appropriate when considering certain low-level indicators, but that partially-compensatory performed better in higher levels of the model hierarchy.

Weighting is a third factor in the model structure. Five out of the twelve cases appear to have implemented some type of weighting (beyond the default weighting inherent in the model structure) [1, 2, 4, 5, 9]. The USFS R6 case [4] included weighting at the indicator level that was tied to their use of context variables: roads along low gradient streams and on sensitive soils received higher weights. The NWFP and NW Anchor Habitats cases [1, 9] included weighting at the indicator level and also further up in the hierarchy. Major differences between assessments of the same watersheds by the NWFP and USFS national [12] assessments were traced back to differences in the weighting of roads, both explicitly and through its averaging with a generally highly rated "soils" attribute in the latter. Clearly, care must be taken in the effects of model weighting and structuring. Steele et al. (2009) make the further point that relationships between criteria weights and evaluation criteria should also be considered. Few of the cases used the Priority Analyst portion of EMDS, but the USFS R1 case [5] demonstrated how models with different weightings can be used to prioritize for different objectives (improve watershed condition, reduce hazardous fuels in wildland urban interface, protect developed recreation values).

5 Conclusion

Watershed assessment is one of the oldest and most frequent application areas for EMDS. Watershed condition is a complex concept, and cases have struggled with representing dynamic processes with indicators based on ecosystem states. Structuring models more according to processes has faced the additional hurdle of overcoming flawed assumptions about multi-collinearity from standard statistical training.

The models built in the cases reviewed here varied considerably, depending on the purposes of the study (e.g., assessment versus prioritization), but also on available data (particularly fish habitat versus fish populations). In contrast, broad consistency in the basic unit of analysis (HUC6) is a benefit for data sharing and comparison of results, although more attention is needed to the versioning of these boundaries. Various efforts have come up with differing assessments for the same watersheds (e.g., cases 1 and 12), apparently more due to unintentional model structuring differences than divergent purpose. Care must be taken in the effects of model weighting and structuring, down to the bottom level of indicators and their evaluation criteria.

The complexity of the contributions to watershed condition means that data for key attributes are often not available. A number of the cases filled in these gaps with expert knowledge., EMDS is well-suited to integration of qualitative and quantitative data sources. Development of key sets of evaluation criteria is always a major challenge, and most models rely heavily on expert knowledge and literature assessments. For reliability and replicability, good documentation of these judgments is essential. While I have suggested a few areas where EMDS might provide further support in terms of software functionality, such as with temporal and spatial relations, many challenges lie in the broader domains of conceptual development and use of the results. Such needs may be best addressed through better sharing of information among a community of practitioners, and I would suggest that facilitating such ongoing dialog is a high priority for the future of the software.

References

Bleier C, Downie S, Cannata S, Henly R, Walker R, Keithley C, Scruggs M, Custis K, Clements J, Klamt R (2003) North coast watershed assessment program methods manual. California Resources Agency and California Environmental Protection Agency, Sacramento

CWA [Clean Water Act] (1977) 33 U.S.C. s/s 1251 et seq

Dai JJ, Lorenzato S, Rocke DM (2004) A knowledge-based model of watershed assessment for sediment. Environ Model Software 19(4):423–433

Dunham J, Gallo K, Shively D, Allen C, Goehring B (2011) Assessing the feasibility of native fish reintroductions: a framework applied to threatened bull trout. N Am J Fish Manage 31(1):106–115

ESA [Endangered Species Act] (1973) 16 U.S.C. 1531–1536, 1538–1540

Gallo K, Lanigan SH, Eldred P, Gordon SN, Moyer C (2005) Northwest Forest Plan—the first 10 years (1994–2003): preliminary assessment of the condition of watersheds. PNW-GTR-647. USDA Forest Service Pacific Northwest Research Station, Portland

Gordon SN (2006) Decision support systems for forest biodiversity management. In: a review of tools and an analytical-deliberative framework for understanding their successful application. Dissertation, Oregon State University, Corvallis

Gordon SN, Gallo K (2011) Structuring expert input for a knowledge-based approach to watershed condition assessment for the Northwest Forest Plan. USA. Environ Monit Assess 172(1):643–661

Hayslip GA, Herger LG, Leinenbach PT (2004) Ecological condition of western cascades ecoregion streams. U.S. Environmental Protection Agency, Region 10, Seattle

Hubler S, Miller S, Merrick L, Leferink R, Borisenko A (2009) High level indicators of Oregon's forested streams. DEQ09-LAB-0041-TR. Oregon Dept. Environmental Quality, Laboratory and Environmental Assessment Division, Hillsboro

Jensen M, Reynolds K, Langner U, Hart M (2009) Application of logic and decision models in sustainable ecosystem management. Paper presented at the 42nd Hawaii international conference on system sciences, Hawaii, HI, 5–8 Jan 2009

Kaufmann PR, Hughes RM (2006) Geomorphic and anthropogenic influences on fish and amphibians in Pacific Northwest coastal streams. In: Hughes RM, Wang L, Seelbach PW (eds) Landscape influences on stream habitats and biological assemblages, American Fisheries Society symposium 48, Bethesda, MD

Lanigan SH, Gordon SN, Eldred P, Isley M, Wilcox S, Moyer C, Andersen H (2012) Northwest Forest Plan—the first 15 years (1994–2008): status and trend of watershed condition. PNW-GTR-856 USDA Forest Service, Pacific Northwest Research Station, Portland, OR

Miewald T, Van Dyk B, Reeves G (2008) Oregon North Coast Salmon conservation assessment. Wild Salmon Center, Portland. http://www.wildsalmoncenter.org/pdf/WSC_OR_NCSalmon Assmnt08.pdf. Accessed 28 Aug 2013

Naiman RJ, Beechie TJ, Benda LE, Berg DR, Bisson PA, MacDonald LH, O'Connor MD, Olson PL, Steel EA (1992) Fundamental elements of ecologically healthy watersheds in the Pacific Northwest coastal ecoregion. In: Naiman RJ (ed) Watershed management: balancing sustainability and environmental change. Springer, New York

NFMA [National Forest Management Act] (1976) Act of October 22; 16 U.S.C. 1600

NOAA Fisheries (1996) Making endangered species act determinations of effect for individual or grouped actions at the watershed scale. National Oceanic and Atmospheric Agency, Fisheries Division. http://www.nwr.noaa.gov/publications/reference_documents/esa_refs/matrix_1996.pdf. Accessed 28 Aug 2013

Pess GR, Montgomery DR, Steel EA, Bilby RE, Feist BE, Greenberg HM (2002) Landscape characteristics, land use, and coho salmon (*Oncorhynchus kisutch*) abundance, Snohomish River, wash. USA Can J Fish Aquat Sci 59:613–623

Potyondy JP, Geier TW (2011) Forest Service watershed condition classification technical guide. USDA Forest Service, Washington. http://www.fs.fed.us/publications/watershed/watershed_classification_guide.pdf. Accessed 28 Aug 2013

Reeves GH, Benda LE, Burnett KM, Bisson PA, Sedell JR (1995) A disturbance-based ecosystem approach to maintaining and restoring freshwater habitats of evolutionarily significant units of anadromous salmonids in the Pacific Northwest. In: Nielson JL, Powers DA (eds) Evolution and the aquatic ecosystem: defining unique units in population conservation, vol 17. American Fisheries Society, Bethesda, pp 334–349

Reeves GH, Hohler DB, Larsen DP, Busch DE, Kratz K, Reynolds K, Stein KF, Atzet T, Hays P, Tehan M (2004) Aquatic and riparian effectiveness monitoring plan for the Northwest Forest Plan. PNW-GTR-577 USDA Forest Service, Pacific Northwest Research Station, Portland

Reiss KY, Gallo K, Konnoff D, Dawson P (2010) Process for evaluating the contribution of national forest system lands to aquatic ecological sustainability: a regional pilot process conducted on the Okanogan-Wenatchee and Colville National Forests. USDA Forest Service, Region 6 (Unpublished draft)
Reynolds K, Peets S (2001) Integrated assessment and priorities for protection and restoration of watersheds. In: Proceedings of the IUFRO 4.11 conference on forest biometry, modeling and information science, Greenwich, 26–29 June 2001
Reynolds K, Reeves GH (2003) Role of knowledge-based systems in analysis and communication of monitoring data. In: Busch DE, Trexler JC (eds) Interdisciplinary approaches to ecological monitoring of ecosystem initiatives. Island Press, New York
Reynolds KM, Saunders MC, Foster M, Olson R, Schmoldt D, Latham D, Miller BJ, Steffenson J (1996) A knowledge-based information management system for watershed analysis in the Pacific Northwest US. AI Appl 10(2):9–22
Reynolds KM, Jensen ME, Andreasen J, Goodman I (2000) Knowledge-based assessment of watershed condition. Comput Electron Agric 27:315–333
Reynolds KM, Rodriguez S, Bevans K (2002) Ecosystem management decision support 3.0 user guide. USDA Forest Service, Pacific Northwest Research Station, Corvallis
SBRWG [Sandy River Basin Working Group] (2006) Salmon and steelhead conservation: an assessment of anchor habitat on the Sandy River, Oregon Trout, Portland. http://www.thefreshwatertrust.org/sites/fwt.ortrout.org/files/pdf/SandyHabRpt.pdf. Accessed 28 Aug 2013
SBRWG [Sandy River Basin Working Group] (2007) Sandy River basin aquatic habitat restoration strategy: an anchor habitat-based prioritization of restoration opportunities. Oregon Trout, Portland. http://www.sandyriverpartners.org/pdfs/SandRestStrategWEB.pdf. Accessed 28 Aug 2013
Seaber, Paul R, Paul Kapanos F, Knapp GL (1987) Hydrologic unit maps. United States Geological Survey Water-supply Papers 2294
Steele K, Carmel Y, Cross J, Wilcox C (2009) Uses and misuses of multi-criteria decision analysis (MCDA) in environmental decision making. Risk Anal 29(1):26–33
Trout Unlimited (2007) Trout unlimited's conservation success index user guide 3.0. http://www.tu.org/science/conservation-success-index. Accessed 28 Aug 2013
US OMB [U.S. Office of Management and Budget] (2006) Forest Service watershed program assessment. Washington. http://georgewbush-whitehouse.archives.gov/omb/expectmore/detail/10003029.2006.html. Accessed 8 Aug 2013
USDA FS [U.S. Department of Agriculture Forest Service] (2008) Aquatic and riparian conservation strategy. Pacific Northwest Region, Portland. http://www.fs.usda.gov/Internet/FSE_DOCUMENTS/stelprdb5316591.pdf. Accessed 28 Aug 2013
USDA FS [U.S. Department of Agriculture Forest Service] (2011) Watershed condition framework: a framework for assessing and tracking changes to watershed condition. Report FS-977, U.S. Department of Agriculture, Forest Service, Washington
USDA FS [U.S. Department of Agriculture Forest Service] (2009) Stream inventory handbook: level I and II. USDA Forest Service, Region 6, Portland
USDA NRCS [U.S. Department of Agriculture, Natural Resource Conservation Service] (2012) What is the WBD? http://nhd.usgs.gov/wbd.html. Accessed 28 Aug 2013
USDA, USDI (1994) Record of decision for amendments to forest service and bureau of land management planning documents within the range of the northern spotted owl. U.S. Department of Agriculture Forest Service and U.S. Department of Interior Bureau of Land Management, Washington
USFWS [U.S. Fish and Wildlife Service] (1990) Endangered and threatened wildlife and plants: determination of threatened status for the northern spotted owl. Fed Reg 55:26114–26194
USGS [U.S. Geological Survey] (2012) National hydrography dataset. http://nhd.usgs.gov/index.html. Accessed 28 Aug 2013

Walker R, Keithley C, Henly R, Downie S, Cannata S (2007) Ecosystem Management Decision Support (EMDS) applied to watershed assessment on California's north coast. In: Standiford RB, Giusti GA, Valachovic Y, Zielinski WJ, Furniss MJ (eds) Proceedings of the Redwood Region Forest Science Symposium: What Does the Future Hold? PSW-GTR-194. USDA Forest Service Pacific Southwest Research Station, Albany CA

Williams JE, Haak AL, Gillespie NG, Colyer WT (2007) The Conservation Success Index: synthesizing and communicating salmonid condition and management needs. Fisheries 32(10):477–493

The Integrated Restoration and Protection Strategy of USDA Forest Service Region 1: A Road Map to Improved Planning

Patrick Bourgeron, Hope Humphries, Chip Fisher, Barry Bollenbacher and Keith Reynolds

Abstract Core design components of the Ecosystem Management Decision Support system were used to develop and implement the integrated restoration and protection strategy of the Northern Region of the U.S. Department of Agriculture Forest Service. Scenarios that spatially optimized hazardous fuel reduction, protected developed recreation values, and improved watershed conditions are presented to illustrate how the evaluation and decision modeling capabilities of the decision support system can be used sequentially in both strategic and tactical planning.

Keywords Landscape analysis · Restoration planning · Departure analysis · Vegetation pattern · Vegetation structure · Spatial decision support · EMDS

P. Bourgeron (✉) · H. Humphries
Institute of Arctic and Alpine Research, University of Colorado, 1560 30th Street, Boulder, CO 80309, USA
e-mail: patrick.bourgeron@colorado.edu

H. Humphries
e-mail: hope.humphries@colorado.edu

C. Fisher · B. Bollenbacher
U.S. Department of Agriculture, Forest Service, Northern Region, 200 East Broadway Street, Missoula, MT 59802, USA
e-mail: cfisher@fs.fed.us

B. Bollenbacher
e-mail: bbollenbacher@fs.fed.us

K. Reynolds
Corvallis Forestry Sciences Laboratory, U.S. Department of Agriculture, Forest Service, Pacific Northwest Research Station, 3200 SW Jefferson Way, Corvallis, OR 97331, USA
e-mail: kreynolds@fs.fed.us

K. M. Reynolds et al. (eds.), *Making Transparent Environmental Management Decisions*, Environmental Science and Engineering,
DOI: 10.1007/978-3-642-32000-2_5,

1 Introduction

The U.S. Department of Agriculture, Forest Service (USFS) embraced the philosophy of "ecosystem management" in its 2008 Planning Rule direction concerning the multiple-use, sustained-yield management of its National Forest System lands (USDA Forest Service 2007). According to Christensen et al. (1996), ecosystem management is "driven by explicit goals, executed by policies, protocols, and practices, and made adaptable by monitoring and research based on our best understanding of the ecological interactions and processes necessary to sustain ecosystem composition, structure, and function." In this light, a major requirement of the 2008 Planning Rule was that forest plans (the primary land and resource management plans of the USFS) provide a strategic vision for maintaining the sustainability[1] of ecological, economic, and social systems across USFS lands. Sustainability consists of realizing desired social, economic, and ecological conditions and trends that interact at varying spatial and temporal scales, and embody the principles of multiple-use and sustained yield.

There is increasing evidence that ecosystem resilience[2] is needed to reach the goal of sustainable management and sustainable ecosystems (Walker et al. 2002; Brand 2009). Indeed, resilience has been described as one of the core underpinnings of sustainable states.[3] Thus, identifying resilience mechanisms must be a primary objective of integrated ecological assessments (Bourgeron et al. 2009), the results of which can then be used to frame and focus ecosystem management.

In forest planning, the strategic vision for ecosystem management is articulated by identifying desired conditions for key ecosystem components that are to be achieved over a 50- to 100-year planning horizon. For example, maintenance of terrestrial ecosystem sustainability involves two primary components: ecosystem diversity and species diversity—central elements of ecosystem management and stewardship (Chapin et al. 2010). Forest plans also include objectives that provide measurable and time-specific (5- to 10-year reporting cycles) projections of management activities and related product flows needed to achieve desired conditions. In this context, the Integrated Restoration and Protection Strategy (IRPS) of the Northern Region (USDA Forest Service) assists with tactical planning for the implementation of strategic forest plan objectives.

The Northern Region of the USFS (including the States of Montana, North Dakota, northern Idaho, and small portions of South Dakota and Wyoming, Fig. 1) recently updated its IRPS (www.fs.usda.gov/goto/r1/irps) using the framework of

---

[1] Sustainability is defined here as meeting the needs of the present generation without compromising the ability of future generations to meet their needs.

[2] The capacity of a system to absorb disturbance and reorganize while undergoing change so as to still retain essentially the same function, structure, identity, and feedbacks; Holling (2001), Walker et al. (2004).

[3] A sustainable state is one which satisfies minimum conditions for ecosystem resilience through time (Perman et al. 2003).

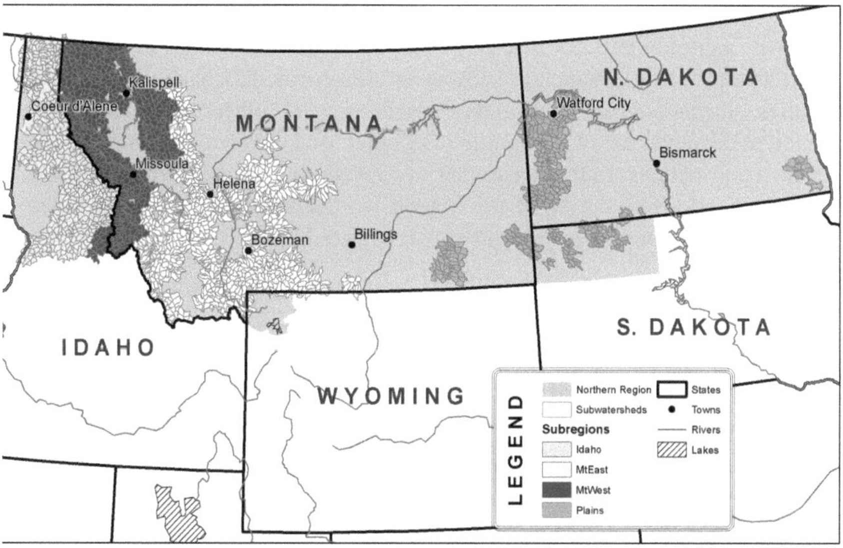

Fig. 1 Location map for the Northern Region of the USDA Forest Service. The figure shows the subwatersheds within the Region that were included in the IRPS assessment (all subwatersheds with at least 1 % USFS ownership) as well as the subregions used for subregional analyses

the Ecosystem Management Decision Support (EMDS) system (Reynolds et al. 2003). Use of the decision support system (DSS) provided a consistent, transparent, and reproducible approach to identifying and prioritizing restoration opportunities, while setting the context for collaboration among all stakeholders in an all-lands approach. The Northern Region IRPS provided information to help local units identify and prioritize potential watersheds for accomplishing forest and grassland plan goals and objectives. It was also intended to assist local units in developing and ranking integrated projects addressing land and water restoration; community wildfire protection plans; and sustainable, resilient, and desired conditions as described in forest and grassland management plans. It provides resource information on values that may be vulnerable or at risk to specific agents of change, including disturbance agents, to help units develop integrated projects. Planning processes such as IRPS are inherently complex. They require the participation of numerous actors, involve decisions within and across spatiotemporal scales and administrative boundaries, and are subject to rapid potential changes in short-term objectives.

The primary goal of this chapter is to provide an example of a regional IRPS implementation that used a flexible yet internally consistent DSS framework. The specific objective is to present its use in the second phase of the IRPS to guide the actual planning process when further plan simplification was required. EMDS was used to develop a prototype knowledge-based system for evaluating ecosystem sustainability (the desired conditions) and decision models to identify priority areas for integrated landscape restoration (the objectives) (Jensen et al. 2009). The result was a prototype DSS that addressed a subset of management issues, and served as a proof of concept for subsequent development. Changes in short term priorities, technological constraints, timelines, and other unforeseen factors resulted in development of a simplified DSS that used the core EMDS components and design principles. In the discussion, we revisit the potential role of EMDS in future applications of the Northern Region IRPS DSS, considering advances in technology since the current project was completed.

2 The Northern Region Integrated Restoration and Protection Strategy

The Northern Region IRPS assessed several key planning questions and identified opportunities and potential priorities for 19 key single resource values that may be at risk to current or projected disturbance and other agents of change, stratified into six individual themes (Table 1). These individual key resource scenarios became the resource objectives for which the DSS helped provide a potential spatial opportunity solution. The individual assessments considered the same three components as the EMDS prototype (Jensen et al. 2009): values, risks, and feasibility. The assessment identified a value (a key resource component), assessed current and projected risks or hazards associated with the value, and then by assigning a weight to these factors, determined the relative opportunity to minimize or reduce the risk factors to restore a more sustainable and resilient condition. In application at the regional scale, feasibility information was not readily available at the broad scale, but rather became a very important factor at finer scales for locating actual project areas. There are many examples of feasibility factors, including but not limited to Forest Plan Standards, Management Area Direction, access, partnership opportunities, and collaborative interest. Greater details of the Northern Region IRPS process, the DSS, and results are discussed in Reynolds et al. (2013).

2.1 Decision Model Design

The overall architecture of the Northern Region IRPS DSS was essentially a three-tiered decision model that retained the core design of the prototype EMDS while simplifying the process in response to technological constraints, changes in short-term priorities, concerns about the complexity of the DSS, and the need for a faster implementation of the strategy update.

At level 1 (the lowest level), the assessment data for each of the 19 scenarios were evaluated by a scenario-specific decision model. An example of this is the

Table 1 Primary resource objectives (scenarios) and associated theme areas in the IRPS

Theme 1: Restoration of forests, grasslands, and human communities to a more resilient condition
Scenario 1A: Community fire resilience
Question: Which human community areas are most critical for improving fire resilience due to burn probability or insect and disease risk?
Scenario 1B: Vegetation resilience and current departure from desired conditions in forested areas
Question: Which forest areas offer the best opportunities to improve vegetation resilience to meet forest plan desired condition due to high departure from historic conditions, crown fire burn probability, or insect and disease risk?
Scenario 1C: Ecosystem resilience and vulnerability in non-forested areas
Question: Which non-forest areas offer the best opportunities to improve non-forest ecosystem resilience due to composition of non-forest types with noxious weed risk and/or departure from historic fire regime (lack of fire) or grazing risk?
Theme 2: Restoration and maintenance of wildlife habitats, including restoration of more resilient vegetation conditions where appropriate, to meet ecological and social goals
Scenario 2A: Whitebark pine ecosystems
Question: Which forest areas offer the best opportunities to restore whitebark pine and associated habitats, considering the high level of mortality from white pine blister rust, mountain pine beetle, and high levels of wildfire burn probability due to succession to spruce-subalpine fir vegetation?
Scenario 2B: Low elevation dry forest communities
Question: Which forest areas offer the best opportunities to restore resiliency of composition and density of dry forest communities and associated habitats, given current and projected insect and disease and high levels of wildfire burn probability due to uncharacteristic high forest density?
Scenario 2C: Dry shrublands (low elevation sagebrush)
Question: Which low-elevation sagebrush areas offer the best opportunities to restore resilience of composition and density and associated habitats, given current levels of conifer encroachment and high levels of wildfire burn probability?
Scenario 2D: Aspen communities
Question: Which aspen habitat areas offer the best opportunities to restore resilience of composition and density and associated habitats, given current levels of conifer encroachment and high levels of wildfire burn probability?
Scenario 2E: Woody draw communities
Question: Which woody draw habitat areas offer the best opportunities to restore resilience of composition and density and associated habitats, given current levels of conifer encroachment, grazing, and high levels of wildfire burn probability?
Scenario 2F: Mixed grass prairie
Question: Which mixed grass prairie habitat areas offer the best opportunities to restore resilience of composition and density and associated habitats, given current levels of conifer encroachment, grazing, and high levels of wildfire burn probability?
Scenario 2G: Riparian areas, wetlands, and seeps
Question: Which riparian and wetland areas offer the best opportunities to restore resilience of composition and density and associated habitats, given current levels of noxious weed hazard, grazing, motorized access, and high levels of wildfire burn probability?
Scenario 2H: Big game winter range
Question: Which big game winter range areas offer the best opportunities to restore resilience of composition and density and associated habitats, given current levels of noxious weed hazard, current vegetation composition and structure vulnerability to disturbance agents, grazing, motorized access, and high levels of wildfire burn probability?

(continued)

Table 1 (continued)

Scenario 2I: Threatened and endangered core grizzly bear habitat
Question: Which core grizzly bear habitat areas offer the best opportunities to provide increased security, considering current open road and motorized trail access and other human disturbance potential?
Theme 3: Restoration and maintenance of resilient, high-value watersheds
Scenario 3: Watershed quality (sediment)
Question: Which subwatersheds are best for restoration due to municipal watershed use, section 303(d) listings, and/or presence of multiple risk factors?
Theme 4: Restoration of high-value fisheries streams—developing more resilient habitat
Scenario 4: Threatened, endangered, and sensitive fish species
Question: Which watersheds are best for restoration due to Forest Plan revision, aquatic species priority, or have multiple risk factors?
Theme 5: Restoration and protection of recreation sites and scenic vistas
Scenario 5A: Safety
Question: Which areas are most important to protect or restore due to high concentrations of use with existing or potential hazard trees due to insects and disease and in areas with high burn probability?
Scenario 5B: Investment protection
Question: Which areas are most important to protect from an investment perspective (e.g., high investment areas that are at risk of damage)?
Scenario 5C: Recreation setting restoration
Question: Which recreation settings are priority areas for restoration (e.g., high-use dispersed recreation areas in vulnerable subwatersheds)?
Scenario 5D: Scenic integrity restoration
Question: Which areas are most important to restore or enhance (e.g., high visibility areas with low or very low scenic integrity)?
Scenario 5E: Scenic integrity protection
Question: Which areas are most important to protect from degradation (e.g., highly visible areas with very high or high existing scenic integrity)?
Theme 6: Protection of people, structures and community infrastructure (roads, trails, bridges, power corridors, recreational developments, etc.) highlighting current and projected mountain pine beetle and wildfire effects
To consider public safety and protection of infrastructure, this theme uses scenarios 1A, 5A, and 5B a second time, based on current regional priorities
Question: Which community areas are best to improve fire resilience due to burn probability or insect and disease risk? (Scenario 1A)
Question: Which areas are most important to protect or restore due to high concentrations of use? (Scenario 5A)
Question: Which areas are most important to protect from an investment perspective? (Scenario 5B)

Within each theme, resource values or scenarios are assessed via a planning question

scenario that addressed resilient forest vegetation condition relative to Desired Condition (DC), S1B (Fig. 2a). It included an assessment of departure of dominance type (similar to forest cover type), contrasting existing condition as a percentage of area versus DC, departure of size class relative to DC, and departure of forest density relative to DC. Examples of risks include loss of western white pine type to root disease and homogenization of forest size classes that leads to susceptibility to disturbance agents such as mountain pine beetle.

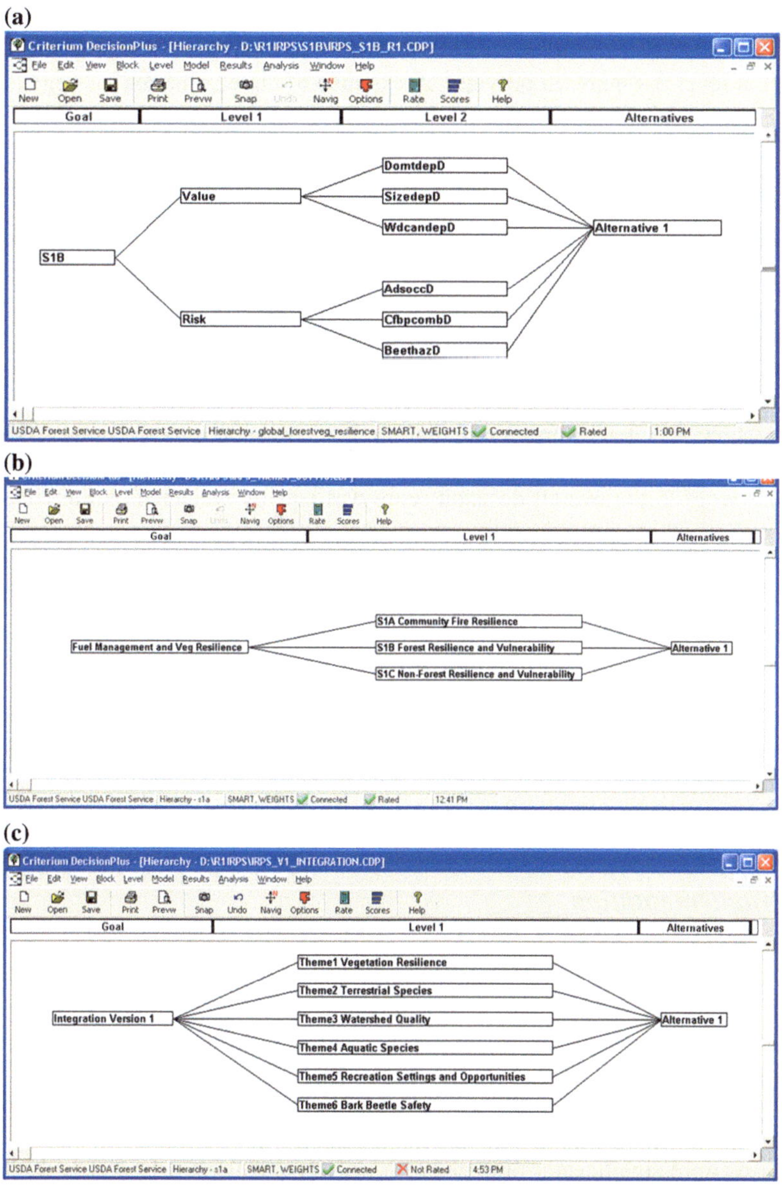

Fig. 2 Decision models used in the IRPS decision support system. **a** An example of the decision model for scenario S1b that addresses resilient forest vegetation condition relative to desired condition. Subcriteria considered under the value criterion include departure of dominance type (similar to forest cover type) relative to desired conditions (DomtdepD), departure of size class relative to desired conditions (SizedepD), and departure of forest density relative to desired condition (WdcandemD). **b** Decision model for priorities under theme 1 (restoration of forests and grasslands, directly adjacent human communities, to a more resilient condition). The goal object refers to the theme itself. Criteria at level 1 of the model, and prefaced by S1A, S1B, and S1C, refer to the three scenarios under the theme. **c** Top-level IRPS decision model, integrating across all themes. See Table 1 for definitions of all scenarios under each theme

At level 2, the 19 scenarios were organized into six broad themes (Table 1). A priority score for each theme was calculated for each theme-specific decision model at level 2. Figure 2b provides an example of three scenarios for vegetation resilience (theme 1). In the current version of the DSS, the priority score for a theme was simply calculated as the average priority score over all scenario components of the theme, meaning that all scenario priorities in a theme were equally weighted, and therefore contributed equally to the priority score for the theme. However, more generic decision models for theme priorities could easily be designed, allowing for differential weights on component scenario priorities, if desired by managers. Finally, at level 3 (Fig. 2c), an overall IRPS priority score was calculated, considering the contributions of the priorities of the six themes.

The overall DSS addressed the same question as the EMDS prototype (Jensen et al. 2009): Where in the Northern Region do all of the identified multiple values at risk show potential priority opportunities for restoration or protection of values to identified risk factors? At level 3, as in level 2, the contribution of each theme priority score to the overall decision score was equally weighted. An alternative approach might have weighted the priorities of contributing themes by the number of scenarios in the theme, a technique known as structural adjustment (Saaty 1992). The effect of structural adjustment is to ensure that all scenarios contribute equally to the final overall priority score. In other words, the contribution of any particular scenario is not diluted by belonging to a theme with a large number of scenarios. The decision to weight themes equally, rather than preserving the equality of scenario contributions, was made by Northern Region leadership, who felt it was preferable to maintain the equality of theme contributions to the overall priority score.

2.2 Implementation

Subwatersheds were used as the unit of analysis. During the implementation phase, restricting the analysis to subwatersheds with 10, 5, and 1 % USFS lands was considered. The Regional leadership team decided to include subwatersheds with at least 1 % USFS lands, because (1) the National Watershed Condition Assessment (Potyondy and Geier 2011) was being performed at that level of National Forest ownership, and (2) the Community Fire Resilience (S1A) and Safety (S5A) scenarios were significantly different with inclusion of these subwatersheds.

Selection of scenario (input) data sources was done with Regional and Forest specialists to identify potential value, risk, and feasibility data sources, and then mapping them to evaluate coverage and consistency issues. Initially, over 120 scenario data sources were identified. During the design phase, several scenario data sources were identified for use in multiple scenarios. For example, a bark beetle risk input was identified for use in six scenarios. In the implementation phase, scenarios with very similar inputs from different data sources were evaluated to simplify them to a single data source when possible. A few scenario inputs had consistency problems across the region because data were compiled from different sources, some regional and others local. An attempt was made to normalize local data sets when they were compiled into regional layers, but there were still noticeable differences when mapped. In this case, local spatial or thematic accuracy was gained at the expense of regional consistency.

A wide range of data types was proposed as inputs to the scenarios, including single feature GIS layers (vector and raster), modeled raster data from multiple sources, modeled vector data from multiple sources, data summaries from the USFS Forest Inventory and Analysis (FIA) program, tabular data summaries, and categorical or binary data. The wide range of data types presented unique challenges for characterizing values of scenario inputs to analysis units. A subwatershed GIS layer with all subwatersheds meeting the 1 % USFS lands requirement was created. A summary table was created from the GIS layer and an attribute was added for every scenario input summary value. As each scenario input summary was calculated, the table was filled in for every subwatershed. This created a simple spreadsheet approach, in which all scenario inputs were located together. Scenario opportunity scores could be calculated from the table, an important consideration for later processing steps.

Several scenario inputs were originally derived from NetWeaver logic models that interpreted and synthesized information from multiple data sources to produce a composite result ("NetWeaver"). However, concerns were raised about the complexity of the logic models and the software. Scenario inputs using NetWeaver logic models were simplified from multi-data-source logic models to single data sources. Scenario inputs that could not be simplified to single data sources were simplified to summaries of multiple data sources. An example is the *road risk* criterion in the Watershed Quality Scenario (S3A) that had three inputs: miles of roads, miles of road within 60 m of streams, and number of stream crossings. To simplify the process, an attribute was added to the value and risk input table for each of the road risk components and then a formula was used to combine the three inputs into a single road risk input:

$$\begin{aligned} \mathit{road\,risk} = {} & (\mathit{road\,density} * 0.2) + (\mathit{riparian\,road\,density} * 0.4) \\ & + (\mathit{stream\,crossings} * 0.4) \end{aligned}$$

in which the three input fields were each first normalized to a standard [0, 1] range (see below). All scenario inputs with multiple components similarly had their attributes added to the value and risk input table so they could be recalculated if necessary.

Many decision models employ a general formula of normalizing inputs, multiplying each input by a weight, and then summing the results to obtain an overall decision score. Because the value and risk inputs had different data types and ranges, they were normalized to a [0, 1] range so they could be summed. A minimum–maximum normalization process was used for all scenario inputs. The minimum and maximum range for each scenario input was based on the range of data values over all subwatersheds in the analysis. To meet time constraints and to simplify the process, the scenario opportunity scores were calculated in ArcMap using the following general formula:

$$\begin{aligned} \textit{Scenario Score} = {} & ((\textit{value}1 - \textit{value}1_{\min}) / (\textit{value}1_{\max} - \textit{value}1_{\min}) * \textit{weight}_{\text{value1}}) \\ & + ((\textit{value}2\ldots) + (\textit{risk}1\ldots) + \ldots) \end{aligned}$$

Decision-model weights for scenario inputs could be changed, and the results viewed in ArcMap in a manner analogous to EMDS. This allowed Regional and National Forest staff to try several versions of scenario inputs and decision-model weights to test the reasonableness of model outputs.

After the scenarios and themes were finalized, four subregional areas were analyzed separately. It was noted that several scenarios had significant differences across the Northern Region due to ecological or resource factors. The Northern Region was spatially partitioned into the Northern Idaho (Idaho Panhandle, Clearwater, and Nez Perce Forests), Western Montana (Kootenai, Flathead, Lolo, and Bitterroot Forests), Eastern Montana (Beaverhead-Deerlodge, Lewis-Clark, Helena, Gallatin Forests, and Beartooth Ranger District [RD] of the Custer Forest), and Plains (Ashland and Sioux RD of the Custer Forest and the Dakota Prairie Grasslands) subregions (Fig. 1). In this case, only subwatersheds within a subregion were used to generate the minimums and maximums for each value and risk input during normalization. In the wildlife theme, some scenarios were excluded from the theme score if the resource did not occur in the subregion.

2.3 IRPS Products

For each scenario, values, risks, and feasibility were assessed for every subwatershed in the Northern Region analysis area (n = 2132). Formulas (Reynolds et al. 2013) were used to calculate opportunity scores for each subwatershed. Higher opportunity scores indicate greater potential opportunity for restoration or protection of a given resource.

Some key findings from the assessment for restoration and protection are:

1. Significant departure from desired forest conditions has resulted in less than desired resilience of forest vegetation as identified in theme 1, which emphasizes the need to:
    a. prioritize restoration of tree composition of western white pine, whitebark pine, western larch, aspen, and ponderosa pine;
    b. reduce forest density on dry forest types (ponderosa pine and Douglas-fir); and
    c. reduce invasive species affecting native ecosystems.
2. Theme 2 indicates priority restoration of wildlife habitat in short supply.

3. Theme 3 indicates restoration of watershed function, including reduction of sedimentation and chemical contamination, and protection or restoration of municipal and watershed water quality.
4. Theme 4 indicates restoration of key fish species habitat.
5. Theme 5 emphasizes restoration and protection of recreation facilities and scenic landscapes.
6. Theme 6 emphasizes protection of people associated with social infrastructure.

Mapped solutions of each of the 19 management concerns indicated the relative priority of potential opportunities to restore aquatic and terrestrial ecosystems, and to protect or sustain many ecosystem services. When scenario assessments were aggregated into the six themes, and then integrated into a single IRPS model across all themes, potential watersheds with multiple resource priorities in the same areas were identified, suggesting areas where the agency could pursue actions that can meet multiple objectives.

Maps (http://www.fs.usda.gov/goto/r1/irps) and histograms of opportunity scores by subwatershed were produced for each of the IRPS themes and scenarios. These maps provide a spatial representation of the key findings across the Northern Region. Analogous map products for each of the four subregions were also created to show how opportunity scores change when evaluating against only those subwatersheds in a geographic subset of the region; Fig. 3 presents mapped opportunity scores for the six themes in northern Idaho and western Montana.

The map for theme 1 (Fig. 3a) summarizes which vegetation communities are most vulnerable, due to their present condition, to disturbance risk agents such as severe fire and potential and actual bark beetle outbreaks. Included in the map for theme 2 (Fig. 3b) are key wildlife habitats that are most vulnerable to disturbance risk factors such as severe fire, bark beetle potential, noxious weeds, and increased forest density. The map for theme 3 (Fig. 3c) includes watersheds that have relative opportunities to reverse trends from risk factors such as too many stream crossings, abandoned mines leaking toxic chemicals, grazing in riparian areas, and high probabilities of severe insect and fire disturbances in the future. The map for theme 4 (Fig. 3d) shows relative opportunities to address key fish species habitat with risk factors that include fish passage problems, road crossings, grazing in riparian areas, abandoned mine sites near streams, dispersed recreation sites next to streams, and water diversions such as dams. In the map for theme 5 (Fig. 3e), relative opportunities are indicated for improving conditions associated with both developed and dispersed recreation sites, as well as opportunities to improve or protect scenery. The map for theme 6 (Fig. 3f) shows relative opportunities to address public safety issues within social infrastructure developments such as in the "wildland-urban interface", developed recreation sites, roads, and power lines that have risk factors caused by the potential for severe fire or bark beetle outbreaks. A map integrating all of the themes illustrates potential locations for addressing multiple management objectives for restoration or protection in a subregional area (Fig. 4).

The DSS products are currently being used as a starting point, combined with local site-specific information such as input from partnerships and collaborative

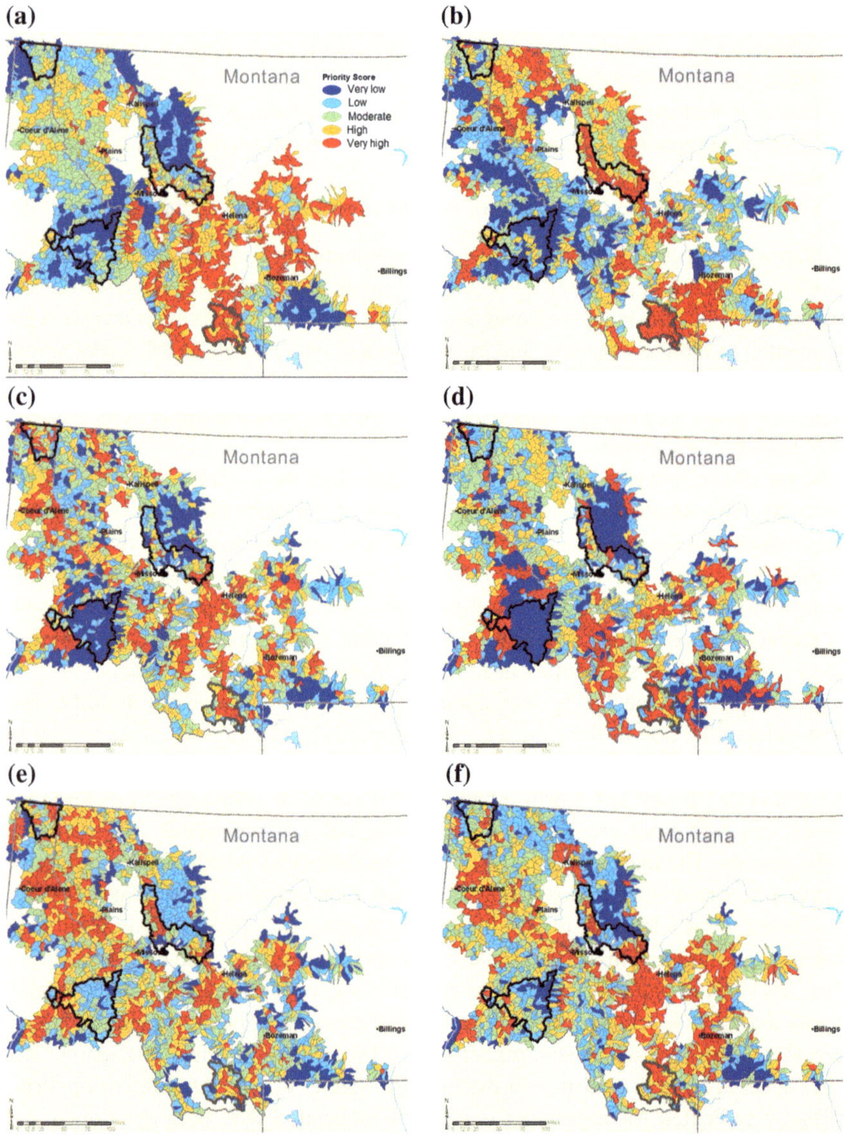

Fig. 3 Northern Idaho and western Montana opportunity scores for **a** restoration of vegetation composition and structure that is vulnerable to uncharacteristic disturbances due to departure from desired conditions; **b** restoration of wildlife habitat vulnerable to multiple risk factors; **c** watershed management and water quality restoration; **d** aquatic species habitat restoration; **e** restoration and protection of recreation facilities; and **f** public safety and infrastructure protection. *Black boundaries* indicate Collaborative Forest Landscape Restoration Act areas

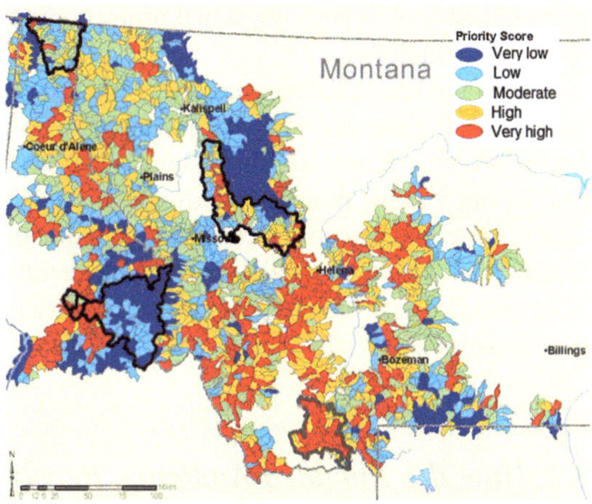

Fig. 4 Theme integration to achieve multiple objectives in the same area at the same time

groups, to identify and sequence priority integrated restoration proposals. This has occurred within Collaborative Forest Landscape Restoration Program (CFLRP) areas on multiple Forests, and has occurred Forest-wide on several Forest Units.

3 Discussion

3.1 What Worked Well

The overall approach in the design of the EMDS prototype and subsequent IRPS application required identification of key resource objectives (as reflected in scenario values) by resource specialists. However, it is important to note that managers (line officers, in the case of the Forest Service) also were critical participants in the process to validate that these were, in fact, the important issues to address in the Northern Region IRPS. The overall assessment of priorities was designed to respond to planning questions related to particular resource values. Identification of the associated objectives gave the strategy team a more integrated perspective of restoration and protection objectives.

Both the EMDS prototype and the subsequent IRPS application required a large volume of data and GIS layers due to the number of planning questions addressed in the assessments. This situation had both positive and negative implications. On the negative side, the data sets were time-consuming and expensive to develop. However, once developed, the data sets in aggregate were seen as a very valuable and powerful asset for the Northern Region, providing context for finer scale evaluations. These default data sets were highly consistent across the entire regional landscape, and can subsequently be enhanced with local data to be much more effective at identifying actual project opportunities in the context of regional landscape conditions.

One of the key findings, after completing the IRPS assessment and evaluation at the regional scale, was that the potential opportunity areas not only provided a useful starting point for discussions, but also afforded a useful context for developing feasible project-opportunity areas. When locally determined feasibility factors are combined with a consistent identification of value and risk factors, local project areas ripe for consideration can be identified. In addition to the 19 regional-scale resource assessments, others may be added at the local level to better address questions such as "why here?" and "why now?"

3.2 How the Intended Audience Received the IRPS DSS

The effectiveness of any assessment depends equally on the scientific methodology and the participation of key actors. At least initially, there has been mixed reaction to the IRPS DSS and subsequent assessment by Forest and Grassland Units. This mixed reaction was based on differences in local interpretation of how the assessment was intended to be used. The USFS units that perceived it as a consistent methodology, to which additional items (such as feasibilities factors) could be added, found it a useful starting point and a consistent framework that can be applied at finer scales. This was indeed a primary purpose of the application. In addition, the overall approach presented here offers the opportunity to integrate other assessments related to major national initiatives at regional and finer scales, for example, by integrating those assessments as new scenarios to be included in an overall opportunity assessment to support forest planning or identification of project-opportunity areas. On the other hand, the USFS units that interpreted the IRPS as a final solution for priority opportunities, or were concerned that this framework had direct and immediate implications for budget allocation to units, were intent on showing why it was not, or should not be, a final solution.

4 Conclusions

Use of EMDS-based analysis methods enabled evaluation of multiple resource values in a transparent manner and produced output maps that displayed high-priority treatment areas. The weighting used for the scenarios can be modified to meet changing needs or modified as information and knowledge increase over time. A by-product of this work was the development of a consistent set of region-wide data themes that have been added to the Northern Region's spatial data library for subsequent use in planning efforts. The result of this work provides consistent interpretations of ecosystem status for future monitoring of current and desired conditions.

Results from the IRPS process in the Northern Region and its associated EMDS prototype (Jensen et al. 2009) suggest that knowledge-based systems such as EMDS are well suited to both strategic and tactical planning, and the following points merit consideration in future National Forest (and other land management) planning efforts:

- Logic models provide a consistent, transparent, and reproducible method for evaluating broad propositions about ecosystem sustainability and resilience. For example: are watershed integrity, ecosystem and species diversity, social opportunities, and economic integrity in good shape across a planning area? The ability to evaluate such propositions in a formal logic framework also allows users the opportunity to determine statistical changes in outcomes over time, which could be very useful for regional and national reporting purposes and for addressing litigation.
- The use of logic and decision models in strategic and tactical Forest planning provides a repository for expert knowledge (corporate memory) that is critical to evaluation and management of ecosystem sustainability and resilience over time. This is especially true for the USFS and other federal resource agencies, which are likely to experience rapid turnover in resource specialist positions within the next several years due to retirements.
- Use of NetWeaver scores in decision models is an efficient and effective method for synthesizing the typically large amounts of information needed to support integrated landscape restoration (Jensen et al. 2009). Moreover, use of logic and decision models to design customized scenarios for integrated landscape restoration offers substantial improvements to traditional GIS-based procedures such as suitability analysis. In particular, the approach demonstrated by Jensen et al. (2009) is not only much more flexible, but also can more easily accommodate much greater complexity than traditional approaches.

4.1 Opportunities for Improvement

The following recommendations would improve various dimensions of the IRPS DSS and its implementation during the assessment phase:

- *Capturing locally available data.* We anticipate the datasets and model will be improved and modified with locally available data.
- *Sensitivity analysis.* The IRPS can be analyzed to determine the importance of the information datasets included in the analysis as well as model weights. Mathematically, the influence of each resource information dataset included in the analysis can be analyzed to determine how much it contributes to the overall score of the resulting prioritization. The most influential datasets should be reviewed to determine whether they accurately represent the management situation on the ground. Concurrently, the weights of the individual datasets or scenarios can be quickly and easily changed to determine the sensitivity of

weights to the different scenarios. More time and attention should be given to those weights that have the greatest impact on overall scenario scores. Additionally, this type of analysis can be used in a collaborative setting to allow collaborators to explore and understand the implications of the assumptions in the model on the resulting opportunity scores.

- *Continuous feedback*. Most datasets used in the Northern Region IRPS analysis are in a constant state of flux. New information becomes available to bolster the existing data, catastrophic disturbance events occur, roads are built and decommissioned, and watershed restoration work is implemented. These are but a few of the changes that occur and cause the data to become obsolete shortly after they are obtained. Therefore, if one could capture this information in an efficient manner, the datasets and model process could be updated periodically to assist with making current, informed decisions.
- *Future use of EMDS*. The regional IRPS DSS closely replicates data processing methods in EMDS, without using the EMDS software to calculate scenario values, partly in response to real and perceived technological constraints. Future versions of EMDS should be able to provide sufficient technological benefits to warrant full or partial adoption for the IRPS analysis process. For example, the next planned release of EMDS will facilitate development of web-based applications ("EMDS 5.0 and Beyond") that would allow internal users and external partners easy access to the datasets and assumptions of the IRPS product, or allow them the ability to quickly and transparently modify some of the assumptions to assess the implications of alternate management schemes.

4.2 Future Applications

The IRPS datasets are region-wide in spatial extent, and include subwatersheds in which at least 1 % of the land area is managed by the Forest Service. This results in a database that includes a significant portion of the landmass in Montana and northern Idaho. Within these subwatersheds, the condition of the land owned and administered by the USFS is known, and in some cases, the condition of non-USFS land is known as well. Most appropriately, analyses can be conducted at the USFS ownership level and on a scale that covers the entire Northern Region. Issues common to all ownerships across the entire Region, such as water quality, can be assessed to determine the key subwatersheds within the Region that could benefit from concentrated restoration activity.

The IRPS framework creates a platform on which future-year planning can occur. Three additional steps should be taken for this to occur. First, the datasets should be updated to include better locally-maintained data where appropriate. Second, the scenarios may need to be reformulated to better describe the issues of more local planning efforts. For instance, if the primary goal of a local plan is to schedule timber activities from which the revenues will be used to improve deteriorated stream crossings, it may not make sense to include information about oil and gas wells. Third, additional feasibility criteria should be developed by the Forests to describe, for example, where it is possible for activities to occur. Such criteria could include timber management feasibility, or opportunities for prescribed burning, among others. Ultimately they should be aligned with the goals of the planning exercise.

Finally, the IRPS product has significant potential to facilitate interactions among partners and collaborators, especially in working meetings. The product can be rapidly modified for real-time, in-person updates that can be displayed in live meetings. This process facilitates immediate analysis of a number of different ideas, for which the effects of different weighting schemes and scenario compositions, for example, can be interactively displayed and evaluated with groups of collaborators.

Acknowledgments We would like to thank the following groups and individuals for contributing to the development and implementation of this project:

- The Fire Laboratory in Missoula, for developing and supplying the burn probability layers.
- The Geospatial Group in the Regional Office Engineering Department, for critiquing the IRPS product to identify areas for needed improvement and ongoing maintenance.
- The Regional Board of Directors, for their input on weighting the themes, which resulted in a useful product for Regional level decisions.

Patrick Bourgeron and Hope Humphries acknowledge support from the National Science Foundation's Dynamics of Coupled Natural and Human Systems program (DEB-1115068: "Dynamics of Coupled Natural and Human Systems in the Colorado Front Range Wildland/ Urban Interface: Causes and Consequences").

We would also like to extend special recognition to Dr. Mark E. Jensen, who initiated and supported this project until his untimely death in 2009. The current state of the Northern Region IRPS assessment owes much to the vision, expertise, and determination that Mark brought to the challenge of creating a rigorous and transparent DSS process for the Northern Region.

References

Bourgeron PS, Humphries HC, Riboli-Sasco L (2009) Regional analysis of socio-ecological systems. Nat Sci Sociétés 17:185–193

Brand F (2009) Critical natural capital revisited: ecological resilience and sustainable development. Ecol Econ 68(3):605–612

Chapin FS III, Carpenter SR, Kofinas GP, Folke C, Abel N, Clark WC, Olsson P, Stafford Smith DM, Walker B, Young OR, Berkes F, Biggs R, Grove JM, Naylor RL, Pinkerton E, Steffen W, Swanson FJ (2010) Ecosystem stewardship: sustainability strategies for a rapidly changing planet. Trends Ecol Evol 25:241–249

Christensen NL, Bartuska AM, Brown JH, Carpenter S, D'Antonio C, Francis R, Franklin JF, MacMahon JA, Noss RF, Parsons DJ, Peterson CH, Turner MG, Woodmansee RG (1996) The scientific basis for ecosystem management. Ecol Appl 6(3):665–691

Holling CS (2001) Understanding the complexity of economic, ecological, and social systems. Ecosystems 4:390–405

Jensen M, Reynolds K, Langner U, Hart M (2009) Application of logic and decision models in sustainable ecosystem management, 2009. In: Proceedings of the 42nd Hawaii international conference on systems sciences, Waikoloa, Hawaii, pp 5–8 Jan 2009

Perman R, Ma Y, McGilvray J, Common M (2003) Natural resource and environmental economics. Addison Wesley Longman, USA

Potyondy J, Geier T (2011) Watershed condition classification technical guide U.S. Department of Agriculture Forest Service FS-978. www.fs.fed.us/publications/watershed/watershed_classification_guide.pdf

Reynolds K, Bollenbacher B, Fisher C, Hart M, Manning M, Henderson E, Sims B (2013) Decision support for the integrated restoration and protection strategy for forest service region one, WO FS Series (in preparation)

Reynolds KM, Rodriguez S, Bevans K (2003) User guide for the ecosystem management decision support system, version 3.0. Environmental Systems Research Institute, Redlands, p 82

Saaty TL (1992) Multicriteria decision making: the analytic hierarchy process. RWS Publications, Pittsburgh

USDA Forest Service (2007) Forest service manual 1900 land management planning, Washington DC

USDA Forest Service (2010) LMP monitoring and evaluation: a monitoring framework to support land management planning northern region supplement to Washington office level M&E framework. Northern Regional Office, Missoula, Montana. Unpublished planning M&E framework document

Walker B, Holling CS, Carpenter SR, Kinzig A (2004) Resilience, adaptability and transformability in social-ecological systems. Ecol Soc 9:5. http://www.ecologyandsociety.org/vol9/iss2/art5

Walker B, Carpenter S, Anderies J, Abel N, Cumming GS, Janssen M, Lebel L, Norberg J, Peterson GD, Pritchard R (2002) Resilience management in social-ecological systems: a working hypothesis for a participatory approach. Conserv Ecol 6(1):14. http://www.consecol.org/vol6/iss1/art14/

Evaluating Wildfire Hazard and Risk for Fire Management Applications

Robert E. Keane, James P. Menakis, Paul F. Hessburg, Keith M. Reynolds and James D. Dickinson

Abstract At several spatial scales, fire managers need accurate and comprehensive assessments of wildfire hazard and risk. Assessments are needed to plan, prioritize, and implement management actions, which can range from pro-active prescribed burning to real-time fire suppression. They can be complex, taking many forms and using few to many variables. Past fire hazard and risk assessment projects often lacked a proper decision support platform (see "An Overview of the Ecosystem Management Decision-Support System") on which to objectively evaluate decision costs, benefits, and trade-offs. However, decision support systems (DSS), like the Ecosystem Management Decision Support System (EMDS), now make such analysis readily within reach. Future fire management decisions will no doubt make increasing use of DSSs in decision-making. This chapter summarizes various methods of computing fire hazard and risk. Then, we present

R. E. Keane (✉) · J. P. Menakis
U.S. Department of Agriculture, Forest Service, Rocky Mountain Research Station, Fire Sciences Laboratory, 5775 Hwy 10 West, Missoula, MT 59807, USA
e-mail: rkeane@fs.fed.us

J. P. Menakis
e-mail: jmenakis@fs.fed.us

P. F. Hessburg
U.S. Department of Agriculture, Forest Service, Pacific Northwest Research Station, Wenatchee Forestry Sciences Laboratory, 1133 North Western Avenue, Wenatchee, WA 98801, USA
e-mail: phessburg@fs.fed.us

K. M. Reynolds
U.S. Department of Agriculture, Forest Service, Pacific Northwest Research Station, Corvallis Forestry Sciences Laboratory, 3200 SW Jefferson Way, Corvallis, OR 97331, USA
e-mail: kreynolds@fs.fed.us

J. D. Dickinson
U.S. Department of Agriculture, Forest Service, Okanogan-Wenatchee National Forest, 215 Melody Lane, Wenatchee, WA 98801, USA
e-mail: jddickinson@fs.fed.us

K. M. Reynolds et al. (eds.), *Making Transparent Environmental Management Decisions*, Environmental Science and Engineering,
DOI: 10.1007/978-3-642-32000-2_6,

two projects that use EMDS to prioritize resources for regional and national fire management application. Finally, we discuss additional fire hazard and risk modeling research needs.

Keywords Wildland fire · Forest fuels management · Wildfire hazard · Wildfire risk · Spatial decision support · EMDS

1 Introduction

Fire management in the United States faces some remarkably challenging issues in the coming years. Decades of fire exclusion policies and successful fire suppression programs have resulted in increased canopy and surface fuel loadings that often foster large wildfires, which can potentially damage ecosystems, burn property, and harm people (Laverty and Williams 2000; Keane et al. 2002b). Proposed fuel treatments designed to mitigate these conditions are becoming risky and expensive to implement, and the cost of fighting large wildfires is rapidly growing (Calkin and Gebert 2006; Venn and Calkin 2007). To make matters worse, many people are moving into the nation's fire-prone wildlands, further escalating wildfire-caused risk to life and property. In addition, predicted changes in climate point to increased fire season length, burned area, and fire severity (Brown et al. 2004; Running 2006; Westerling et al. 2006). To face some of these issues, government fire management agencies have implemented the National Fire Plan (NFP). The NFP is a comprehensive strategy to return the landscape to healthy, sustainable, and fire-resilient ecosystems that will provide protection to those who live in the wildland urban interface (GAO/RCED 1999).

One tool to aid fire management in solving complex future issues and implementing the NFP is fire hazard and risk modeling (Hann and Bunnell 2001). Assessing potential damage and benefits of wildland fire to ecosystems, structures, and people will provide fire management with the critical information they need to decide (1) where to allocate funds and fire fighting resources, (2) which landscapes are in need of fuel or ecosystem restoration treatments, and (3) where and how to implement possible mitigation treatments. Fire hazard and risk assessment products can be used to plan, design, prioritize, and implement fire management strategies at several spatial scales, and across many organizational boundaries (Keane et al. 2010). Assessment products can be incorporated into decision support tools, such as the EMDS system, to facilitate fire management planning, and increase transparency of decision-making. Fire management will find EMDS to be a critical decision tool in the future because evaluations of fire hazard and risk will need a generalized, consistent, and transparent context in which to assess possible fire management decisions (Hessburg et al. 2007).

Fire hazard and risk modeling can be accomplished using diverse methods and approaches, each of which has unique limitations. Advances in computer technology, fire behavior modeling, and fire ecology simulation have greatly improved fire hazard and risk prediction, but further advances in these fields are needed to ensure that fire hazard and risk maps portray the most appropriate and consistent variables for evaluating management concerns. This chapter discusses the use of fire hazard and risk modeling as input in decision support tools. First we define and summarize risk and hazard terminology and methods, and then we describe the use of fire hazard and risk analysis in fire management. The challenges facing fire hazard and risk modeling for decision support are discussed next, along with two detailed examples that illustrates both limitations and benefits of this type of modeling at two spatial scales. Finally, we discuss ongoing research needs.

1.1 Background

In this chapter, "hazard" is considered an act or phenomenon with the potential to do harm (NRC 1989; Hardy 2005). The notion of fire hazard represents the potential susceptibility or vulnerability of existing conditions to wildfires. Fire hazard is typically described independent of weather, using surface and canopy fuel characteristics. It can be either expressed as potential fire behavior (e.g., flame length or fireline intensity) arising from existing fuels, or as a critical fuel property (e.g., loading or biomass) (Hogenbirk and Sarrazin-Delay 1995). The term "risk" is used to describe the probability that a fire event might occur, as affected by other causative agents and interacting factors (Bachmann and Allgower 1999; Bachmann and Allgower 2001). We amend this definition to also include the subsequent ignition of the adjacent fuels (i.e., fire spread) and the potential for that ignition to create a specific fire event. We likewise adopt the Bachmann and Allgower (1999) definition of fire risk as the likelihood a specified event will occur within a specific time period or from the realization of a specified hazard.

Fire hazard has been described using a variety of approaches and variables including expected fire behavior (Hardwick et al. 1998; Hessburg et al. 2007), fuel characteristics (Hogenbirk and Sarrazin-Delay 1995), satellite image classifications (Cohen 1989; Jain et al. 1996; Ercanoglu et al. 2006), topography analysis (Yool et al. 1985), expert knowledge (Gonzalez et al. 2007), socio-economic analysis (Bonazountas et al. 2005), and crown fire index calculations (Fiedler et al. 2001). Fire risk, on the other hand, has been described as the probability of fire weather occurrence (Gill et al. 1987), the frequency of rare fire events (Neuenschwander et al. 2000), the probability of a wildfire causing tree mortality or loss of wildlife habitat (Ager et al. 2011, 2007, respectively), and the probability distribution of ignitions, fire sizes, and burning conditions (Parisien et al. 2005). The diversity of fire risk analysis approaches arises from specialized management objectives, while fire hazard assessment projects are generally designed around the availability of the spatial data layers used to represent hazard. Fire risk, on the other hand, has been described as the probability of fire weather occurrence (Gill et al. 1987), the frequency of rare fire events (Neuenschwander et al. 2000), the probability of a wildfire causing tree mortality or loss of wildlife habitat

(Ager et al. 2011, 2007, respectively), and the probability distribution of ignitions, fire sizes, and burning conditions (Parisien et al. 2005). The diversity of fire risk analysis approaches arises from specialized management objectives, while fire hazard assessment projects are generally designed around the availability of the spatial data layers used to represent hazard.

An adequately comprehensive assessment of either fire hazard or risk is difficult because it requires extensive knowledge across a wide variety of disciplines that includes simulation modeling, geographic information system (GIS) analysis, advanced statistics, fire behavior and effects, and fuels sampling and map prediction (Keane et al. 2010, 2013; but see Thompson and Calkin 2011). As a result of limited data and experience, many fire managers find it difficult to create those layers that can be directly used in hazard and risk assessment. Additionally, the ability of managers and scientists to adequately sample fuels and predict much needed fuels maps may be much poorer than originally thought (Keane et al. 2013).

2 Challenges of Fire Hazard and Risk Mapping

2.1 Selected Variables

Most analyses attempt to describe a full range of hazard and risk using multiple variables because of the astounding complexity and variability of wildland fire. These variables are computed using a variety of methods, protocols, and computer programs (Neuenschwander et al. 2000; Sampson and Sampson 2005). For example, to define fire danger in subwatersheds ($\sim$10,000 ha), Hessburg et al. (2007) evaluated three primary topics—fire hazard, fire behavior, and ignition risk. Fire hazard evaluated conditions for surface and canopy fuels; fire behavior evaluated probable spread rate, flame length, fireline intensity, and crownfire potential; ignition risk evaluated relative plant greenness index (NDVI), lightning strike potential, and the Keetch-Byram and Palmer drought indices. In their application, they were working to explain all of the variance in fire danger expressed at the subwatershed scale.

In some applications, use of various combinations of fire hazard measures together may be inappropriate because they are highly correlated, contradictory, or unsuited for the fire management issue being addressed. For example, one hazard analysis effort may use high fireline intensity to connote high fire hazard, while another may use fast fire spread rates (Hardy 2005; Sampson and Sampson 2005). A high intensity fire may be a hazardous fire, but this type of fire can be appropriate and desirable in stand-replacement fire regime ecosystems, such as lodgepole pine (Heinselman 1981; Romme and Knight 1981; Agee 1998). It is clear that the selection of variables and the logic for rating fire hazard and risk is ultimately governed by the objectives of the analysis. A concise statement of objectives is vitally important before any hazard analysis is initiated, and the variables selected for the analysis must be matched to these objectives.

There is both art and science in deciding which variables to include in the fire hazard and risk analysis. Complexity, uncertainty, and errors tend to increase as more variables are added to hazard analysis, yet too few variables may not adequately answer the hazard analysis objective. A good approach is to select a small set of variables that integrate the behavior of other associated variables. Fireline intensity, for example, is better at describing fire hazard than fire behavior fuel model because it integrates temperature, relative humidity, wind, and slope, along with fuel model, into its calculations (Hough 1968). Scorch height might be even better because it could also be used to evaluate potential tree mortality if that fit the analysis objective. To eliminate potential bias though, it is critical that selected variables are not correlated to each other within the analysis, unless the objective is to explain as much of the variance as is possible.

It is fundamentally important that the origin, temporal domain, spatial scale and resolution, data limitations, and uncertainty of fire hazard variables and maps be recognized when interpreting them (Cohan et al. 1984; Tainaka 1996). Complete metadata on the selection, computation, and units of fire hazard variables is critical for understanding and interpreting analysis results in the decision making process. The quality, type, and source of the data input used to compute fire hazard variables must also be known to facilitate appropriate fire hazard analysis. Finally, the accuracy and error associated with each input, intermediate, and derived variable used in analyses should be evaluated in fire hazard interpretation. Unfortunately, this step is rarely accomplished because it is difficult and expensive to quantify error and uncertainty in most fire hazard and risk variables and maps.

2.2 Analysis

An enormous challenge in fire hazard and risk modeling is adequately translating potential fire behavior and effects from terrain, slope, fuel, and weather characteristics to the appropriate spatial and temporal contexts. One estimate of fire behavior for one point in time and space may over-generalize the complex interactions that occur during a fire, or within a fire regime (Morgan et al. 2001). A common assumption in many fire hazard analyses is that the fire that occurs in a pixel is a head fire and therefore, represents a worst case scenario (Keane et al. 2008). However, that pixel might be in a topographic position where head fires are rare, or that pixel may rarely experience the weather needed to sustain a head fire. A more accurate representation of fire hazard would be to quantify the distribution of probable fire intensities and spread rates at each pixel as would occur under a wide variety of weather conditions, wind directions, fire ignition locations, spread patterns, and suppression strategies. Measures of hazard and risk could then be derived from the distributions, such as the probability of wildfire occurring above a threshold fireline intensity (Farris et al. 2000).

Finney et al. (2011) have implemented this strategy in the FSPro simulation package, which simulates the probability of fire spread based on thousands of FARSITE[1] runs with multiple weather scenarios and random ignition locations for real-time, operational wildfire application. The down-side of this approach is that it is computationally demanding, making it difficult to complete for large regions, at resolutions that are often required of hazard analysis (Keane et al. 2008). Moreover, it is somewhat problematic to implement a temporal component because the fuels are considered static for the entire simulation, and a finite set of historical weather scenarios are used.

Some efforts at describing fire hazard have taken disparate GIS layers and merged them together to create a final layer (Sampson and Sampson 2005). A typical example would be merging the flame length, surface fuel model, and canopy bulk density map layers to create a fire hazard map: two layers describe continuous variables with differing units, while the third is a categorical variable. This approach has its own problems since each layer has a unique error distribution, mapping resolution, analysis scale, and computational detail, all of which are compromised when merged. That is, the solution acquires the lowest significant figure of the components. An approach that might result in lower error propagation might be to explicitly set a threshold value for continuous maps or a set of values for categorical maps, above which fire hazard is high and below which hazard is low. This could be used to create a binary variable data layer that can then be merged with other binary maps (Hessburg et al. 2007). Threshold values could be based on a theoretical or physical context, and take into account the sensitivity and error of the parameters that were used to create the continuous data layer or to compute the behavior from the fire model.

2.3 Weather

The use of weather in fire analyses is also quite challenging. Many efforts assume severe fire weather (e.g., 90 or 99th percentile temperature, high wind, low fuel moistures) to compute the fire characteristics that describe "worst-case" hazard or risk, in the context of a specific management objective. These analyses, however, rarely describe the frequency of that severe weather event for all ecosystems within the analysis area. Extremely dry conditions, for example, may occur frequently in low-elevation pinyon-juniper patches, but they may be relatively rare in high-elevation lodgepole pine ecosystems, yet both may have the same hazard value. It is important that analysis weights the frequency of the fire event with the severity of impacts when describing hazard and risk.

[1] FARSITE is a Windows-based fire behavior and growth simulator used widely by USDA and US Department of Interior agencies (Finney 1998).

A similar problem with percentile weather variables arises when assessing fire hazard across large landscapes, especially in mountainous terrain, as 90, 95, and 99th percentile fire weather differs across diverse topography. Severe fire weather at low-elevation, south-facing, and ridgetop sites may be quite different from severe fire weather in high-elevation, north-facing, or valley bottom settings. Some hazard analyses use only one fire weather scenario across the analysis landscapes taken from a single weather station, which can grossly underestimate weather variability.

Some fire management analyses include weather variables as a rating of fire hazard or risk (Gill et al. 1987). While this provides a useful representation of the climate that fosters high fire hazard, the rating is not useful as a measure to detect changes in hazard as a result of fire management activities. Weather is the one factor in fire dynamics over which managers have the least control. The challenge then is to select a fire hazard or risk variable that integrates the weather with fuel conditions, such that it can be used to monitor changes in fire hazard as fuel treatments are implemented, and as wildfires burn. Smoke emissions, for example, is a valuable variable for predicting the likelihood of fire effects because it integrates fuel loadings by size class with weather to compute consumption and then emissions (Lutes et al. 2009).

Wind also presents a special challenge in simulating fire hazard and risk even though it is a major factor governing fire behavior, because it is difficult to quantify in a spatial and temporal domain for many fire analyses (Forthofer et al. 2003). Wind is complex because it has a magnitude (wind speed), duration (gusts), and direction, all of which can heavily influence predicted fire behavior and subsequent fire hazard. Moreover, these three properties are correlated with each other and often vary by geography and with the terrain (Rehm and Mell 2009). Therefore, it may be difficult to derive wind scenarios that have adequate spatial resolution, and are applicable across large regions and long time periods. It is critical that wind scenarios in fire hazard analysis also match management objectives and be described in sufficient detail so that results are interpreted in the proper context.

2.4 Scale

Many fire hazard analyses tend to concentrate on stand-level fuels and their characteristics without recognizing the spatial influence of topography, winds, and adjacent fuels on wildland fire (Finney 2005). The spatial patterns of landscape composition and structure are important to fire hazard because fuel pattern will influence fire spread and intensity (Loehle 2004). Spatial patterns of fuels can also dictate the design and placement of fuel treatments on the landscape (Agee et al. 2000; Finney 2001). However, as spatial resolution decreases and the extent of the analysis area increases, spatial relationships may become less important. Regional and national evaluations of fire hazard and risk to prioritize watersheds for fuels treatments, for example, may not require detailed analysis of spatial pattern as much as project-level analyses conducted to optimize fuel treatment locations (Hessburg et al. 2007).

Scale is also an important consideration in determining the size of the fire hazard analysis area, especially if spatial interactions of fire spread are included (Bourgeron and Jensen 1994; Tang and Gustafson 1997; White et al. 2000). If the analysis area is too small, the spatial representation of simulated fires will be truncated because most fires will encounter the defined landscape border before they reach their full size (Keane et al. 2002a). However, if the landscape is too big, the high data and computational requirements may prohibit a statistically sufficient number of fire simulations. The challenge is to select a landscape extent that adequately represents the spatial dynamics of disturbance, while still being appropriate for the management objective. Karau and Keane (2007) found that approximately 100 km^2 is adequate for many spatially explicit simulations of fire spread, but this varies by topography, ecosystem, and geography. They also found that the simulation area must contain sufficient buffer around the analysis landscape to minimize bias in fire spread. Keane et al. (2002a) found that this bias becomes negligible when the size of the surrounding buffer area is roughly eight to ten times the size of the focal landscape.

The resolution and detail of the primary data layers used to create hazard maps are also important. The fire behavior fuel models that are used in many fire hazard analyses, for example, are simplified classifications of fire observations that result in a decreased resolution of output (Scott and Burgan 2005). And while the LANDFIRE project[2] represents significant progress in providing the fine-scale spatial data needed in fire management across the conterminous US (Rollins et al. 2009), its national scope demands a coarser treatment of vegetation and fuels descriptions that sometimes result in moderate or lower map accuracy at local scales (Keane et al. 2013). Fuel characteristics are notoriously variable and scale-dependent, making them difficult to sample and map (Keane et al. 2013), and few fire behavior and fire growth models have sufficient resolution and detail to adequately represent actual fuel load distributions (Keane et al. 2001, 2013; Keane 2013). It is important that users and decision-makers recognize the scale and resolution limitations of the input spatial data when interpreting the fire hazard and risk assessment outputs. Elaborate and detailed analysis may actually expand rather than contract modeling error and uncertainty in comparison to more simplified analysis.

3 The PNW EMDS Fuels Management Project

A regional-scale decision support system (DSS) was developed for the Pacific Northwest Region (PNW) of the US Forest Service based on wildland fire hazard and risk analysis. The PNW fire danger model was designed to aid fire managers,

---

[2] A nation-wide geospatial mapping project offering data layers at the landscape-scale. For more information go to http://www.landfire.gov.

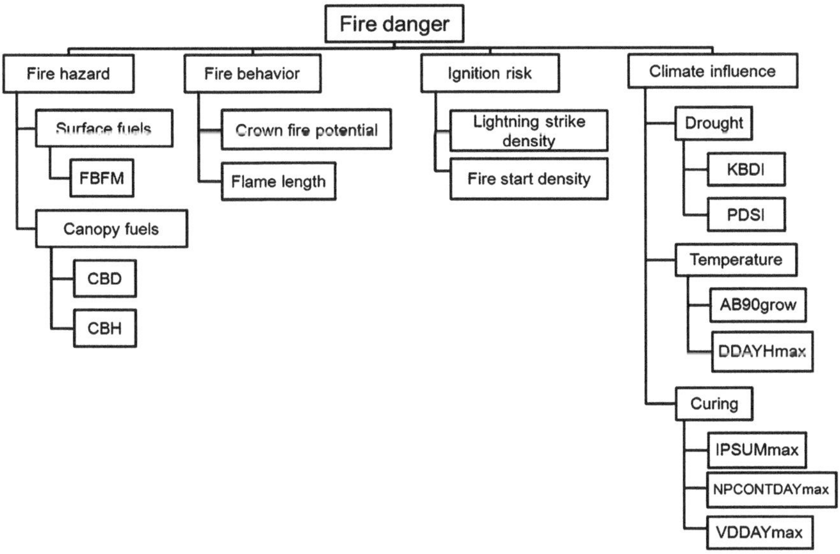

Fig. 1 Dendrogram showing organization of the PNW EMDS fire danger model. The evaluation of fire danger is composed of four primary topics—fire hazard, fire behavior, ignition risk, and climate influence. Under each of these four primary topics are secondary and elementary topics, where data are evaluated. Abbreviations are: FBFM = the fire behavior fuel models of Scott and Burgan (2004), CBD = canopy bulk density (kg·m^{-3}), CBH = canopy base height (m), KBDI = the Keetch Byram Drought Index, PDSI = Palmer Drought Severity Index, AB90grow = the number of days above 90 °F (32.2 °C) during the local growing season, DDAYHmax = the mean penultimate maximum degree-days heating during the local growing season while temperatures were >18 °C, IPSUMmax = an inverted precipitation index for the local growing season, NPCONTDAYmax = the mean penultimate maximum consecutive days without measurable precipitation (<0.3 cm), while temperatures were ≥10 °C, and VDDAYmax = the mean penultimate maximum days with a vapor pressure deficit <1000 Pa while temperatures are ≥10 °C during the local growing season

who were interested in preventing large wildfires, with prioritizing watersheds for possible ecosystem restoration and fuel reduction treatments. This is a prevention-oriented model that is designed to detect changes in an analysis area as treatments are implemented, and as controlled and uncontrolled wildfires modify the landscape. In contrast, a suppression-oriented model would be designed to aid in predicting which subwatersheds of a Region might experience a large-scale and severe wildfire in a given fire season. Suppression-oriented models are often used to strategically allocate fire preparedness and suppression resources where the threat of a large and rapidly expanding fire are likely.

The PNW fire danger model consists of logic and decision models. The logic model (which we discuss here) evaluates the state of each landscape (in this example, each subwatershed) as a function of four primary topics: fire hazard, fire behavior, ignition risk, and climate influence (Fig. 1) (Hessburg et al. 2007). Each primary topic has secondary topics under which data are evaluated. Surface

(FBFM, fire behavior fuel model) and canopy fuels (crown base height and crown bulk density) were included under fire hazard. Under fire behavior, flame length and crown fire potential were computed assuming an escaped wildfire burn scenario. Lightning strike and fire start density were computed under ignition risk. Climate influence evaluated three secondary subtopics: drought likelihood (drought), fire season temperature (temperature), and fuel curing likelihood (curing) topic. The long and short term drought descriptors of Palmer drought severity index (PDSI) and the Keetch-Byram drought index (KBDI), respectively, were evaluated under the drought subtopic. Under the temperature subtopic, the effects of two temperature variables on fire danger were considered; "the number of hot days during the growing season", which was computed as the number of days above 90 °F (32.2 °C) during the local growing season (AB90grow), and "growing season degree-day heat sums", which was computed as the mean penultimate maximum degree-days heating during the local growing season while temperatures were >18 °C (DDAYHmax). These data were obtained from Oak Ridge National Laboratory (ORNL, Hargrove et al. 2004). Subwatershed values were calculated as zonal means.

Three precipitation/moisture variables were considered for their effects on fuel curing; "seasonal dryness"—an inverted precipitation index for the local growing season (IPSUMmax), "mild to hot days without rain"—the mean penultimate maximum consecutive days without measurable precipitation (<0.3 cm), while temperatures were ≥10 °C (NPCONTDAYmax), and "mild to hot days with low humidity"—the mean penultimate maximum days with a vapor pressure deficit <1000 Pa while temperatures are ≥10 °C during the local growing season (VDDAYmax). These data were also obtained from ORNL, (Hargrove et al. 2004). Subwatershed values were calculated as zonal means.

All mapped continuous and ordinal fire hazard variables were evaluated based on a threshold value and then summarized to subwatersheds (12-digit Hydrologic Unit Code or HUC), to create new, spatially-descriptive variables. For example, pixels with canopy bulk density (CBD) values exceeding 0.15 $kg{\cdot}m^{-3}$ were analyzed to compute area (CBDarea) and level of aggregation (CBDaggregation) of high crown fuels for the subwatershed. Values of these new subwatershed variables were then evaluated based on a threshold value and a fuzzy membership function that computed a value between zero (meaning no strength of evidence that the watershed is above the threshold) and 1.0 (maximum evidence that the watershed is above threshold) (Hessburg et al. 2007).

The FIRE HAzard and Risk Model (FIREHARM) was used to compute all pixel variables for the fire behavior topics and KBDI (Fig. 2) (Keane et al. 2010). FIREHARM is a C++ program that computes fire characteristics over time using spatially-explicit daily climate data to simulate fuel moisture and a host of embedded routines that calculate commonly used measures of fire behavior, fire danger, and fire effects (Fig. 2). FIREHARM is a modeling platform that integrates previously developed fire simulation models into its structure and contains no new fire behavior or effects simulation methods. Although FIREHARM's input and output are spatial, the model is not spatially explicit because it does not simulate

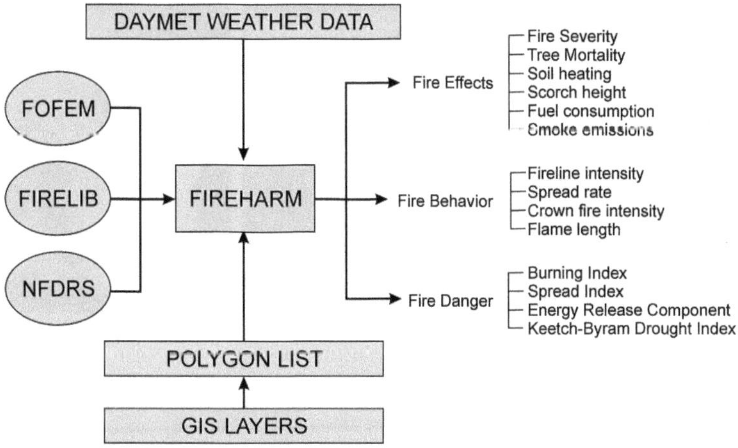

Fig. 2 Compartment diagram of the FIREHARM model showing input requirements and output data. FOFEM is the First Order Fire Effects model (Reinhardt and Keane 1998), FIRELIB is a set of functions for predicting fire behavior (Bevins 1996), and NFDRS is the National Fire Danger Rating System (Deeming et al. 1977)

interactions across pixels, such as fire spread. Instead, the model assumes that pixels (or polygons) experience head fires, and then simulates the fire characteristics from antecedent weather. FIREHARM does not simulate crown fires directly, but it does calculate crown fire intensity (Rothermel 1991; Finney 1998).

FIREHARM can be run in two modes. In the event mode, which is used for most fire hazard mapping, the user enters fuel moistures and ambient weather conditions for a given situation or event, such as a worst case wildfire, and the program will calculate all fire variables for this specified situation. In the temporal mode, FIREHARM simulates fuel moistures from daily weather to compute daily fire characteristics over 18 years. The program then calculates the probability of a user-specified event occurring during the 18-year weather record and this probability value is used to describe potential risk.

Most of the data used to support this project came from the National LANDFIRE mapping project (www.landfire.gov). Seven LANDFIRE map zones (1, 2, 3, 7, 8, 9, 10, 18) defined the project area (www.nationalmap.gov). Map zones are broad biophysical land units represented by similar surface landforms, land-cover conditions, and natural resources (Fig. 3). Subwatersheds, defined by the US Geological Survey (Seaber et al. 1987, http://nhd.usgs.gov/wbd.html) as nationally consistent, 12-digit hydrologic units, were the smallest subunits within map zones, and all data were summarized to them. All spatial data representing the fire hazard topic (surface and canopy fuels) were taken directly from LANDFIRE for the analysis area. These same data were also used as input to the FIREHARM program. Weather data are input into FIREHARM using the DAYMET US database (http://ww.daymet.org) developed by Thornton et al. (1997). DAYMET is a computer model that was used to generate daily spatial surfaces of temperature,

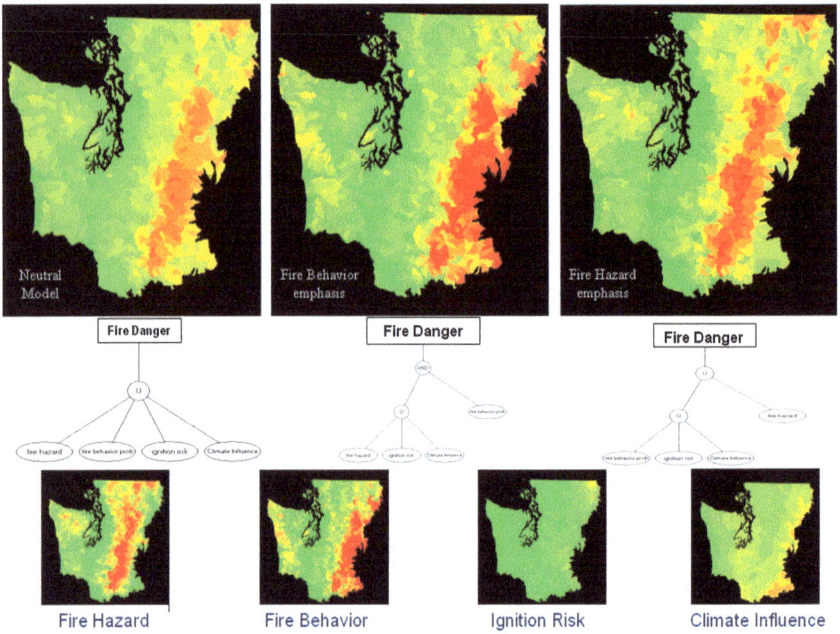

Fig. 3 Example EMDS fire danger model outputs for map zone 1 of the PNW study area. Three sets of outputs are shown for a single map zone (zone 1, western Washington State). The *top row* of maps shows differences in *fire danger* when changing the logical operators used. The *middle panel* shows the differences in the logical operators used to evaluate the data in the NetWeaver models. The *bottom row* of maps shows the results after evaluating each primary topic—*fire hazard*, *fire behavior*, *ignition risk*, and *climate influence*. The results of primary topic evaluations do not vary among the three examples, only the use of operators. The union (U) operator in the left example treats the data associated with each primary topic as equivalently compensating factors. This is the neutral model. The middle example uses the AND operator to set off fire behavior as the limiting factor (*fire behavior emphasis*). The right example uses the union operator to evaluate fire hazard after the other three topic evaluation results have been considered as equal and compensating, effectively weighting fire hazard threefold over each of the other topics (*fire hazard emphasis*)

precipitation, humidity, and radiation over large regions of complex terrain (Thornton et al. 1997, 2000). Additional spatial data comes from other established sources such as PDSI (http://www.ncdc.noaa.gov/data-access), the National Climate Data Center, or derived using existing, published, and documented models and modeling procedures.

In Fig. 3, we show three fire danger evaluation outputs for a single map zone (zone 1, western Washington State). The results of primary topic evaluations do not vary among the three examples, only the use of operators. These input data layers for the map zone 1 evaluations were created by running FIREHARM in the temporal or probabilistic mode. The example in Fig. 3 shows how decision-makers and analysts can "game" the fire danger model to develop alternative ways of emphasizing or considering the data. The fire danger output layers can be then used to create ordered lists of watersheds for use in prioritizing areas geographic areas for fuel treatments.

Independent of the DSS model development, there were issues that had to be addressed during development. First, LANDFIRE base layers could not be edge-matched across zones to create seamless regional or national layers. This was due to the well-known and pre-existing problem of satellite imagery radiometric readings being calibrated by map zone, rather across map zones; an intractable task due to the larger extent of the area and the variability represented across satellite scenes. The result of calibration was high integrity of satellite sensor values within a map zone, but somewhat differing numerical values for the same conditions across map zone boundaries. That is, map zones could not be edge-matched in terms of their numerical data. This affected all vegetation, fire effects, and fire behavior base layers that were based on satellite imagery, or their derivatives. To resolve this problem, all fire danger calculations were relativized within each map zone so that map zone outputs could be evaluated side-by-side, on a common footing. Second, regional fire managers were more interested in a simple suppression-oriented than prevention-oriented model for operational use, because of upper level management emphasis on wildfire protection rather than pro-active fuel treatments. Their selected variables mostly concerned fire weather rather than fuels conditions. This would make the task of detecting change in fire hazard and risk as a result of fuel treatments problematic.

4 National EMDS Fuels Treatment Prioritization Project

A national-scale DSS was developed across multiple land management agencies (Department of Agriculture's Forest Service; and Department of the Interior's Bureau of Land Management, Bureau of Indian Affairs, National Park Service, and Fish and Wildlife Service) to allocate the substantial fuel-treatment budget on federal lands within and across agencies. To address the concerns by Congress and the General Accountability Office for a rational, transparent, and reproducible process for allocating the fuels-treatment budget (GAO 2002, 2003, 2004, 2007), the National EMDS Fuels Budget decision support model was designed to evaluate wildfire potential across all administrative units in the continental US, and to establish priorities for allocating the fuel-treatment budgets within the agencies.

The National EMDS Fuels Budget model consists of both logic and decision models (Reynolds et al. 2009). The logic model delineates wildland fire potential in terms of fire behavior and fire probability of occurrence under extreme fire weather conditions. Fire behavior is represented by crown fire potential and surface fire potential, in terms of fireline intensity, rate of spread, and flame length. Crown fire potential was delineated by assigning relative classes (such as very low to extreme) to spatial data of existing vegetation types based on expert knowledge (Menakis et al. 2003; Menakis 2008). Surface fire potential was represented by surface fuels, which were calculated either using the Fuels Characteristic

Classification System (Ottmar et al. 2007), or the BehavePlus fire behavior model (Andrews 2008) and the Scott and Burgan (2005) fire behavior fuel models mapped by LANDFIRE.

Fire probability was represented by both fire weather and fire occurrence. Fire weather included problem fire days and length of fire seasons, with problem fire days computed as an average number of days a year from 1982 to 1997 that a fire could experience extreme fire weather, based on thresholds of high temperature, high wind, and low humidity (Menakis et al. 2003). Fire season length was based on the average number of days per year that the relative energy release component (RERC) was above 95 % based on daily RERC maps from 1980 to 2005 (Jolly 2008). Fire occurrence was based on small and large fire occurrence from 1980 to 2003. Small fires were defined as any fire greater than 0.1 acre and large fires were any fire greater than 500 acres.

The criteria in the decision model for assessing priorities were the wildland fire potential from the logic model, and its consequences, along with performance and opportunity criteria (Reynolds et al. 2009). The consequences criterion described the negative effects of wildland fire associated with untreated fuels, and included measures of proximity to wildland urban interface (WUI), ecosystem vulnerability, potential emissions, and potential water quality impacts. WUI data (Radeloff et al. 2005) were used to measure the number of structures that could be affected from a wildfire on federal lands. Ecosystem vulnerability measured the risk of losing key ecosystem components from a wildfire, as measured by the FRCC (Schmidt et al. 2002, http://www.frames.gov/partner-sites/frcc/frcc-home). Emissions measured the amount of smoke that would likely be produced by a wildfire and its impact on people. Lastly, water quality measured the impact of wildland fire to mapped municipal watersheds. The performance criterion was designed into the model to gauge administrative-unit performance in terms of efficiency and effectiveness of fuel treatments. Unfortunately, since data were unavailable in the agency databases for this criterion, it remained unimplemented in the decision model (Reynolds et al. 2009). However, the percent of land with minimal departure from frequent, low-severity fires that occurred historically was included as proxy for efficiency, since these lands are typically more economically efficient to maintain than to rehabilitate once fire has been removed for several years (Schmidt et al. 2002; Reynolds et al. 2009).

The opportunities criterion described other resource benefits that could be accomplished with fuels management. The two major measures included in this criterion were ecosystem restoration and biomass opportunities. Ecosystem restoration focused on fuel treatments that would restore ecosystems back to their historic condition, and was based on the FRCC metric (Schmidt et al. 2002). Biomass opportunity focused on the availability of biomass to produce bio-energy and bio-based products (such as lumber, composites, paper and pulp, furniture), and included both the availability of biomass and its proximity to processing plants. For information security reasons, in these analyses we only had the amount of biomass available on federal lands (Blackard et al. 2008).

The National EMDS Fuels Budget decision support model has vastly improved since it was first implemented in 2006. Further improvements could be gained with respect to the extent of data coverage, data quality, and the sophistication of the available models (Reynolds et al. 2009). As national, consistent data become available, more refined methods should be incorporated into this decision model (such as the methods used in the aforementioned PNW fire danger model). This is especially true with the simplistic methods used to calculate wildland fire potential. Nevertheless, this approach has proved to be successful in providing an accountable and transparent methodology in setting budget priorities for fuel-treatment nationally. This success can be attributed to the effective engagement of scientists, managers, and technical specialists at various levels of the agencies.

5 Research and Management Needs

5.1 Fire Simulation Models

The single biggest need for future fire hazard and risk projects is the development of new fire behavior and effects models that are spatially explicit and integrate cross-scale dynamic factors into their design. While contemporary one-dimensional point models of fire behavior (e.g., BEHAVE) represented a major advancement in wildland fire science in the early 1970s, they are now over 30 years old, and fail to fully account for the multitude of three-dimensional spatial interactions of fire spread and intensity. Crown fires, for example, are poorly simulated in current fire behavior models because heat transfer properties of convection and radiation are not directly simulated in all three dimensions. Matching the scale of fire behavior and effects with the scale of simulation is also a research need. Smoldering combustion, for example, must be simulated at finer scales than the scales used to simulate crown fires.

It is also critical that these new fire models use inputs and outputs that are measurable on the landscape. Many inputs and most outputs of the Rothermel (1972) spread model are difficult, if not impossible, to measure with any degree of precision or accuracy. Designing models around output variables that can be precisely measured in the field using standardized protocols is important because it will allow the manager to evaluate the performance of fire models with confidence. It is also important that the inputs are easily and accurately measured so that high quality digital input maps can be created and used in hazard and risk projects.

Fire management will probably not be able to use the new and improved three-dimensional spatially-explicit fire behavior models in the near future because the complexity of the models will make them unwieldy to use. Therefore, a critical research need is the proper synthesis of complex fire behavior model results into simplified management-oriented models for easily accomplished fire hazard analysis. For example, results from complex fire models that simulated fire behavior for many environmental conditions (weather, fuels, and topography) can then be used to develop empirical models, following the physical relationships used in Rothermel (1972), to make a more simplified fire behavior model for managers. Or, complex models can be run using different parameterizations to determine thresholds of response that identify when the Rothermel (1972) algorithms are weak and should be modified to correct limitations. Or, scale adjustment factors could be approximated for the Rothermel (1972) model from response surfaces of the complex fire models.

5.2 Fuels

Considerable research in the field of wildland fuel science is needed to meet the burgeoning needs of fire management in the coming years. Most fire behavior models use simplistic fuel models to represent complex fuelbeds (Burgan and Rothermal 1984). Fire managers often used the standard 13 Northern Forest Fire Laboratory (NFFL) fuels models (Anderson 1982) until Scott and Burgan (2005) created a new set of 40+ models. These fuel models do not describe actual fuelbed loadings on the ground, rather they contain fuel characteristics that are calibrated to agree with observed fire behavior. Therefore, fuel models can never be used to quantitatively describe fuelbed characteristics (e.g., fuel loadings, the key fuelbed property) because they are abstract representations of fire behavior. These coarse fuel models represent the finest resolution that can be used currently to describe fire hazard and they are too broad to adequately represent significant changes in the fuelbed. Effective fire analyses will depend on the subtle differences in fuels to characterize relative differences in fire hazard and risk across the landscape.

Accurate fuel sampling techniques must be developed to quantify key fuel characteristics (e.g., loading, specific density, surface-area-to-volume ratio) across multiple new and existing fuel components (e.g., logs, duff, litter) (Keane et al. 2013). These sampling techniques must be easy for management to learn and implement, and they must also provide for accurate assessments at the scale of evaluation. Canopy fuels represent a special challenge because it is the three-dimensional distribution of canopy fuels that ultimately governs crown fire behavior (Reinhardt et al. 2006). Ideally, researchers should develop a variety of sampling techniques and methods so that managers can select the most appropriate technique for their application. Moreover, the next generation of fuel models must describe real fuels across multiple fuel components (e.g., size classes), and also the spatial variability of these fuels as they are distributed across the simulation landscape, both within and among stands (Keane et al. 2001).

5.3 Weather

Creation of comprehensive weather and climate databases is a critical need to support wildland fire hazard and risk analysis. Spatially-explicit, long-term historical databases are needed that ensure consistent hazard and risk analysis over large regions, and across diverse weather conditions. These data should be readily accessible and contain understandable formats with complete metadata documentation. Hourly weather data are needed to support spatial fire spread modeling, but daily, monthy, and annual summaries can facilitate coarser-scale fire analyses.

Future weather and climate databases should also be developed for use in fire hazard simulations and quantifying conditions under climate change (Millar et al. 2007). Historical and future climate variables should be the same and should be derived for comparable spatial scales and domains to facilitate direct comparison of fire hazard and risk analysis outputs, and statistical modeling of changes in key fire behavior covariates.

We also suggest the development of several comprehensive, standardized climate and weather scenarios for comparative fire hazard analysis that encompass conditions of interest to fire management. A diverse set of weather scenarios that specify weather-related inputs such as fuel moistures, maximum temperature, wind, and relative humidity, can be developed for various fire management applications. For example, weather scenarios are needed for prescribed burning, wildfire, and wildland fire use conditions. A more representative set of weather scenarios is also needed to represent expected variation in climatic severity, such as moist, dry, very dry, and extreme weather conditions; examples of which can be seen in the First-Order Fire Effects Model (FOFEM) fire effects program (Reinhardt and Keane 1998). These scenarios could be hierarchically nested by geographical area using climate severity designations appropriate to the areas. The advantage of common weather scenarios is that they would aid in standardizing fire hazard and risk analysis across agencies, geographical areas, and management applications.

5.4 *Analysis and Decision Support*

One problem with many hazard and risk analyses is that there are few statistical techniques that can be used to compute the significance or uncertainty in the selected variables and final assessment. New statistical methods are needed to aid managers in the evaluations of uncertainty associated with results, and these results need to be factored into overall decision-making. These multivariate statistical methods need to be easy to implement and interpret for fire managers. Since many variables are computed from simulation models, it is critical that the models themselves be evaluated for accuracy, precision, and error rate. This will require novel statistical analysis techniques because they must assess accuracy across models, input map layers, and input parameters. Advanced sensitivity analysis, for example, could be used to assess the accuracy and significance of variables simulated from a model. Map accuracies for fire analysis variables must be assessed at multiple levels such as the geograpical accuracy, model accuracy, and input data accuracy.

Managers and researchers will also require significant increased computing resources and expertise to conduct future fire hazard and risk analysis projects. Spatial simulations of fire spread for multiple weather and fuel scenarios requires thousands of simulations using multi-processor computers, and complex computer programs that rely on high quality, high resolution, and spatially consistent input data for good results (Finney et al. 2011). It would be even more computationally demanding to perform these simulations for all possible future fuel conditions resulting from vegetation development, disturbance, climate change, and management policies (Keane and Finney 2003). Currently, these computationally intensive techniques are beyond the technical and computing resources available to fire management, so any quantification of fire hazard probably will require a compromise between the management objective, available computing resources, modeling expertise, and time. Both scientists and managers will need extensive training and background in fire modeling to understand and interpret future fire hazard analyses, which may be impractical to implement across fire management agencies. Moreover, developing management-friendly, easy-to-use software tools for complex hazard analysis may be difficult because of the complicated set of inputs, parameters, and outputs involved in fire behavior and effects simulation. We feel the best route might be the creation of computing centers with highly trained experts in the fields of fire behavior modeling, fuels, GIS, and landscape ecology who could run these models and train people on how to use and interpret the results. It is usually the availability of computational resources, input data, and modeling expertise that dictate the rigor of most hazard assessments for fire management (Keane et al. 2008).

Finally, flexibility is required in decision support tools to accommodate results from multiple model run scenarios and multiple analyses. Results from a single EMDS run, for example, do not fully describe the decision space for fire managers because the full range, dimensionality, and behavior of fire hazard predictions may not be fully represented across the landscape. Varying logical operators within a given logic model in EMDS, while using the same input data, will yield highly differing landscape prioritizations (Fig. 3). Moreover, these disparate results provide critical information on the sensitivity and importance of the key input variables.

Reassigning weights and importance in the EMDS decision model will also expand the decision space and account for possible uncertainty in allocating weights across hazard variables. For example, programming a Monte Carlo simulation across a characteristic distribution of weights might clearly illustrate uncertainty in the decision space. Adding results from multiple weather scenarios for fire hazard variables into the EMDS logic model will also provide greater depth of insight into the uncertainties associated with fire management decision-making.

Synthesizing critical landscape variables from pixel-level fire hazard predictions to local landscape predictions (e.g., Hessburg et al. 2007, summarize the area and aggregation of area of key fire danger variables within subwatersheds) is also important for future analyses, so as to integrate spatial contagion considerations into hazard evaluations. Ultimately, it is critically important to have standardized models using defined sets of fire hazard variables, so that comparative analyses can be done across multiple scales and geographical areas by different agencies (e.g., Reynolds et al. 2009; Hessburg et al. 2007).

References

Agee JK (1998) The landscape ecology of western forest fire regimes. Northwest Sci 72:24–34

Agee JK, Bahro B, Finney MA, Omi PN, Sapsis DB, Skinner CN, van Wagtendonk JW, Weatherspoon CW (2000) The use of shaded fuelbreaks in landscape fire management. For Ecol Manage 127:55–66

Ager AA, Finney MA, Kerns BK, Maffei H (2007) Modeling wildfire risk to northern spotted owl (*Strix occidentalis* caurina) habitat in Central Oregon, USA. For Ecol Manage 246:45–56

Ager AA, Vaillant NM, Finney MA (2011) Integrating fire behavior models and geospatial analysis for wildland fire risk assessment and fuel management planning. J Combust Article ID 572452:19. doi:10.1155/2011/572452

Anderson HE (1982) Aids to determining fuel models for estimating fire behavior. General Technical report INT-122. USDA Forest Service Intermountain Research Station, Ogden

Andrews PLB (2008). f. m. s., version 4.0: Variables. Gen. Tech. Rep. RMRS-GTR-213WWW. Fort Collins, CO: Department of Agriculture, Forest Service, Rocky Mountain Research Station, p 107. http://www.fs.fed.us/rm/pubs/rmrs_gtr213.pdf. BehavePlus fire modeling system version 4.0: Variables. General Technical Report RMRS-GTR-213WWW, USDA Forest Service Rocky Mountain Research Station, Fort Collins CO

Bachmann A, Allgower B (1999) Framework for wildfire risk analysis. In: International conference on forest fire research; Conference on fire and forest meteorology, Lusa, pp 2177–2190

Bachmann A, Allgower B (2001) A consistent wildland fire risk terminology is needed! Fire Manag Today 61:28–33

Bevins CD (1996) Firelib: user manual and technical reference. SEM Technical Document, Systems for Environmental Management, Missoula

Blackard JA, Finco MV, Helmer EH, Holden GR, Hoppus ML, Jacobs DM, Lister AJ, Moisen GG, Nelson MD, Riemann R, Ruefenacht B, Salajanu D, Weyermann DL, Winterberger KC, Brandeis TJ, Czaplewski RL, McRoberts RE, Patterson PL, Tymcio RP (2008) Mapping U.S. Forest biomass using nationwide forest inventory data and moderate resolution information. Remote Sens Environ 112:1658–1677

Bonazountas M, Kallidromitou D, Kassomenos PA, Passas N (2005) For fire risk analysis. Human Ecol Risk Assess 11:617–626

Bourgeron PS, Jensen ME (1994) An overview of ecological principles for ecosystem management. In: Jensen ME, Bourgeron PS (eds) Ecosystem management: principles and applications. U.S. Department of Agriculture, Forest Service, Pacific Northwest Research Station, Portland, pp 45–57

Brown TJ, Hall BL, Westerling AL (2004) The impact of twenty-first century climate change on wildland fire danger in the western United States: an applications perspective. Clim Change 62:365–388

Burgan RE, Rothermal RC (1984) BEHAVE: fire behavior prediction and fuel modeling system–FUEL subsystem. General Technical report INT-167, USDA Forest Service

Calkin DE, Gebert KM (2006) Modeling fuel treatment costs on forest service lands in the western United States. W J Appl For 21:217–231

Cohan D, Haas SM, Radloff DL, Yancik RF (1984) Using fire in forest management: decision making under uncertainty. Interfaces 14:8–19

Cohen WB (1989) Potential utility of the TM tasseled cap multispectral data transformation for crown fire hazard assessment. In: ASPRS/ACSM annual convention proceedings: agenda for the 90s, Baltimore, pp 118–127

Deeming JE, Burgan RE, Cohen JD (1977) The National fire danger rating system—1978. General Technical report INT-39. USDA Forest Service Intermountain Forest and Range Experiment Station, Ogden

Ercanoglu M, Weber KT, Langille J, Neves R (2006) Modeling wildland fire susceptibility using fuzzy systems. GISci Remote Sens 43:268–282

Farris CA, Pezeshki C, Neuenschwander LF (2000) A comparison of fire probability maps derived from GIS modeling and direct simulation techniques. In: The joint fire science conference and workshop: crossing the millenium: integrating spatial technologies and ecological principles for a new age in fire management. University of Idaho, Moscow, pp 130–137

Fiedler CE, Keegan CE, Woodall CW, Morgan TA, Robertson SH, Chmelik JT (2001) A strategic assessment of fire hazard in Montana. Submitted to joint fire sciences program. University of Montana, Missoula

Finney MA (1998) FARSITE: fire area simulator—model development and evaluation. Research Paper RMRS-RP-4. United States Department of Agriculture, Forest Service Rocky Mountain Research Station, Ft. Collins

Finney MA (2001) Design of regular landscape fuel treatment patterns for modifying fire growth and behavior. Forest Sci 47:219–228

Finney MA (2005) The challenge of quantitative risk analysis for wildland fire. For Ecol Manage 211:97–108

Finney M, Grenfell I, McHugh C, Seli R, Trethewey D, Stratton R, Brittain S (2011) A method for ensemble Wildland Fire simulation. Environ Model Assess 16:153–167

Forthofer JM, Butler BW, Shannon KS, Finney MA, Bradshaw LS, Stratton R (2003) Predicting surface winds in complex terrain for use in fire spread models. In: Fifth symposium on fire and forest meteorology and second Wildland fire ecology and fire management congress. American Meterological Society, Orlando, pp J2.6

GAO (2002) Severe wildland fires: leadership and accountability needed to reduce risks to communities and resources. Report to congressional requesters GAO-02-259. United States General Accounting Office, Washington, DC

GAO (2003) Additional actions required to better identify and prioritize lands needing fuels reduction. Report to congressional requesters GAO-03-805. United States General Accounting Office, Washington, DC

GAO (2004) Forest service and BLM need better information and a systematic approach for assessing risks of environmental effects, GAO-04-705. United States General Accounting Office, Washington, DC

GAO (2007) Wildland fire management: better information and a systematic process could improve agencies approach to allocating fuel reduction funds and selecting projects, GAO-07-1168. United States General Accounting Office, Washington, DC

GAO/RCED (1999) Western National forests–a cohesive strategy is needed to address catastrophic wildfire threats, GAO report GAO/RCED-99-65. United States General Accounting Office, Washington, DC

Gill AM, Christian KR, Moore PHR, Forrester RI (1987) Bushfire incidence, fire hazard, and fuel reduction burning. Aust J Ecol 12:299–306

Gonzalez JR, Kolehmainen O, Pukkala T (2007) Using expert knowledge to model forest stand vulnerability to fire. Comput Electron Agric 55:107–114

Hann WJ, Bunnell DL (2001) Fire and land management planning and implementation across multiple scales. Int J Wildland Fire 10:389–403

Hardwick PE, Lachowski H, Forbes J, Olson RJ, Roby K, Fites J (1998) Fuel loading and risk assessment Lassen National Forest. In: Proceedings of the seventh forest service remote sensing applications conference. American Society for Photogrammetry and Remote Sensing, Bethesda, pp 328–339

Hardy CC (2005) Wildland fire hazard and risk: problems, definitions, and context. For Ecol Manage 211:73–82
Hargrove WW, Hoffman FM (2004) A flux atlas for representativeness and statistical extrapolation of the AmeriFlux network. Oak Ridge National Laboratory Technical Memorandum ORNL/TM–2004/112. http://www.geobabble.ornl.gov/flux-ecoregions
Heinselman ML (1981) Fire intensity and frequency as factors in the distribution and structure of northern ecosystems. In: Proceedings of the conference fire regimes and ecosystem properties. USDA Forest Service General Technical report WO-26, pp 7–55
Hessburg PF, Reynolds KM, Keane RE, James KM, Salter RB (2007) Evaluating wildland fire danger and prioritizing vegetation and fuels treatments. For Ecol Manage 247:1–17
Hogenbirk JC, Sarrazin-Delay CL (1995) Using fuel characteristics to estimate plant ignitability for fire hazard reduction. Water Air Soil Pollut 82:161–170
Hough WA (1968) Fuel consumption and fire behavior of hazard reduction burns. Research Paper SE-36, USDA Forest Service, Southeastern Forest Experiment Station, Ashville
Jain A, Ravan SA, Singh RK, Das KK, Roy PS (1996) Forest fire risk modelling using remote sensing and geographic information systems. Curr Sci 70:928–933
Jolly M (2008) US National fire danger rating system: past, present and future. In: International workshop on advances in operational weather systems for fire danger rating, Edmonton
Karau EC, Keane RE (2007) Determining landscape extent for succession and disturbance simulation modeling. Landscape Ecol 22:993–1006
Keane RE, Burgan RE, Wagtendonk JV (2001) Mapping wildland fuels for fire management across multiple scales: integrating remote sensing, GIS, and biophysical modeling. Int J Wildland Fire 10:301–319
Keane RE, Drury SA, Karau E, Hessburg PF, Reynolds KM (2008) A method for mapping fire hazard and risk across multiple scales and its application in fire management. Ecol Model
Keane RE, Drury SA, Karau EC, Hessburg PF, Reynolds KM (2010) A method for mapping fire hazard and risk across multiple scales and its application in fire management. Ecol Model 221:2–18
Keane RE (2013) Describing wildland surface fuel loading for fire management: a review of approaches, methods and systems. Int J Wildland Fire
Keane RE, Finney MA (2003) The simulation of landscape fire, climate, and ecosystem dynamics. In: Veblen TT, Baker WL, Montenegro G, Swetnam TW (eds) Fire and global change in temperate ecosystems of the Western Americas. Springer, New York, pp 32–68
Keane RE, Herynk JM, Toney C, Urbanski SP, Lutes DC, Ottmar RD (2013) Evaluating the performance and mapping of three fuel classification systems using forest inventory and analysis surface fuel measurements. For Ecol Manage 305:248–263
Keane RE, Parsons R, Hessburg P (2002a) Estimating historical range and variation of landscape patch dynamics: limitations of the simulation approach. Ecol Model 151:29–49
Keane RE, Veblen T, Ryan KC, Logan J, Allen C, Hawkes B (2002b) The cascading effects of fire exclusion in the Rocky Mountains. In: Baron J (ed) Rocky mountain futures: an ecological perspective. Island Press, Washington, DC, pp 133–153
Klaver JM, Klaver RW, Burgan RE (1998) Using GIS to assess forest fire hazard in the Mediterranean region of the United States
Laverty L, Williams J (2000) Protecting people and sustaining resources in fire-adapted ecosystems—a cohesive strategy. Forest Service response to GAO report GAO/RCED 99-65, USDA Forest Service, Washington, DC
Loehle C (2004) Applying landscape principles to fire hazard reduction. For Ecol Manage 198:261–267
Lutes DC, Keane RE, Caratti JF (2009) A surface fuels classification for estimating fire effects. Int J Wildland Fire 18:802–814
Menakis JP (2008) Mapping wildland fire potential for the conterminous United States. In: USDA forest service Twelfth biennial remote sensing conference—supporting resource management with technology—RS-2008. USDA Forest Service, Salt Lake City

Menakis JP, Cohen J, Bradshaw L (2003) Mapping wildland fire risk to flammable structures for the conterminous United States. In: Fire conference 2000: the first national congress on fire ecology, prevention, and management. Tall Timbers Research Station, Tallahassee

Millar CI, Stephenson NL, Stephens SL (2007) Climate change and forests of the future: managing in the face of uncertainty. Ecol Appl 17:2145–2151

Morgan P, Hardy CC, Swetnam TW, Rollins MG, Long DG (2001) Mapping fire regimes across time and space: understanding coarse and fine-scale fire patterns. Int J Wildland Fire 10:329–342

Neuenschwander LF, Menakis J, Miller M, Sampson RN, Hardy CC, Averill R, Mask R (2000) Indexing Colorado watersheds to risk of wildfire. J Sustain For 11:35–56

Nrc NRC (1989) Improving risk communication. National Academy Press, Washington, DC

Ottmar RD, Sandberg DV, Riccardi CL, Prichard SJ (2007) An overview of the fuel characteristic classification system—quantifying, classifying, and creating fuelbeds for resource planning. Can J For Res 37:2383–2393

Parisien MA, Kafka VG, Hirsch KG, Todd JB, Lavoie SG, Maczek PD (2005) Mapping wildfire susceptibility with the BURN-P3 simulation model. Informational report NOR-X-405. Canadian Forest Service, Northern Forestry Centre, Edmonton

Radeloff VC, Hammer RB, Stewart SI, Fried JS, Holcomb SS, McKeefry JF (2005) The wildland-urban interface in the United States. Ecol Appl 15:799–805

Rehm RG, Mell WE (2009) A simplified model for wind effects of burning structures and topography on WUI surface fire propagation. Int J Wildland Fire 18:290–301

Reinhardt E, Keane RE (1998) FOFEM—a first order fire effects model. Fire Manage Notes 58:25–28

Reinhardt E, Scott JH, Gray KL, Keane RE (2006) Estimating canopy fuel characteristics in five conifer stands in the western United States using tree and stand measurements. Can J For Res 36:1–12

Reynolds KM, Hessburg P, Keane RE, Menakis JP (2009) National fuel treatment budgeting in US federal agencies: capturing opportunities for transparent decision-making. For Ecol Manage 258:2373–2381

Rollins MG (2009) LANDFIRE: a nationally consistent vegetation, wildland fire, and fuel assessment. Int J Wildland Fire 18:235–249

Romme WH, Knight DH (1981) Fire frequency and subalpine forest succession along a topographic gradient in Wyoming. Ecology 62:319–326

Rothermel RC (1972) A mathematical model for predicting fire spread in wildland fuels. Research Paper INT-115, United States Department of Agriculture, Forest Service, Intermountain Forest and Range Experiment Station, Ogden

Rothermel RC (1991) Predicting behavior and size of crown fires in the Northern Rocky Mountains. Research Paper INT-438, United States Department of Agriculture, Forest Service Intermountain Forest and Range Experiment Station, Ogden

Running SW (2006) Is global warming causing more, larger wildfires. Science 313:927–928

Sampson RN, Sampson RW (2005) Application of hazard and risk analysis at the project level to assess ecologic impact. For Ecol Manage 211:109–116

Schmidt KM, Menakis JP, Hardy CC, Hann WJ, Bunnell DL (2002) Development of coarse-scale spatial data for wildland fire and fuel management. General Technical report RMRS-GTR-87. USDA Forest Service Rocky Mountain Research Station, Fort Collins

Scott J, Burgan RE (2005) A new set of standard fire behavior fuel models for use with Rothermel's surface fire spread model. General Technical report RMRS-GTR-153. USDA Forest Service Rocky Mountain Research Station, Fort Collins

Tainaka K (1996) Uncertainty in ecological catastrophe. Ecol Model 86:125–128

Tang SM, Gustafson EJ (1997) Perception of scale in forest management planning: challenges and implications. Landscape Urban Plan 39:1–9

Thompson MP, Calkin DE (2011) Uncertainty and risk in wildland fire management: a review. J Environ Manage 92(8):1895–1909. doi:10.1016/J.JENVMAN.2011.03.015

Thornton PE, Hasenauer H, White MA (2000) Simultaneous estimation of daily solar radiation and humidity from observed temperature and precipitation: an application over complex terrain in Austria. Agric For Meteorol 104:255–271

Thornton PE, Running SW, White MA (1997) Generating surfaces of daily meteorological variables over large regions of complex terrain. J Hydrol 190:214–251

Venn T, Calkin D (2007) Challenges of socio-economincally evaluation wildfire management in and adjacent to non-industrial private forestland in the western United States. In: Improving the triple bottom line returns from small-scale forestry. The University of Queensland, Ormoc, pp 383–393

Westerling AL, Hidalgo HG, Cayan DR, Swetnam TW (2006) Warming and earlier spring increase in western US forest wildfire activity. Science 313:940–943

White PS, Harrod J, Walker JL, Jentsch A (2000) Disturbance, scale, and boundary in wilderness management. In: Wilderness science in a time of change. USD AForest Service Rocky Mountain Research Station RMRS-P-15-VOL-2, Missoula, pp 27–42

Yool SR, Eckhardt DW, Estes JE, Cosentino MJ (1985) Describing the brushfire hazard in Southern California. Annu Assoc Am Geogr 75:417–430

Landscape Evaluation and Restoration Planning

**Paul F. Hessburg, R. Brion Salter, Keith M. Reynolds,
James D. Dickinson, William L. Gaines and Richy J. Harrod**

Abstract Contemporary land managers are beginning to understand that landscapes of the early 20th century exhibited complex patterns of compositional and structural conditions at several different scales, and that there was interplay between patterns and processes within and across scales. Further, they understand that restoring integrity of these conditions has broad implications for the future sustainability of native species, ecosystem services, and ecological processes. Many too are hungry for methods to restore more natural landscape patterns of habitats and more naturally functioning disturbance regimes; all in the context of a warming climate. Attention is turning to evaluating whole landscapes at local and regional scales, deciphering their changes and trajectories, and formulating scale-appropriate landscape prescriptions that will methodically restore ecological functionality and improve landscape resilience. Here, we review published landscape evaluation and planning applications designed in EMDS. We show the utility of EMDS for designing transparent local landscape evaluations, and we reveal approaches that have been used thus far. We begin by briefly reviewing six projects from a global sample, and then review in greater depth four projects we have developed with our collaborators. We discuss the goals and design of each project, its methods and utilities, what worked well, what could be improved and

P. F. Hessburg (✉) · R. Brion Salter
U.S. Department of Agriculture, Forest Service, Pacific Northwest Research Station, Wenatchee Forestry Sciences Laboratory, 1133 North Western Ave., Wenatchee, WA 98801, USA
e-mail: phessburg@fs.fed.us

K. M. Reynolds
U.S. Department of Agriculture, Forest Service, Pacific Northwest Research Station, Corvallis Forestry Sciences Laboratory, 3200 SW Jefferson Way, Corvallis, OR 97331, USA
e-mail: kreynolds@fs.fed.us

J. D. Dickinson · W. L. Gaines · R. J. Harrod
U.S. Department of Agriculture, Forest Service, Okanogan-Wenatchee National Forest, 215 Melody Lane, Wenatchee, WA 98801, USA
e-mail: jddickinson@fs.fed.us

K. M. Reynolds et al. (eds.), *Making Transparent Environmental Management Decisions*, Environmental Science and Engineering,
DOI: 10.1007/978-3-642-32000-2_7,

related research opportunities. It is our hope that this review will provide helpful insights into how spatial decision support technologies may be used to evaluate and plan for local and perhaps larger-scale landscape restoration projects.

Keywords Landscape analysis · Restoration planning · Reference variation · Departure analysis · Future range of variability · Historical range of variability · Vegetation pattern · Vegetation structure · Climate change · Spatial decision support · EMDS

1 Introduction

Over the last several centuries, human settlement, development, and management have altered the ecological patterns and processes of forested landscapes across the U.S. such that every ecosystem has been touched by at least one of these influences. Wildfire suppression, management practices that excluded wildfire (e.g., road and rail construction), and domestic livestock grazing have altered even wilderness and roadless areas. In the western U.S., these influences occurred in the late 19th and 20th centuries. Today, few forests on public lands fully support their native flora and fauna, and wildfires and insect outbreaks are especially unprecedented in their periodic severity and spatial extent. In response, there is public mistrust of foresters and land managers, and a succession of environmental laws has ensued to constrain forest management. Additionally, there is little shared insight as to methods or philosophies that could guide landscape restoration and maintenance in a manner that cooperates with native ecosystem structure and function.

Toward development of a shared vision and goals, scientists, public land managers, and citizens are beginning collaborative partnerships to develop a common understanding of the causes and consequences of past management on national forests, and of their possible future trajectories with climatic warming. Here, we review several applications developed with EMDS for landscape evaluation and restoration planning in forests of the Inland Northwest U.S. These applications can be used to:

(1) strategically and tactically plan for landscape restoration,
(2) evaluate the status and trends of forest-landscape vegetation and habitat conditions,
(3) evaluate the vulnerability of conditions to wildfire, insect, and pathogen disturbances, and
(4) conduct these evaluations in the context of recent historical and likely future climates.

Our goal in presenting these examples is to show the broad utility of EMDS in landscape evaluation and planning environments. We refer readers to the references for additional details of the projects. First though, we begin by providing a brief background on the settlement and management history of this region and related effects. This context clarifies the origins of specific restoration goals and potential pathways to achieve them.

1.1 Background

Subsistence agriculture, hunting, and burning activities dominated early aboriginal management of the Holocene North American landscape. These activities enabled colonization of the continent and cultural development over thousands of years, but not without attendant landscape impacts associated with hunter-gatherer, nomadic, and subsistence lifestyles (Pyne 1982; Sauer 1971; White 1991, 1992, 1999). Burning by Native Americans created new and expanded existing herblands, meadows, and open wooded expanses, thereby enhancing harvest of edible plants, nuts, and berries. It also increased sighting distances in the event of sneak-attacks by marauding tribes, and improved forage for wild ungulates, which enhanced hunting both near and away from encampments. Intentional burning was also employed along major travel routes to improve food supplies while traveling and to increase travel ease and safety. Indian burning lacked direct spatial controls on burned area or fire effects, and burns often travelled farther and killed more forest than intended. Nonetheless, Native Americans were the first fire managers, and their use of intentionally lighted fires greatly aided their travels and lifestyles.

In the mid-19th century, settlement and management of the Great Plains, and the Pacific, Rocky Mountain, and Intermountain West by Euro-American settlers accelerated to a fever pitch with the discovery of lush and productive prairies on the plains, and in the intermountain valleys, rich gold and silver ore deposits, and abundant acres for homesteading and a fresh start (Pyne 1982; Robbins 1994, 1997, 1999; White 1991). Westward migration, Native American expatriation from ancestral homelands by the U.S. cavalry, and forced settlement onto reservations produced the final downfall of the indigenous population. However, the major depredation to aboriginal populations had already been done via the introduction of exotic diseases by trappers and fur traders in the late-1700s (Hunn 1990; Langston 1995; Robbins 1994, 1997, 1999).

With settlement came land clearing for homesteads, expansion of agriculture, timber harvesting (Hessburg and Agee 2003; Langston 1995; Robbins 1997, 1999), and early attempts at wildfire suppression, which became highly effective only after the 10 a.m. rule was enacted as federal policy between 1934 and 1935 (Pyne 1982; van Wagtendonk 2007). This policy of suppressing fires by 10 a.m. of the next burn period after detection forever changed the role of wildfire, especially as it applied to primeval western landscapes. The rule was removed in the early 1970s, but moderately aggressive wildfire suppression is still practiced in the U.S.

Natural variability in wildfire frequency, duration, severity, seasonality, and extent were unavoidably altered by decades of fire exclusion, wildfire suppression, and broadly-popularized fire-prevention campaigns. Wildfire exclusion by cattle grazing, road and rail construction, successful wildfire prevention and suppression policies, and industrial-strength selective logging, beginning in the 1930s and continuing for more than 50 years, contributed not only to extensive alteration of natural wildfire regimes, but also to changes in forest insect and pathogen disturbance regimes, causing them to shift significantly from historical analogues. For example, the duration, severity, and extent of conifer defoliator and bark beetle outbreaks increased substantially (Hessburg et al. 1994), becoming more chronic and devastating to timber and habitat resources (e.g., see Hummel and Agee 2003).

Selective logging accelerated steadily during and after the Second World War. Fire exclusion and selective logging primarily advanced the seral status and reduced fire tolerance of affected forests with the removal of fire-tolerant species and the largest size and age classes (Hessburg and Agee 2003). It also increased the density and layering of the forests that remained because selection cutting favored the regeneration and release of shade-tolerant and fire-intolerant tree species such as Douglas-fir, grand fir, and white fir (Hessburg et al. 2005). Recent warming and drying of the western U.S. climate has exacerbated these changes (McKenzie et al. 2004; Westerling and Swetnam 2003; Westerling et al. 2006), and will continue to do so.

Changes from pre-settlement era variability of structural and compositional conditions affected regional landscapes as well. Prior to the era of management, regional landscape resilience to wildfires naturally derived from mosaics of previously burned and recovering vegetation patches from prior wildfire events, and a predictable distribution of prior fire event sizes (Moritz et al. 2011). This resilience yielded a finite and semi-predictable array of pattern conditions (Hessburg et al. 1999a, b, c, 2000a) that supported other ecological processes at several scales of observation.

As a result of these many changes, land managers faced substantial societal and scientific pressure to improve habitat conditions and viability of native species, and the food webs that support them. Because alternatives to managing for historical analogue or related conditions are untested or untestable (Millar et al. 2007; Stephens et al. 2010), public land managers have been required to restore a semblance of the natural abundance and spatial variability of habitats. This has largely been reinforced by endangered species and environmental laws.

On the other hand, public mistrust over decades of commodity-driven management on public lands paralyzes most attempts at large-scale landscape restoration, and with some good reason. Restoration prescriptions for thinning, underburning, and slash disposal are often seen as blanket remedies, and another form of landscape oversimplification by management, which is the current problem. The time is ripe for more transparent evaluation of landscape patterns, processes, changes in their interactions and associated restoration planning, and even riper for management applications to be conducted experimentally and transparently, with full access to scientific methods and adaptive learning.

Below, we briefly highlight several examples in which EMDS was used to conduct landscape evaluations for decision-making in a variety of planning contexts. In these examples, tools within the EMDS modeling framework were used to develop evaluations that considered the effects of various management strategies or tactics on the natural or developed environment, or to select specific lands or man-made features for management, management avoidance, or modification. These examples illustrate how EMDS might be used at a variety of scales with varied goals in mind. Hopefully it becomes apparent that if the management goals and contexts can be clearly articulated, a logical and transparent application can be developed in EMDS to represent it.

1.2 Previous Examples of Evaluations Using EMDS

Stolle et al. (2007) developed an EMDS application to evaluate natural resource impacts that might be caused by conventional management practices (site preparation, planting, and harvesting) in a forest plantation. Using logic networks designed with the NetWeaver developer tool (Miller and Saunders 2002, see also Chapter 2), they evaluated the effects of management activities on ambient soil and site conditions as a means of representing the inherent risks associated with standard management practices of commercial plantation forestry. They mapped *fragility areas* on a forest property that were sensitive to standard forestry practices (according to an established set of criteria), which enabled them to implement low-impact management of the natural resources, while producing an economic return.

Girvetz and Schilling (2003) used EMDS to build a knowledgebase that evaluated the environmental impact of an extensive road network on the Tahoe National Forest, CA, USA. Using spatial data for natural and human processes, the authors evaluated the assertion that any road has a high potential for impacting the environment. They used modeled potential environmental impact to negatively weight roads for a least-cost path network analysis to more than 1500 points of interest in the forest. They were able to make solid recommendations for providing access to key points of interest, while streamlining and reducing the road network and reducing its environmental impacts.

Janssen et al. (2005) developed an EMDS model to provide decision support for wetland management in a highly managed wetland area of the northern Netherlands. Because legislation in the European Union has mandated the importance of preserving wetland ecosystems, they funded development and implementation of an operational wetland evaluation decision-support system to support the European policy objectives of providing ongoing agriculture, expanding recreation opportunities, maintaining residential opportunities, and conserving wetland habitats. They compared three possible management alternatives for their influence on water quality and quantity, the local climate and biodiversity, and social and economic values: (1) modern peat pasture (current), (2) historical peat pasture, and (3) dynamic mire. The model adequately framed management options and provided needed context for decisions about future land allocations.

Wang et al. (2010) developed an integrated assessment framework and a spatial decision-support system in EMDS to support land-use planning and local forestry decisions concerning carbon sequestration. The application integrated two process-based carbon models, a spatial decision module, a spatial cost-benefit analysis module, and an Analytic Hierarchy Process (AHP) module (Saaty 1992, 1994). The integrated model provided spatially-explicit information on carbon sequestration opportunities and sequestration-induced economic benefits under various scenarios of the carbon-credit market. The modeling system is demonstrated for a case study area in Liping County, Guizhou Province, China. The study demonstrated that the tool can be successfully applied to determine where and how forest land uses may be manipulated in favor of carbon sequestration.

Staus et al. (2010) developed an EMDS application to evaluate terrestrial and aquatic habitats across western Oregon, USA, for their suitability of meeting the ecological objectives spelled out in the Northwest Forest Plan (USDI 1992; USDA 1994), which included maintenance of late-successional and old-growth forest, recovery and maintenance of Pacific salmon (*Oncorhynchus* spp.), and restored viability of northern spotted owls (*Strix caurina occidentalis*). Areas of the landscape that contained habitat characteristics supporting these objectives were modeled as having high conservation value. The authors used their model to evaluate the ecological condition of 36,180 township and range sections (~260 ha each) across the study domain. They identified 18 % of study area sections as providing habitats of high conservation value. The model provided information that could be considered in future land management decisions to spatially allocate owl habitats in the western Oregon portion of the Northwest Forest Plan area. Furthermore, their results illustrated how decision-support applications can help land managers develop strategic plans for managing large areas across multiple ownerships.

White et al. (2005) developed an EMDS knowledge base for evaluating the conservation potential of forested sections in the checkerboard ownership area of the central Sierra Nevada in California, USA (see also Chapter "Forest Conservation Planning"). Four primary topics were evaluated including each section's (1) existing and potential terrestrial and aquatic biodiversity value, (2) existing and potential mature forest connectivity, (3) recreation access and passive use resource opportunities, and (4) risks of exurban development, unnatural fire, and management incompatible with mature forest management. Results of evaluations of each primary topic were networked in a summary knowledgebase. The knowledgebase allowed the science team to recommend arrangements of sections within the checkerboard ownership that showed the highest promise of conserving important terrestrial and aquatic species and habitats, in the long term.

2 Four Detailed Examples

In sections that follow, we review in more detail four EMDS model applications that we and our collaborators developed to evaluate landscapes in unique contexts for the purpose of determining restoration needs and treatment priorities.

In Sect. 3, we present an approach to estimating the extent to which present forest landscape patterns in the Inland Northwest have departed from the conditions that existed before the era of modern management (~ 1900). In Sect. 4, we describe the use of EMDS to evaluate existing patterns of forest vegetation in a random sample of watersheds of one ecoregion against a corresponding broad envelope of historical reference conditions for the same ecoregion. In a third application (Sect. 5), changes in spatial patterns of various patch types of forested landscapes were evaluated in two watersheds in eastern Washington, USA, with respect to the patterns of two sets of reference conditions; one representing the broad variability of pre-management era (~ 1900) conditions, and another representing the broad variability associated with one plausible warming and drying climate-change scenario. Finally, in Sect. 6, we present an EMDS application designed to provide decision support for landscape restoration of a managed dry forest area in the eastern Cascade Mountains of Washington State.

3 Evaluating Changes in Landscape-Level Spatial Patterns

In Hessburg et al. (2004), we present a landscape evaluation approach to estimating the extent to which present-day forest landscape patterns have changed from the variety of conditions that existed before the era of modern management (~ 1900). Our goal in this foundational project was to approximate the range and variation of these recent historical patterns, use that knowledge to evaluate present forest conditions, and assess the trajectory and ecological importance of any significant changes. The approach was based on the Wu and Loucks (1995) hierarchical patch dynamics paradigm, which we briefly summarize here because it frames the analytical approach.

The paradigm holds that an ecosystem can be viewed as a multi-level hierarchy of patch mosaics. An ecosystem's overarching dynamics derive from emergent properties of concurrent patch dynamics occurring at each level in a hierarchy. Across the temporal scales of a hierarchy, regional spatial patterns of biota, geology, geomorphic processes, and climate provide top-down constraint on ecological patterns and processes occurring at a meso-scale. Likewise, fine-scale patterns of endemic disturbances, topography, environments, vegetation, and other ecological processes provide critical bottom-up context for patterns and processes occurring at a meso-scale. At all spatial and temporal scales of the hierarchy, ecosystems exhibit transient patch dynamics and non-equilibrium behavior. This is due to a mix of both stochastic and deterministic properties of the supporting land

and climate systems and ecosystem processes at each level. Lower level processes are incorporated into the next higher-level structures and processes, and this happens at all levels.

Landscape patterns at each level in a hierarchy are never the same from year to year, and they never repeat in the same arrangements. However, transient dynamics are manifest as envelopes of pattern conditions at each level (literally, a naturally occurring range of variation), owing to the recurring patterns and interactions of the dominant top-down and bottom-up spatial controls (Hessburg et al. 1999a, c). Thus, patterns don't repeat in the same spatial arrangements, but they do exhibit predictable spatial pattern characteristics, for example, in the percentage area in different cover species, size class, or structural conditions, the range in patch sizes, or the dispersion of unique patch types.

Moreover, because contexts and constraints are non-stationary, the processes and patterns they reflect are non-stationary as well. In a warming climate, for example, the envelope of pattern conditions at each level in a patch dynamics hierarchy may be reshaped by the strength and duration of warming, with the existing patterns as initial context. Reshaping within a level can be figuratively represented as an envelope of conditions that drifts directionally in a hyper-dimensional phase space. Because this is impossible to illustrate, we illustrate a simpler cartoon of conditions shifting in a two-dimensional phase space (Fig. 1). Relatively small amplitude and short-term changes (multi-annual to multi-decadal) in climatic inputs will do little to reshape the envelope, but large amplitude and long-term changes (centenary to multi-centenary and longer) have much greater likelihood of significantly reshaping pattern envelopes.

In this project, we developed an approach to estimating the non-equilibrium conditions associated within a meso-scale landscape in a forest patch dynamics hierarchy. For simplicity, we termed the conditions for the climatic period ending in the early 20th-century "reference conditions." Typical variation in these conditions was termed "reference variation" (RV). For our estimate of RV we chose the median 80 % range of a diagnostic set of five class and nine landscape spatial-pattern metrics (McGarigal and Marks 1995), because most historical observations typically clustered within this middle range. The class metrics were: the percentage of the total landscape area (%LAND), patch density per 10,000 ha (PD), mean patch size (MPS, ha), mean nearest-neighbor distance (MNN, m), and edge density (ED, m·ha-1). The landscape metrics were: patch richness (PR) and relative patch richness (RPR), Shannon's diversity index (SHDI) and Hill's transformation of Shannon's index (N1, Hill 1973), Hill's inverse of Simpson's λ, N2, (Hill 1973; Simpson 1949), Simpson's modified evenness index, and Alatalo's evenness index, R21, (Alatalo 1981), a contagion index (CONTAG); and an interspersion and juxtaposition index (IJI). We supplemented the FRAGSTATS source code (McGarigal and Marks 1995) with the equations for computing the N1, N2, and R21 metrics.

The focal level of the study was forest landscapes of meso-scale watersheds and their spatial patterns of structure, species composition, fuels, and wildfire behavior attributes. Structural classes were an approximation of stand succession and

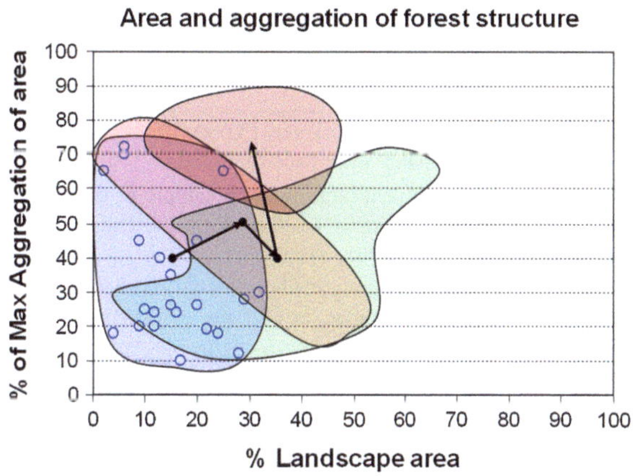

Fig. 1 Graphical representation of how landscape area and aggregation of area of a single forest structural component might vary in phase space (for example, old multilayered forest or stand initiation structure) as the climate of an ecoregion shifts. Within the concept of historical or natural range of variation, clouds or envelopes of conditions exist in phase space, for any number and combination of structural and compositional features, across a broad range of metrics, and no two are alike. The same is true for current and future ranges of variation. This broad dimensionality is readily captured in data space, quantified, and then may be used to detect significant changes in spatial patterns and variability in those patterns

development phases. Cover types reflected forest overstory species and mixes. Estimates of surface and canopy fuels reflected the available fuels to support wildfires and either surface or crownfire behavior. We focused on patterns of living and dead vegetation at this level because many of the most important changes in the dynamics of altered forest ecosystems are reflected in the living and dead structure of the affected structural and compositional landscapes (Spies 1998). We stratified landscapes into ecoregions to reflect top-down biogeoclimatic constraint on forest structural patterns and related disturbances (Hessburg et al. 2000b, 2004). Study landscapes were 4,000–12,000 ha subwatersheds.

We developed a repeatable quantitative method (outlined in Table 1) for estimating RV in historical forest vegetation patterns and of vulnerability to disturbance. The objective was to estimate a RV so that we could evaluate the direction, magnitude, and potential ecological importance of the changes observed in present-day forest landscape patterns (Keane et al. 2002, 2009; Landres et al. 1999). To automate this approach, we programmed a departure analysis application in EMDS that compared the spatial pattern conditions of a test landscape with the estimated RV that would be expected within its ecological subregion (Reynolds 1999a, 2001a). Via automation, this analysis could be repeated for any number of subwatersheds within the same ecoregion. By means of the comparison with RV, we could identify vegetation changes that were beyond the range of the RV estimates. Changes that fell within the range of the RV estimates were assumed to be within the natural variation of the interacting land and climate

Table 1 Outline of methods used in Hessburg et al. (2004) for estimating departure of present forest landscape patterns from historical (circa. 1900) reference conditions

Step	Action	References
1	Stratified Inland Northwest U.S. subwatersheds (5,000–10,000 ha) into ecological subregions using a published hierarchy	Hessburg et al. (2000b)
2	Mapped the historical vegetation of a large random sample of the subwatersheds of one subregion (ESR4 – the Moist and Cold Forests subregion) from 1930–1940s aerial photography	Hessburg et al. (1999a)
3	Statistically reconstructed the vegetation attributes of all patches of sampled historical subwatersheds that showed any evidence of prior timber harvest	Moeur and Stage (1995)
4	Ran spatial pattern analysis on each reconstructed historical subwatershed calculating a finite, descriptive set of class and landscape metrics in a spatial analysis program (FRAGSTATS)	McGarigal and Marks (1995) Hessburg et al. (1999a)
5	Observed the data distributions from the spatial pattern analysis output of the historical subwatersheds and defined reference conditions based on the typical range of the clustered data	Hessburg et al. (1999a, b)
6	Defined reference variation as the median 80 % range of the class and landscape metrics for the sample of historical subwatersheds	Hessburg et al. (1999a, b, c)
7	Estimated ESR4 reference variation for spatial patterns of forest composition (cover types), structure (stand development phases), modeled ground fuel accumulation (loading), and several fire behavior attributes	Hessburg et al. (1999a, b, c) Huff et al. (1995) O'Hara et al. (1996) Hessburg et al. (2000a)
8	Programmed ESR4 reference conditions into a decision support model (EMDS)	Reynolds (1999a, b) Reynolds (2001a, b)
9	Mapped the current vegetation patterns of an example watershed, Wenatchee_13, from the Wenatchee River basin, also from ESR4	Hessburg et al. (1999a)
10	Objectively compared a multi-scale set of vegetation maps of the example watershed with corresponding reference variation estimates in the decision support model	Hessburg et al. (1999a, b)

system, and dominant ecosystem processes. Changes that were beyond the range of RV estimates were termed "departures" that could be explored in more detail for their potential ecological implications.

We also programmed transition analysis on the test landscapes' historical and current maps of cover type and structural class to discover the path of each significant change. To conduct transition analysis, we converted the polygon maps of historical and current cover type and/or structural class to raster format (30-m resolution). These raster maps were combined such that each pixel had a historical and current cover type (and/or structural class) identity. We computed the number of pixels for each unique type of historical-to-current transition, divided this number by the total number of pixels, and multiplied that result by 100 to derive a percentage of the subwatershed area in a transition type.

Using departure and transition analyses, we were able to highlight a variety of important changes to the test landscape. For example, we found that timber harvests had converted much area dominated by the ponderosa pine (*Pinus ponderosa*) cover type to Douglas-fir (*Pseudotsuga menziesii*); regeneration harvest had highly fragmented forest cover; and old forests of the western hemlock (*Tsuga heterophylla*), western redcedar (*Thuja plicata*), Douglas-fir, and ponderosa pine zones had suffered significant depredation from selective and regeneration harvesting (18 % reduction in area).

Departure and transition analyses of fuel loading, wildfire rate of spread, crownfire potential, flame length, and fireline intensity attributes under prescribed and wildfire (90th percentile) burn scenarios depicted an historical landscape that displayed large contiguous areas with very high fuel loading and high potential for crown fires under an average wildfire scenario, typically high to extreme flame lengths, and high to extreme fireline intensities. This ordinarily high fire danger could be accounted for by a preponderance of moist to wet growing environments and a low-frequency, high-severity, stand-replacement fire regime. Large fires were relatively uncommon and they were likely driven by extreme weather or severe climatic events. However, current conditions showed that past management activities in the test landscape had reduced the likelihood of large stand-replacing fires with the introduction of nearly 50 clearcut units.

Departure analysis using landscape metrics showed poor correspondence between the present-day combined cover type-structural class mosaic and the estimates of RV. Timber harvesting had increased patch type richness, diversity, dominance, evenness, interspersion, and juxtaposition of structural class patches, and reduced overall contagion in the cover type-structural class mosaic. The historical landscape was simply patterned, consisting of fairly large patches borne of infrequent, large, high-severity fires. Management had made it more complexly patterned and fragmented.

3.1 What Worked Well?

Overall, this EMDS application did a reasonably good job of evaluating landscape pattern departures. Changes in landscape vegetation patterns were compared to a RV that simultaneously considered 18 vegetation, fuel, and fire-behavior features (e.g., physiognomies, cover types, structural classes, and potential vegetation types, fuel loads, fire rate of spread, and crownfire potential) according to a diagnostic set of 14 class and landscape pattern metrics. Results of departure and transition analyses were intuitive and useful to explaining the ecological effects of 20th-century management and settlement.

Landscape evaluations like this one must examine a host of class and landscape pattern metrics applied across a variety of mapped conditions to accurately infer class and landscape-level changes, and their significance. Evaluation in EMDS enabled analysis across a large number of landscape dimensions, with multiple metrics on each dimension. Structuring evaluations in this manner was useful to inferring change in ecological functionality with change in structure or patterning. This application tackled a hyper-dimensional problem, and it did so with relative ease.

3.2 What Could be Improved?

The EMDS application was a relatively straightforward proof-of-concept. Once it was clear that complex evaluations could be structured for analysis and interpretation, one could clearly see how other important dimensions could be integrated into the evaluations. For example, a wildlife manager could develop RV estimates for a variety of habitat features and networking arrangements. Departure analyses could evaluate and translate changes in landscape vegetation patterns and features into important changes in keystone or focal plant and animal species habitats and those of functional groups of species. A multi-scale habitat analysis would allow managers to directly interpret scaled effects of altered vegetation patterns on species varying by body size, mobility, and home range.

Future evaluations could also include the characterization of RV of landscape vulnerabilities to various insect and pathogen disturbances (e.g., see Hessburg et al. 1999a, d, 2000a). This feature is developed in the fourth application discussed below. A more inclusive application structure might ultimately include the use of other insect, pathogen, or noxious weed modeling platforms for predicting host or habitat contagion, spread potential, and intensification in the context of variables and conditions unrelated to vegetation.

Finally, it would be of theoretical value to represent landscape pattern departures across broad physiographic gradients. An expanded evaluation of this sort would help scientists and managers better understand relative degree and variation in spatial controls contributed by regional biology, geology, geomorphology, and climate. For example, a gradient-oriented analysis would lead to testable hypotheses about the nature, degree, and mechanisms of climatic influence on pattern and process; an especially hot current topic.

3.3 Research Opportunities

Perhaps the greatest opportunity to advance this application would be to develop and incorporate empirical data that adequately represent RV estimates at multiple levels in the patch dynamics hierarchy. Admittedly, this would be a large and costly task, but it would provide an immense payoff. For example, spatial heterogeneity in vegetation and fuels conditions exists in the landscape within patches, multi-patch neighborhoods, and regional landscapes as well. Improved understanding of departures at these added levels would aid understanding of the degree and manner of cross-connections between spatial scales.

Understanding typical within-patch spatial heterogeneity would give scientists and managers better insight into the lower level structures and processes influential to those occurring at a higher level, and it would improve knowledge of the principal underlying mechanisms and pathways that drive changes in vegetation patterns and vulnerability to disturbances. It would also help managers to better achieve their vegetation and disturbance regime restoration goals by helping them to more aptly specify the multi-scale patterns and variability that their patch-level silvicultural and prescribed burning prescriptions can approximate.

The same arguments can be made at other levels as well. For example, one can observe that many present-day regional landscapes are synchronized for broad-scale and damaging biotic and abiotic disturbances that may produce long-term, game-changing effects (Allen et al. 2010). Only by evaluating the spatial and temporal pattern variability of regional and local landscapes and of patches and patch neighborhoods can observers begin to understand the scales of motivating factors, the degrees and patterns of spatial and temporal controls, and primary mechanisms driving these broad-scale and emergent processes.

4 Strategic Planning for Landscape Restoration

In this second example, Reynolds and Hessburg (2005) developed methods and an EMDS decision-support application to strategically plan for landscape restoration. We illustrated a two-phase approach to evaluating departure of present-day pattern conditions of 15 forested landscapes within a single ecoregion from pre-management-era reference conditions (RV), similar to but simpler than that described in the first application above. We then computed the restoration priority among the subwatersheds, in light of the departures and other technical and economic feasibility considerations. Methods for the departure analysis are summarized in the first application above. Here, we briefly summarize methods unique to this application.

To identify sample landscapes constrained by similar environmental contexts, we used the Hessburg et al. (2000b) ecological subregions to stratify subwatersheds (ca. 4,000–12,000 ha) of the eastern Washington Cascades into biogeoclimatic zones (Fig. 2a). Subwatersheds (Fig. 2b) were used as the basic sampling units because they provided a rational means to subdivide land areas that shared similar climate, geology, topography, and hydrology. Subwatersheds compose the (12-digit) 6th level in the established hierarchy of US watersheds (Seaber et al. 1987, National Hydrology dataset available at: http://nhd.usgs.gov/).

We selected ecological subregion #4 (ESR4) as the biogeoclimatic zone in which we sampled and estimated reference conditions (Fig. 2a). Landscapes of this subregion are dominated by moist (67 % of the area) and cold (21 % of the area) forest types, with total annual precipitation of 1100–3000 mm/year, generally warm growing-season temperatures (mean annual daytime temperature, 5–9 °C), and relatively low levels of solar radiation (frequently overcast skies, 200–250 W·m^{-2};

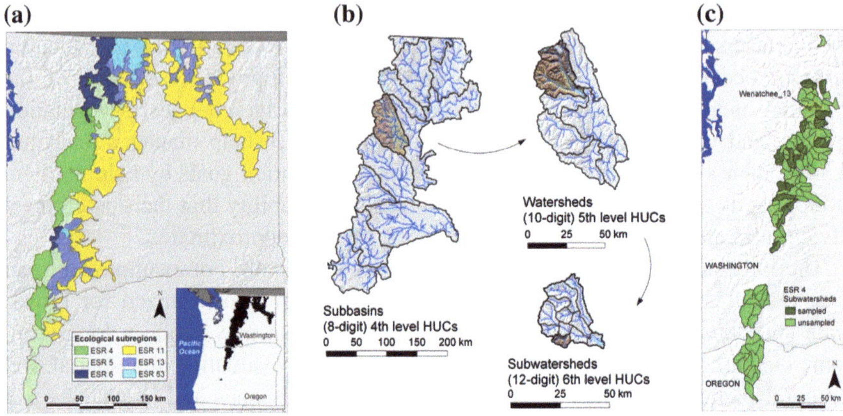

Fig. 2 Ecological subregions of the eastern Washington Cascades in the western United States (adapted from Hessburg et al. 2000b). **a** The ecological subregions (ESR) are defined as follows: *4* Warm/Wet/Low Solar Moist and Cold Forests, *5* Warm/Moist/Moderate Solar Moist and Cold Forests, *6* Cold/Wet/Low and Moderate Solar Cold Forests, *11* Warm/Dry and Moist/Moderate Solar Dry and Moist Forests, *13* Warm and Cold/Moist/Moderate Solar Moist Forests, and *53* Cold/Moist/Moderate Solar Cold Forests. **b** Hierarchical organization of sub-basins (4th level), watersheds (5th level), and subwatersheds (6th level) in the eastern Washington Cascades of the western United States (see also Seaber et al. 1987). The example shows the Wenatchee River sub-basin at the 4th level, the Little Wenatchee River watershed at the 5th level, and subwatershed Wenatchee 13 at the 6th level. **c** Subwatersheds included in this study were randomly selected from ESR 4

Hessburg et al. 2000b). The subregion contained 93 subwatersheds. To map a sample of the historical and current vegetation, we randomly selected 15 of the 93 in order to sample at least 15 % (actual 16.1 %) of the total number of subwatersheds and 15 % (actual 19.2 %) of the subregions' area (Fig. 2c).

Four vegetation features for the historical and current conditions were interpreted from stereo aerial photography and mapped in each subwatershed: physiognomic class, cover class, structure class, and late-successional old-forest class. As in the first application above, present-day maps of each vegetation feature were compared against ecoregion RV estimates developed for each feature. As in the preceding application (Sect. 3), five class metrics and nine landscape metrics were used to compare the current and RV conditions. Using this diagnostic set of metrics, we could (1) detect key changes in landscape patterns that had potential ecological significance and (2) understand the specific class changes that were driving shifts in the mosaics.

The phase 1 objective of designing a NetWeaver knowledgebase for this problem was to assess how well current conditions in the sampled subwatersheds of ESR4 corresponded to pre-management era RV. We used the term integrity to express the degree of correspondence, and departures were integrity departures. Primary topics for evaluation, corresponding to mapped attributes were: physiognomic integrity, cover integrity, structural integrity, cover-structure combined integrity, and late successional/old-growth forest integrity. Class metrics of each attribute class and the landscape metrics were evaluated for the current condition of each landscape. An evaluation for any metric was done by comparing its value for the current condition to a ramp function for the same metric derived from the historical data. The result of an evaluation was an expression of the degree of support for correspondence of the current conditions to the RV encoded in the ramp function.

Phase 2 provided a decision model for assigning restoration priorities to subwatersheds. The model included four primary criteria: compositional integrity, structural integrity, feasibility of management, and fire risk (Table 2). All subcriteria of compositional and structural integrity criteria were measures of support from the landscape analysis. Subcriteria of fire risk and feasibility represented attributes of subwatersheds that were not part of the logic-based evaluation, but were included in the decision model as logistical considerations.

Pair-wise comparisons among primary and secondary criteria using standard methods for the Analytic Hierarchy Process provided weights for the decision model (Table 2). Simple Multi-Attribute Rating Technique (SMART) utility functions for rating criteria at the lowest level of the model were also specified. Utility functions for feasibility subcriteria gave greater preference to subwatersheds with shallow slopes, road access to stands, and satisfactory timber value, which could financially underwrite restoration costs. Utility functions for subcriteria of fire risk gave greater preference to subwatersheds with higher ratings for crown-fire potential and fuel loading, based on a rationale of protecting the existing forest resources. Fuel loading and crown-fire potential were attributed to individual vegetation patches using published methods (Huff et al. 1995; Hessburg et al. 1999b, 1999c).

Priorities for landscape restoration were based on: (1) the assessment of departures in compositional and structural integrity; (2) feasibility of management, which was composed of steepness of the watershed, road access, and value of the timber; and (3) fire risk, which was composed of crown-fire potential under an average wildfire burn scenario, and fuel loading. Feasibility and risk criteria were incorporated to inform the decision-making process with real-world criteria that could influence a manager's ability to make and execute restoration decisions.

4.1 What Worked Well?

The objectives for developing this decision model were to demonstrate that landscapes of any ecoregion could be evaluated on a common basis to determine key ecological departures and then assessed for restoration priority. The application met those objectives and highlighted the importance of using decision-support technology to assess technical and economic feasibility of any proposed restoration, also on a common footing. Because many parameters (844) were used in developing this demonstration model, departure analyses were simplified in comparison with the first application shown in Sect. 3. This was done to make the

Table 2 Structure of an analytic hierarchy process model for determining priorities for restoring subwatersheds in ESR4

Criterion[a]	Weight[b]	Description
Compositional integrity	0.25	Synthesis of cover and physiognomic integrities
Cover integrity	0.67	Strength of evidence for cover integrity from evaluation phase of analysis
Physiognomic integrity	0.33	Strength of evidence for physiognomic integrity from evaluation phase of analysis
Structural integrity	0.25	Synthesis of structural integrities
All forest integrity	0.67	Strength of evidence for structural integrity from evaluation phase of analysis
Late-successional/ old-forest integrity	0.33	Strength of evidence for late-successional/ old-growth integrity from evaluation phase of analysis
Feasibility of management	0.25	Synthesis of feasibility factors
Steepness	0.25	Percent of subwatershed area with slope 30 %
Road access	0.25	Percent of subwatershed within 250 m of any road
Timber value	0.50	Relative measure of timber value in a subwatershed
Fire risk	0.25	Synthesis of fire risks
Crown fire potential	0.75	Percent of subwatershed area with high, very high, or severe crown fire potential rating
Fuel loading	0.25	Percent of subwatershed area with high or very high fuel bed loading

[a] Primary decision criteria were: compositional integrity, structural integrity, feasibility, and fire risk. Secondary decision criteria are shown indented under their primary criteria, and, because they were the lowest criteria in the model, also represent the attributes of subwatersheds that are being evaluated. Each attribute was evaluated against a utility function, specified with the Simple Multi-Attribute Rating Technique. The decision score on each primary criterion was derived as the weighted average of the utility scores of the criterion's subcriteria

[b] Each weight expressed the relative importance of a subcriterion with respect to its parent criterion. In the case of primary criteria, importance was with respect to the overall model goal of assigning restoration priorities

problem more tractable, and the added detail was not relevant to the demonstration. Departure analyses like those described in the first application can be easily incorporated into the second.

4.2 What Could be Improved?

The use of feasibility and effectiveness criteria in the decision model highlighted the need to adequately ground management decisions within their appropriate contexts. Contextual grounding might also include human social and aesthetic values, legal concerns, human safety values, life-cycle costs and benefits, impacts of restoration treatments on terrestrial and aquatic habitats, resources, and species, the period of those effects, and the expected time period of effective restoration. For example, Hessburg et al. (2007a) present a decision-support application that evaluated danger of severe wildland fire and prioritized 575 subwatersheds in the Rocky Mountain region for vegetation and fuels treatments. They showed that many subwatersheds, while in relatively poor condition with respect to fire hazard, expected fire behavior, and ignition risk, were not the best candidates for treatment when considered in the context of the amount of associated wildland–urban interface (WUI). Considering fire danger in the context of the people and structures that might be most impacted by the fires restructured watershed-treatment priority in a useful manner.

4.3 Research Opportunities

This example demonstrates a relatively straightforward application of EMDS in which logic is first used to assess the current state of landscape features with respect to a number of ecosystem properties of interest, and then management priorities are derived, which take into account practical considerations that are important to decision makers. This approach is strategic in the sense that it identifies which landscape features are the priority, but the solution by itself does not necessarily suggest what specific management actions should be implemented in any given feature to produce improved conditions (e.g., a form of operational planning). Consequently, an interesting research and development opportunity exists to expand EMDS functionality to support this type of operational planning.

Another area of system functionality that is ripe for further research and development is more explicit support for the adaptive management process. The current example is typical of most EMDS applications to date in that it assesses current condition. However, federal oversight agencies in the US, and presumably in other countries as well, are demanding that national land-management agencies do a much better job of monitoring project performance and cost effectiveness with respect to issues such as landscape restoration. To that end, one can envision using precisely the same logic specification to reassess a particular spatial extent at one or more points in time in the future, producing a new distribution of modeled outcomes at each point in time. Clearly, given two or more distributions of model outcomes, standard statistical tests could be used to test hypotheses for significant changes in distributions over time. In important respects, such capabilities have always been the holy grail of adaptive management. Because EMDS already supports multiple assessments, and assessments may be temporally defined, integrating hypothesis testing into the EMDS framework to better support adaptive management would be highly desirable.

5 Tactical Planning for Landscape Restoration

In a third application, Gärtner et al. (2008) demonstrated an approach to evaluating current landscape vegetation patterns with reference to two climate scenarios: one was retrospective, representing the pre-management era climate, a second was prospective, representing change to a warmer and drier climate. We used decision-support modeling in EMDS to set treatment priorities among the landscapes and select alternative treatments. The analysis did not seek to accurately predict climate change, but to interpret landscape consequences given a plausible scenario. We used a NetWeaver logic model to assess landscape departure from the two sets of reference conditions and a decision model developed in Criterium DecisionPlus (CDP) to illustrate how various landscape conditions could be prioritized for management treatments in light of two climate scenarios, taking into account not only considerations of landscape departure, but also logistical considerations pertinent to forest managers. Our methods represented a hedging approach managers might use to determine how best to proceed with restorative management in an uncertain climatic future.

The study area encompassed the 6,070 ha Gotchen Late-Successional Reserve (LSR, Hummel et al. 2001 and Hummel and Calkin 2005), and adjacent lands totaling 7,992 ha. The reserve is located east of the crest of the Cascade Mountain Range in Washington State, USA, on the Gifford Pinchot National Forest (Fig. 3). The study area is part of a regional network of LSRs established as one component of the Northwest Forest Plan, which required protection of the northern spotted owl (*Strix occidentalis caurina*) and associated species with an adequate distribution and arrangement of late-successional habitats (ROD 1994).

In this application, we evaluated landscape departure of two landscapes, comprising the bulk of the study area, from RV associated with one historical and one future climate reference condition. As in the prior two applications, the reference conditions represented broad envelopes of vegetation conditions common to an ecoregion. The landscapes were evaluated relative to these reference conditions in EMDS. We evaluated outputs from the decision model to determine which landscape should be treated first, and which landscape treatments might be most effective at favorably altering conditions in light of the two climate references.

The study area fell in ESR 4 as described in the preceding application (Fig. 3, Hessburg et al. 2000b). To consider the natural landscape patterns that might occur under a climate-change scenario, we adopted a change scenario involving a climatic shift to drier and warmer conditions because limiting factors for forest

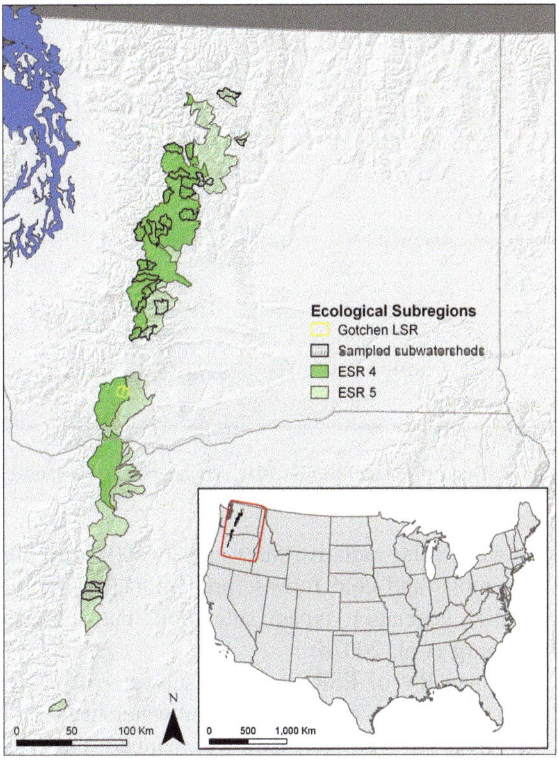

Fig. 3 Location of the Gotchen Late-Successional Reserve (study area) and Ecological subregions (ESR) *4* the subregion of the study area. *ESR 5* is shown as the subregion immediately to the east of *ESR 4* along the west-east temperature and precipitation gradient (Hessburg et al. 2000a)

growth, tree mortality, and high wildfire risk are often associated with protracted warming.

Empirical data from the next drier and warmer ecoregion (ESR 5) were used as a reference set to simulate the climate-change analogue for the study area. We reasoned that use of ESR 5 for these climate-change reference conditions was rational for several reasons: (1) ESR 5 sat adjacent to ESR 4 on the west to east climatic gradient of temperature and precipitation (Fig. 3); (2) ESR 5 received more solar radiation during the growing season and was drier than ESR 4; (3) ESR 5 was composed of the same forest species and structural conditions as were found in ESR 4 and was ordinarily influenced by fire regimes that are more similar to those forecast for a warming and drying climate-change scenario (Gedalof et al. 2005; Littell et al. 2009; McKenzie et al. 2004); and (4) ESR 5 landscapes had existed for a long time under these warmer and drier climatic conditions such that conditions reflected the natural spatio-temporal variation in landscape patterns that would exist under the influences of succession, disturbance, and the local climate.

Climatic conditions in ESR 5 represented a significant difference in total annual precipitation and average growing season daytime solar radiative flux (Hessburg et al. 2000b). ESR 5 was characterized as a warm (5–9 °C annual average temperature), moderate solar (250–300 $W{\cdot}m^{-2}$ annual average daylight incident

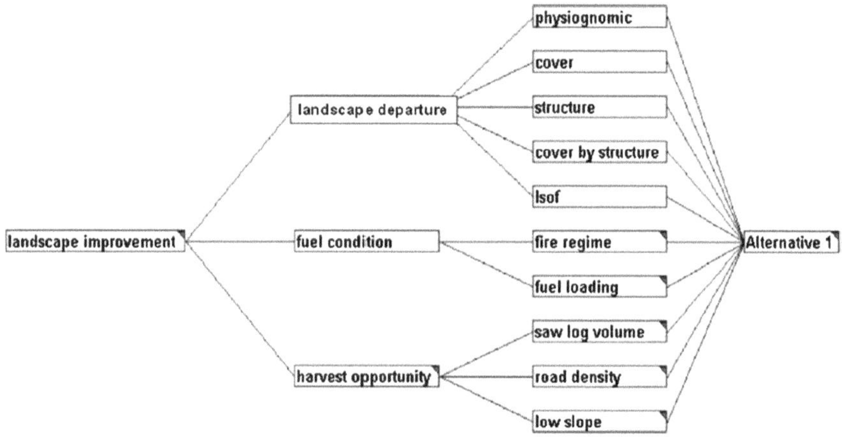

Fig. 4 Decision model to prioritize subwatersheds for landscape restoration

shortwave solar radiative flux), moist (400–1100 mm/year total annual precipitation), moist and cold forests (predominantly occupied by moist and cold forest potential vegetation types) subregion, but subwatersheds included dry forests (Hessburg et al. 2007b).

To map RV of ESRs 4 and 5, subwatersheds were randomly selected to represent at least 10 % of the total subwatersheds and area of each subregion. For each selected subwatershed, we mapped pre-management era vegetation by interpreting representative stereo aerial photographs. The resulting vegetation features enabled us to derive forest cover types (Eyre 1980), and structural classes (O'Hara et al. 1996), using methods detailed in Hessburg et al. (1999b, 1999c). Five different vegetation features were used to characterize the attributes of the historical subwatersheds of ESRs 4 and 5. The five features were the physiognomic condition, the cover-type condition, the structural class condition, the combined cover type by structural class condition, and the late-successional and old-forest condition. Five class and nine landscape metrics generated by FRAGSTATS (McGarigal and Marks 1995) were chosen to display spatial relations within classes and landscapes of these features. The metrics were the same as those outlined in preceding applications.

In a first phase, we evaluated landscape departure of the two subwatersheds in terms of departure of current conditions from the two climatically defined reference conditions. In a second phase, we determined which of the two subwatersheds exhibited a higher priority for restoration. The decision model for assigning restoration priorities included three primary criteria: landscape departure, fuel condition, and harvest opportunity (Fig. 4). All subcriteria of landscape departure were measures of evidence from the landscape analysis performed with the NetWeaver logic engine.

Subcriteria of fuel condition and harvest opportunity represented attributes of subwatersheds that were not part of the logic-based evaluation, but were included in the decision model as logistical considerations for management (Fig. 4). Fuel condition was evaluated in terms of probable fire regime and fuel loading. Harvest opportunity was evaluated in terms of available merchantable volume, road density, and proportion of subwatershed area with slope ≤ 10 %. The slope specification was intended not so much as a feasibility but cost criterion, indicative of areas with easy access for ground-based harvesting and yarding equipment. Road density and slope were calculated from a digital elevation model and map layers provided by national forest staff. Fire regime was calculated as the proportion of the subwatershed that had a fire regime condition class >1. Fire regime condition class depicted the degree of departure from historical fire regimes (Schmidt et al. 2002).

Stand-level tree-inventory data were collected following Hummel and Calkin (2005). From the stand-level data, we estimated fuel load and sawlog volume in each subwatershed using available plot data sets. The proportion of subwatershed area with a high fuel loading was calculated as the proportion of plots with a fuel load class >1, following methods of Ottmar et al. (1998). Sawlog volume (mean $m^3 \cdot ha^{-1}$) in stands was calculated with NED-2 (Twery et al. 2005), based on tree lists from the plot data.

We found little significant change in physiognomic or cover type conditions among the two test subwatersheds; but surprisingly, the evidence for no change was greater in the western subwatershed under the climate-change scenario, indicating that current spatial patterns of cover types, while not departed from ESR 4 historical conditions, would actually be closer to conditions that would be anticipated under the warming/drying climate-change scenario (Fig. 5). Similarly, we found significant evidence for structural class departures in both subwatersheds when historical reference conditions were considered, but departures were less evident in one of the two subwatersheds when the RV for the climate-change scenario was considered. Results for cover type by structure evaluation were analogous (Table 3). Evidence for limited late-successional/old-forest departure was strong in both subwatersheds using the historical RV scenario, but declined in both subwatersheds under the climate-change scenario, indicating that warmer and drier conditions would likely favor expanded area of these structures.

To determine which of the two subwatersheds had the highest priority for landscape restoration, we applied the decision model and its primary criteria to the selection process (Fig. 4). The eastern-most of the two evaluated subwatersheds received a higher priority rating for landscape improvement in the context of both the historical climate and climate-change scenarios. The overall decision score under the historical reference scenario was highest for the eastern subwatershed, but scores were nearly identical for the climate-change scenario. On balance, the two subwatersheds were found to be in relatively good condition, regardless of the climatic reference (Table 3).

Contributions of harvest opportunity and fuel condition to restoration priority were essentially the same for both subwatersheds in either scenario. The only features that changed the overall decision score were related to landscape departure.

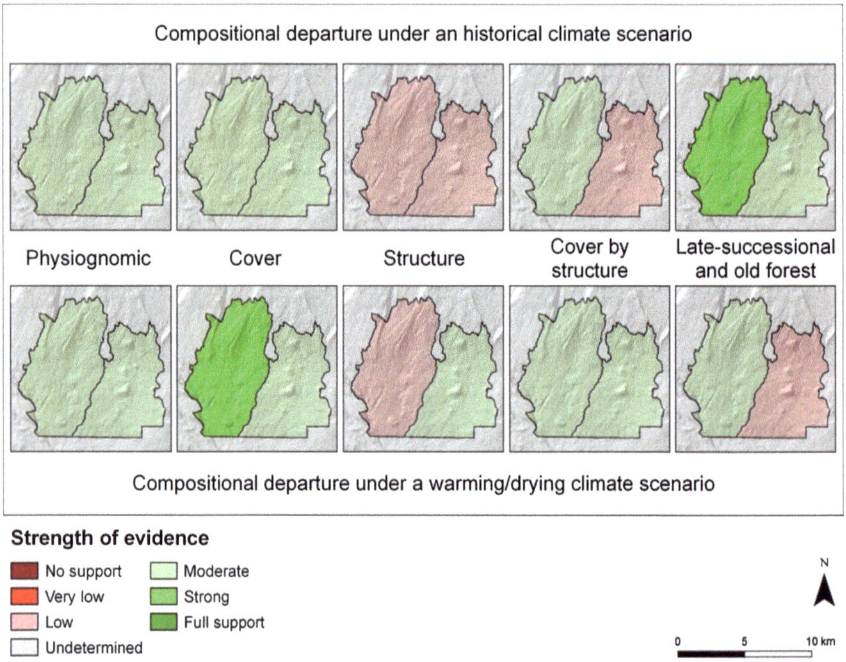

Fig. 5 Illustration of the landscape departure evaluation of the current Gotchen landscape relative to reference conditions representing pre-management era (*above*) and future warming climates (*below*). Each of the small figures shows the two subwatersheds of the Gotchen landscape; the coloration displays the degree of departure under the historical (*upper*) and warming (*lower*) climate conditions

Table 3 Contributions of subcriteria to decision scores of the eastern and western Gotchen watersheds when compared with the historical and future climate reference conditions

Watershed	Historical reference		Climate change reference	
	East	West	East	West
Physiognomic condition	0.037	0.024	0.023	0.012
Structural condition	0.098	0.094	0.073	0.081
Cover type-structural condition	0.039	0.034	0.013	0.01
Late-successional/old-forest condition	0.182	0.087	0.222	0.195
Fire regime condition	0.119	0.119	0.119	0.119
Fuel loading condition	0.089	0.094	0.089	0.094
Harvest opportunity	0.012	0.037	0.012	0.037
Overall decision score	0.576	0.489	0.551	0.548

Scores for landscape pattern departure differed slightly between the historical reference and climate-change scenarios, and in both cases the contributions of late-successional/old forest had the most impact on treatment priority.

5.1 What Worked Well?

The application met its objectives of evaluating the degree of departure in the watersheds relative to retrospective and prospective sets of reference conditions. Addition of the two tactically-oriented criteria to the decision model (vulnerability to severe wildfire and timber harvest opportunity) were helpful to assigning the relative priority of landscape restoration treatments between the two subwatersheds.

We found it noteworthy that the two sets of reference conditions were more similar than different in most aspects. That is, ranges of conditions were mostly overlapping rather than unique. This lends empirical credibility to the notion that envelopes of pattern conditions were historically nudged and reshaped rather than re-invented wholesale by shifting climatic regimes (Keane et al. 2009; Moritz et al. 2011). With the enormous legacy of spatial pattern alteration caused by past fire exclusion and suppression, timber harvest, road development, and livestock grazing elsewhere in the Inland Northwest, this may not be the case in a future climate unless spatial patterns are restored. Applications like that of Gärtner et al. (2008) may become highly useful to designing, evaluation, and comparing alternative recipes in a world of uncertain climatic outcomes.

5.2 What Could be Improved?

A general enhancement of the model would be to include specific threats to resource values—those currently existing as well as those imposed by restoration activities. Across a broad regional landscape, where numerous landscapes may be considered, and especially in the context of the western US, threats to resource values associated with wildfire should to be considered in any decision model of this type. Where the legacy of past management to native species, food webs, and habitats is a concern, models such as this one should evaluate existing threats to species, populations, and habitats, and compare these with any threats derived from restoration treatment intensity and distribution. Such an evaluation would aid manager calibration of treatment scenarios that optimized improvements over deleterious effects.

5.3 Research Opportunities

A novel aspect of this study was that the analysis of vegetation condition, as a prelude to making decisions about investments in restoration, was both retrospective (comparing existing conditions to an envelope of historical reference conditions) and prospective (comparing existing conditions to plausible reference conditions of a future climatic scenario). In light of the current reality of global climate change and its downscaled regional influences (McNulty and Aber 2001; Spittlehouse and Stewart 2003), it is reasonable and perhaps essential to not only consider where a system has come from, but where it may be headed, and the tradeoffs associated with the changes. Logic- and scenario-based modeling, as illustrated in this study, may help surface ramifications of contemporary management that might otherwise be overlooked. The conundrum for forest managers is that the actual conditions and variability of a future climate scenario cannot be predicted with reasonable certainty. However, extending the example offered here, by including multiple plausible climate change scenarios, may help identify management strategies that demonstrate trade-offs associated with each scenario, minimize future risk, and conserve the greatest number of management, species, and process options for the future.

6 Decision Support for Project Planning

In this section, we present a fourth and final EMDS application that provides decision support for restoring a mixed coniferous forest landscape on the Naches Ranger District of the Okanogan-Wenatchee National Forest in eastern Washington, USA. The project (hereafter, "Nile Creek") was the first landscape restoration project developed under a newly minted, peer-reviewed, forest-wide restoration strategy (hereafter, the Strategy, USDA-FS 2010).

Under the Strategy, the objectives of landscape evaluations are to: (1) transparently display how projects move landscapes towards drought, wildfire, and climate resilient conditions; (2) describe and spatially allocate desired ecological outcomes (e.g., adequate habitat networks for focal wildlife species; disturbance regimes consistent with major vegetation types); (3) logically identify project areas, treatment areas, and the associated rationale; and (4) spatially allocate desired ecological outcomes and estimate outputs from implemented projects. Landscape evaluations under the strategy assemble and examine information in five topic areas: (i) patterns of vegetation structure and composition; (ii) potential for spread of large wildfires, insect outbreaks, and disease pandemics across stands and landscapes given local weather, existing fuel and host conditions; (iii) damaging interactions between road, trail, and stream networks; (iv) wildlife habitat networking and sustainability; and (v) minimum roads analysis, (i.e., which of the existing roads are essential and affordable for administrative and recreation access). Over time and as needed, additional topics are being added to this working prototype.

For simplicity, the strategy for landscape evaluation was implemented in approximately eight steps:

Step 1—determine the landscape evaluation area,
Step 2—evaluate landscape patterns and departures,
Step 3—determine landscape and patch scale fire danger,

Step 4—identify key wildlife habitat trends and restoration opportunities,
Step 5—identify aquatic/road interactions,
Step 6—evaluate the existing road network,
Step 7—identify proposed landscape treatment areas (PLTAs), and
Step 8—refine PLTAs and integrate findings from steps 2–6 into landscape restoration prescriptions.

District specialists from multiple disciplinary fields worked in partnership to complete each of the steps. Steps 1–6 occurred concurrently and were completed prior to Steps 7 and 8. These steps were applied in the Nile Creek analysis area; we present the landscape-evaluation model for that area.

6.1 Determining the Landscape Evaluation Area

Determining the size of the evaluated area had implications for ecological and planning efficiency. Evaluating two or more subwatersheds (12-digit, 6th-field hydrologic unit code [HUC], 4,000–12,000 ha each) was recommended by Reynolds and Hessburg (2005) and Hessburg et al. (2005) based on the findings of Lehmkuhl and Raphael (1993), who showed that some attributes of spatial pattern are influenced by the size of the area being analyzed when analysis areas are too small. We used subwatersheds larger than 4000 ha to avoid this bias. Watershed size also coincided with previous watershed assessments, generally providing a range of elevations and forest types, and was useful in evaluating hydrological influences of anticipated forest restoration treatments.

Watershed size was large enough to evaluate many cumulative effects, but wide-ranging wildfires, native carnivore species and most salmonids required analysis of larger areas than subwatersheds (e.g., Ager et al. 2007; Gaines et al. 2003; Reeves et al. 1995).

Several future project areas could be acceptably planned via a single large-scale analysis, thereby reducing paperwork, decreasing planning time and cost, and increasing environmental analysis efficiencies leading to project implementation. The actual project area included three subwatersheds (the Dry-Orr Project) covering an area of ~29,000 ha. For brevity, this paper discusses landscape analysis in just one of these subwatersheds, Nile Creek, which encompasses an area of 8295 ha (Fig. 6, see also Hessburg et al. 2013).

The EMDS application for the Nile Creek project evaluated five primary topics in a NetWeaver logic model. The vegetation pattern departure, major insect and pathogen vulnerabilities, patch level fire attributes, and habitat availability for focal wildlife species topics evaluated how the current landscape compared to the pre-management era and future warming climate reference conditions. The fire movement potential topic was evaluated at a subbasin scale (see Fig. 2b). The aquatic-road interactions and minimum roads analysis required Forest-wide modeling efforts, which were not yet completed in time for this project area, and truncated versions were incorporated in this evaluation.

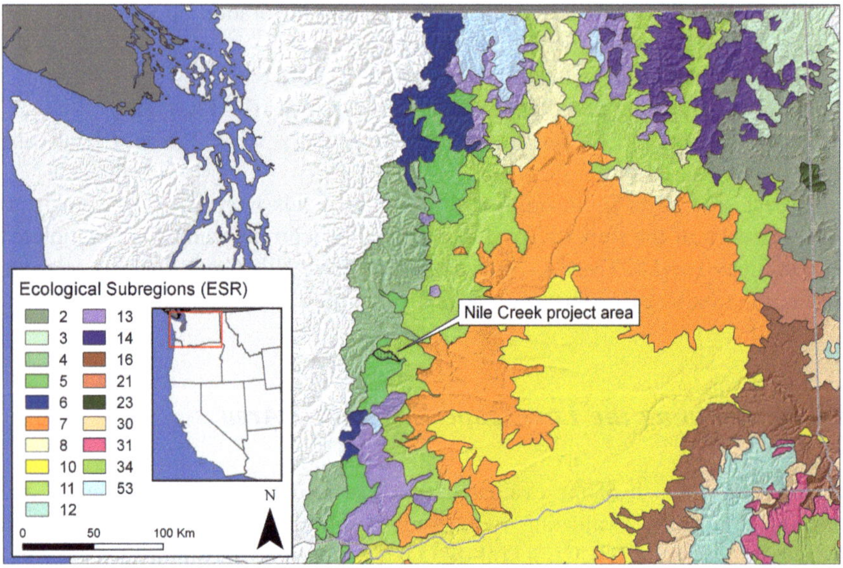

Fig. 6 Ecological subregions in eastern Washington, USA (from Hessburg et al. 2000b). The Nile Creek project area is outlined with stippling in ESR5

6.2 Evaluating Landscape Vegetation Patterns and Departures

As in preceding applications, we evaluated departure of the current vegetation conditions for the Nile Creek subwatershed from RV associated with one historical and one future climate reference condition. The project area fell in ESR 5 as described above (Figs. 3 and 6, Hessburg et al. 2000b), and we used the RV estimates of this ecoregion to represent natural variation in spatial patterns for the pre-management era. To consider the natural landscape patterns that might occur under a climate-change scenario, we adopted a scenario involving a climatic shift to drier and warmer conditions using reasoning described in the prior application (see Sect. 5, Gärtner et al. 2008). Empirical data from the next drier and warmer ecoregion (ESR 11) were used as a reference set to represent RV associated with the climate-change scenario for the project area (Hessburg et al. 2000b).

Two of eight available features–combined cover type-structure class (CTxSC) and combined potential vegetation type-cover type-structural class (PVGxCTxSC)–were subcriteria evaluated under vegetation pattern departure. Five class and nine landscape metrics generated by FRAGSTATS (McGarigal and Marks 1995) were chosen to display spatial relations and RV within classes of the two features, and within entire landscapes of these features. The metrics were the same as those outlined in preceding applications. Departures from the RV estimates of the two climate references across the full suite of metrics and vegetation features formed the basis of vegetation departure analysis (Fig. 7).

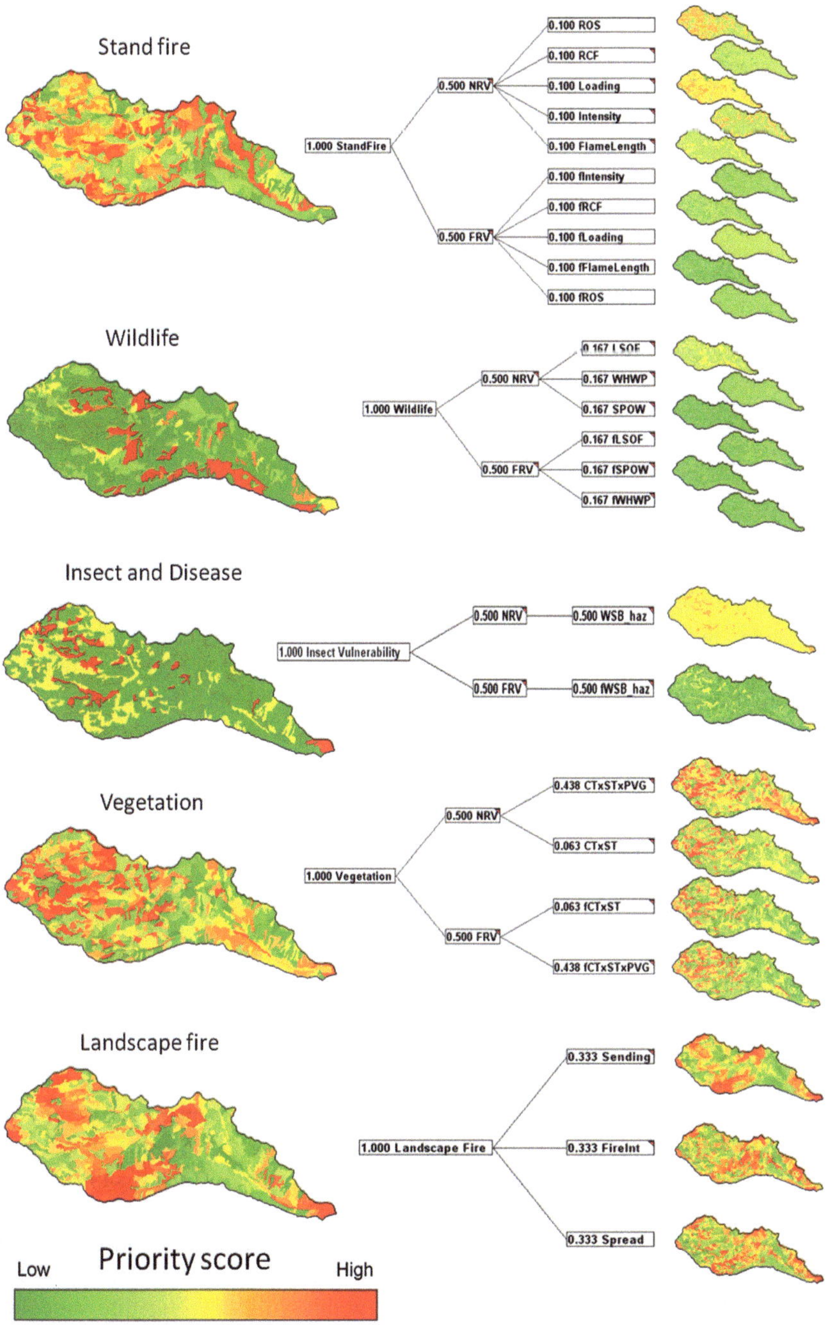
Stand fire
1.000 StandFire
0.500 NRV
0.100 ROS
0.100 RCF
0.100 Loading
0.100 Intensity
0.100 FlameLength
0.500 FRV
0.100 fIntensity
0.100 fRCF
0.100 fLoading
0.100 fFlameLength
0.100 fROS
Wildlife
1.000 Wildlife
0.500 NRV
0.167 LSOF
0.167 WHWP
0.167 SPOW
0.500 FRV
0.167 fLSOF
0.167 fSPOW
0.167 fWHWP
Insect and Disease
1.000 Insect Vulnerability
0.500 NRV
0.500 WSB_haz
0.500 FRV
0.500 fWSB_haz
Vegetation
1.000 Vegetation
0.500 NRV
0.438 CTxSTxPVG
0.063 CTxST
0.500 FRV
0.063 fCTxST
0.438 fCTxSTxPVG
Landscape fire
1.000 Landscape Fire
0.333 Sending
0.333 FireInt
0.333 Spread
Low
Priority score
High

◀ **Fig. 7** Five CDP models representing the contributions of network evaluations to treatment priority scores (range 0 [*darkest green*], to 1 [*darkest red*]) in the Nile subwatershed. Acronyms in the figure are: *Stand fire* the weighted results of subtopic departure analyses (weights are shown with each topic and subtopic); *NRV* the weighted results of all subtopics that evaluate departure from the natural range of variation; *ROS* wildfire rate of spread; *RCF* risk of crownfire, *Loading* surface fuel loading; *intensity* fireline intensity; *Flame length* flame length; *FRV* the weighted results of all subtopics that evaluate departure from the future range of variation; to avoid confusion, an "*f*" is placed immediately before a subtopic acronym to indicate that it is associated with the FRV portion of a departure analysis; *Wildlife* the weighted results of subtopic departure analyses for key wildlife habitat pattern and abundance; *LSOF* late successional and old forest; *WHWP* white-headed woodpecker; *SPOW* northern spotted owl; *WSB haz* western spruce budworm hazard departure; *CTxSC* departure of combined cover type and structural class conditions; *CTxSCxPVG* departure of combined cover type, structural class, and potential vegetation group conditions; *Sending*, *FireInt*, and *Spread* denote the varying degrees of fire sending (node influence), fireline intensity, and wildfire rate of spread occurring during the FlamMap simulations

We evaluated the vulnerability of each landscape and its component patches to a native insect relative to the historical and future climate reference conditions, using methods of Hessburg et al. (1999d, 2000a). Each patch was assigned to a vulnerability class based on vegetation factors that increased patch and vulnerability and landscape contagion with respect to the insect. In Nile Creek, we evaluated landscape vulnerability to the western spruce budworm (*Choristoneura occidentalis*). Damage associated with this insect had increased over the 20th century; District foresters wanted to understand the extent of the vulnerability increase. The product of this step was a map of patch vulnerability to the western spruce budworm for the current landscapes, which were compared against the two reference conditions for the same landscape vulnerability (Fig. 7).

6.3 Determining Patch and Landscape Scale Fire Danger

Patch-level expected wildfire behavior was modeled for all current and reference condition patches using methods detailed in Hessburg et al. (2000a) and Huff et al. (1995). Current conditions of patches were evaluated against reference ranges of conditions to determine departure under either climate scenario.

We modeled expected landscape fire behavior during a typical wildfire (97th-percentile burn conditions) at the scale of the entire subbasin (8-digit) 4th-field HUC. In the case of the Nile Creek project area, the larger Naches subbasin that surrounds Nile Creek was modeled; it encompasses an area of approximately 180,000 ha. Available forest-wide fuels layers were resampled to 90 m-resolution rasters and 97th-percentile fuel moistures and weather conditions were used to condition fuels for fire behavior modeling within the *FlamMap* fire modeling framework (Finney et al. 2007, and references therein).

Custom wind grids, created using *WindNinja* modeling software (Forthofer et al. 2009), were derived for the five most likely prevailing wind directions and used as input to the *FlamMap* model. For each of the wind directions, the subbasin landscape was ignited with 1000 randomly distributed fire starts one hundred times each, and fires were allowed to burn for six hours each until all of the landscape was exposed to multiple fires ($\sim$100,000 ignitions). Each model run created several raster outputs that were stored for further analysis, including: fireline intensity, active and passive crown fire activity, rate of spread, flame length, and node influence. The node influence is a value assigned to a given pixel in *FlamMap* that represents the number of pixels that burn during the simulation as a result of that pixel burning. Node influence is highly variable, depending on ignition location, fuel arrangement, simulation edge effect, and simulation duration. To create a meaningful node influence grid, all node influence outputs were composited from all ignitions, and from each wind direction. We created an additional composite layer, using all fires from each of the five wind directions that represented how similarly fires spread considering slope and fuel interactions. We termed this layer the congruence (of fire spread direction) layer. The flame length layer was also composited across the five different wind directions.

Finally, the composited node influence was combined with flame length and the congruence layer to create an index that showed the relative contribution of each pixel to the spread and intensification of fire. Areas with large clusters of high fire danger pixels (i.e., $\geq$80th-percentile scores for combined flame length, node influence, and congruence) were identified as priority treatment areas to interrupt the flow of wildfire across large landscapes.

6.4 Identifying Wildlife Habitats and Restoration Opportunities for Focal Species

In this evaluation, we: (1) determined the location and amount of habitat for focal wildlife species present within the landscape-evaluation area, (2) compared the current amount and configuration of habitats for focal species to historical and future climate reference conditions, and (3) identified habitat restoration opportunities and priorities that could be integrated with other resource priorities and carried forward into project planning.

Focal wildlife species were selected because they are either federally listed or identified as focal species by the USDA Forest Service, Pacific Northwest Region (USFS 2006) and their life-history requirements were appropriately assessed at the scale of our evaluation. The selected focal species are closely associated with forested habitats, and their populations are influenced by changes to forest structure, among other factors. Focal species included the northern spotted owl, northern goshawk (*Accipiter gentilis*), white-headed woodpecker (*Picoides albolarvatus*), American marten (*Martes americana*), pileated woodpecker (*Dryocopus pileatus*), Lewis's woodpecker (*Melanerpes lewis*), and black-backed woodpecker (*Picoides arcticus*). The habitat definitions that were used in the landscape evaluation for these species are described in USDA-FS (2010) and in Gaines et al. (2010). The products of this evaluation step were maps showing the location and amount of habitat for each of the focal species and maps and tabular data showing the degree of departure in habitat amounts and configuration between current and the two reference conditions. The applied class and landscape metrics used to estimate departures in the amount of habitats were those described above in preceding sections.

6.5 Evaluating Aquatic Ecosystem and Road Interactions

In this step, we identified the road segments that had the greatest impacts on streams, channel features and migration, and in-stream habitats to determine restoration priorities. The components of the aquatic/road interactions evaluation were hydrologic connectivity of roads and streams, fish distribution, slope stability and soil properties, and stream channel confinement. These components were evaluated using *NetMap* (Benda et al. 2007) and results were incorporated into project planning and alternative comparison, but outside of EMDS, due to timing issues. The hydrologic connectivity evaluation ranked the relative importance of flow routes connecting the road system to streams by combining a georeferenced roads layer with a flow-accumulation file generated from a 10-m digital elevation model (DEM). The evaluation of fish distribution linked current in-stream and other survey data with a current streams layer. This would enable later integration in EMDS of potential treatment areas with current fish distributions for listed and sensitive fish species. Slope and soil stability was modeled by combining an existing soils layer (SSURGO, USDA-NRCS 2005, 2006) with the DEM, and assigning slope breaks of 0–34.9, 35–60, and >60 %. Unstable soils and steeper slopes were used to identify slope and soil related hazards. Stream-channel confinement was evaluated using a layer developed by the local Forest that identified stream channels with <3 % gradient within 30-m feet of a road.

6.6 Integrating Landscape Evaluation Results in EMDS

Each of the described primary topics was evaluated using a relatively simple logic model (five networks total) that related class metrics of each primary topic (Fig. 7). The results were then combined in the single CDP decision model as a network of networks, as illustrated in Fig. 8. Results of landscape evaluation enabled the District planning team to attach a treatment priority to all patches in a subwatershed and to identify areas with clusters of high-priority patches, termed potential landscape treatment areas (PLTAs) that could form the nucleus of several project areas. In Fig. 9, we illustrate mapped PLTAs in the Nile subwatershed. The circled areas represent likely PLTAs emerging from the landscape evaluation.

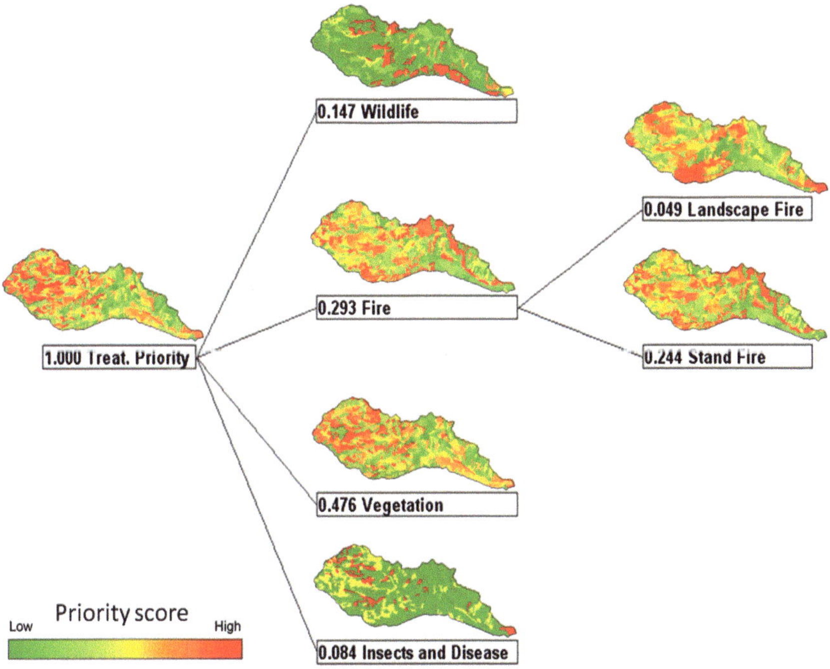

Fig. 8 Second stage CDP decision model for the Nile Creek project area. Landscape treatment priority scores of polygons within the Nile subwatershed (range 0 [*darkest green*], to 1 [*darkest red*]) were derived from primary criteria associated with four major topics (see also Fig. 7): *Wildlife, Fire, Vegetation,* and *Insects and Diseases.* Primary criteria were weighted by managers using the SMART utility in EMDS. Under the *Fire* criterion, the *Landscape Fire* and *Stand Fire* networks (Fig. 7) were evaluated as subcriteria, and weighted by District managers. The map on the far left shows the results of the entire CDP evaluation of priority treatment scores assigned to patches. These scores are later used for identifying proposed landscape treatment areas (PLTAs, Fig. 9) and potential restoration treatment locations

The results of evaluations of each primary topic provided information that could be used by all members of the interdisciplinary planning team to develop a prescription for each landscape (i.e., a landscape-level prescription). For example, results generated from the landscape pattern, fire, and habitat evaluations allowed the interdisciplinary team to quantify the amount, types, and spatial locations of treatments to accomplish multiple restoration objectives. These objectives included strategically altering large-scale fire behavior, increasing the amount and improving the networking of key wildlife habitats, restoring ecosystem functions by restoring landscape pattern and process interactions, reducing risk to human communities, and minimizing the road network needed to access treatment areas, provide access for fire protection, and provide for other administrative uses.

Upon completion of the initial landscape evaluation, identification of the PLTAs, and proposal of preferred landscape treatment options, the vegetation data was edited to reflect the effects of treatment. These edited landscapes were then

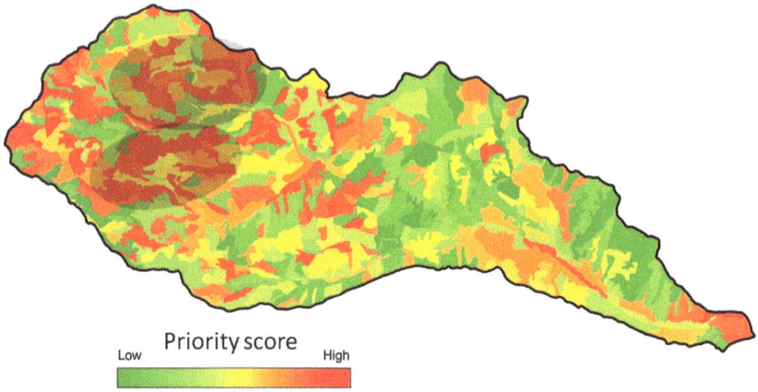

Fig. 9 Nile subwatershed patch-level priority scores resulting from CDP evaluation of subcriteria and criteria in EMDS. Landscape treatment priority scores of polygons within the Nile subwatershed (range 0 [*darkest green*], to 1 [*darkest red*]) were derived from primary criteria associated with four major topics (see also Figs. 7, 8): *Wildlife, Fire, Vegetation,* and *Insects and Diseases.* Circles show example potential landscape treatment areas (PLTAs) where restoration projects (*shaded areas*) might focus treatments appropriate to the need, to achieve multi-way and multi-level restoration goals

re-evaluated by the EMDS application, and managers were able to determine the degree to which progress was made toward restoration goals with regard to both climate scenarios. Using EMDS, the interdisciplinary team was then able to evaluate a variety of landscape prescriptions and treatment options, and to assess how the various options would affect fish habitats, insect and disease risks, landscape patterns, and the flammability of the larger landscape. The final product was a refined map of PLTAs and preferred options for landscape treatment for the Nile Creek project area. This process of landscape evaluation provided important advantages to the environmental analysis that followed in terms of transparency, efficiency, and credibility.

6.7 What Worked Well?

First and foremost, the development of the EMDS application improved communication within the interdisciplinary team, as it gave the members a concrete framework for organizing the analytical and decision space necessary for exploring restoration management opportunities. Resource managers were able to organize the logic and analysis needs for their area of expertise and share their submodels with the interdisciplinary team as primary topics that can feed into the overall application structure.

The use of EMDS in this application allowed for much better integration across resource disciplines and yielded transparent and repeatable landscape evaluation and decision-making processes. The alternative development portion of the process allowed the planning team to identify priority areas for restoration treatments that could achieve multiple objectives. The comparison of current conditions to historical range of variation (HRV) and the expected future climate range of variation (FRV) conditions in EMDS enabled the planning team to develop objective measures that could be used to describe resilient landscapes and measure progress towards achieving the restoration goals. Integration of a climate change scenario into EMDS allowed the incorporation of current climate-change science into the landscape evaluation process and informed project-level planning and decision-making.

The landscape evaluation allowed the interdisciplinary team and the decision-maker to strategically locate project areas to meet multiple restoration objectives. In addition, EMDS provided a mechanism to transparently display how emphasizing a certain resource more than another influenced prioritization and the spatial allocation of treatments.

To date, no other planning process has allowed managers on the Okanogan-Wenatchee NF to strategically and spatially locate treatments based on the complex and simultaneous interactions of multiple landscape conditions and resource variables. Managers were better able to describe restoration needs at a landscape scale rather than stand by stand. As a result, new opportunities for restoration treatments were discovered. For example, the District interdisciplinary team chose a PLTA in mesic forests to address patch types and arrangements rather than solely focused on thinning in dry forests, which had occupied much of the Forest focus in preceding years.

In comparison with previous planning efforts, the interdisciplinary team was better able to truly integrate concerns for multiple resources. Prior projects were largely driven by the need to manipulate vegetation for forest health improvement and wildfire mitigation. The landscape evaluation process more fully integrated planning, simultaneously emphasizing wildlife and aquatic habitat conditions, landscape and patch-scale fire behavior, vegetation and fuels patterns, and the pros and cons of continued road access, leading to restoration opportunities for a multiplicity of resources. The Nile Creek Project became a good example of simultaneous problem-solving rather than an exercise in trade-off analysis.

6.8 What Could be Improved?

A simple CDP decision model was developed in this EMDS application for want of time and additional resources. Alongside information reflecting knowledge about the state of the system, other criteria might have been included, such as those reflecting social and economic values, and other feasibility and efficacy criteria. Examples might include consideration of fire risks to human developments, effects on meeting other resource objectives where restoration is not the primary goal, matters of technical and economic feasibility and social acceptability, relationships to life-cycle costs and benefits, retreatment frequency and the duration of positive treatment effects, uncertainties associated with management outcomes and data quality, and trade-offs associated with more or less strategic placement of treatments.

6.9 Research Opportunities

Two opportunities for increasing the research and heuristic value of this project-level planning tool would include adding stochastic succession and disturbance dynamics to modeled landscape treatment prescriptions and to evaluate alternative landscape prescriptions against FRV conditions representing several plausible future climate scenarios. In the first instance, stochastic behavior could be added to modeled landscape-treatment scenarios by simulating them spatially in models such as the Landscape Succession and Disturbance Model–LANDSUM (see Keane et al. 2002; Barrett 2001) or many others. LANDSUM provides state-transition models for the potential vegetation types of a study area. Within each state-transition model (STM) are successional states defined by cover types and structural classes, a complete set of transition pathways that show all potential succession paths between states, and transition times related to each potential path. Initialized disturbance probabilities by disturbance severity determine the likelihood that any state will transition to any other state. In this context, landscape treatments would occur as prescribed, but other unplanned disturbances caused by wildfires, forest insects, and forest pathogens could occur as well. The net result would be annualized depictions of planned and unplanned vegetation outcomes, which would be a more accurate depiction of likely outcomes of implemented scenarios.

In a related manner, a range of climatic futures could also be simulated using a "climatized" version of LANDSUM (e.g., Cary et al. 2006) or other STM. Simulations would occur as described above, but in this case the conditioning climate would influence fire probabilities by means of a scalar applied to historical fire probabilities assigned from the climate change and area burned literature. The advantage of this sort of approach would be in developing hedging strategies for landscape management in an uncertain climatic future.

7 Final Thoughts

First, some readers will no doubt be curious about the level of effort needed to fully implement decision-support applications for landscape analysis such as those presented in our four detailed examples. Put simply, the effort can be daunting if the process must begin with collection of new field or satellite data. As a very rough guide, we suggest that each day of modeling and analysis is supported by

10 days of geoprocessing, and each day of geoprocessing is supported by 10 days of collecting and processing field data. In other words, designing, implementing, and running a landscape DSS typically represents a very modest fraction of the overall effort. Developing good quality, map-based information about the landscape(s) of interest, for each of the dimensions that may be co-considered is what takes the time and effort. If the needed data are already at hand, additional investment in DSS development can return disproportionately large value added relative to the investment.

There are at least a few strategic lessons to be gleaned from the four examples that have been presented. Addressing questions about ecosystem integrity or landscape departure with respect to vegetation required very high-dimensional logic representations in order to adequately address the facets of structure and composition. Indeed, all the logic models discussed evaluated 100s of input variables and 1000s of parameters. Contributing to the very large size of these models, five class metrics were used to evaluate each patch type, and nine landscape metrics were used to evaluate the spatial properties of patch types. Notice also that the same set of 14 metrics was used across all four examples for simplicity. As a practical matter, we consider that the utility of the metrics chosen is entirely dependent upon the questions being addressed, and there are over 100 to choose from. The last three examples demonstrated how decision models can usefully augment the logic-based analysis, thereby introducing practical management issues into the priority setting process, while simplifying the analysis by decomposing it into two relatively simpler problems—understanding the status of the systems in question, and then asking what might be done given the condition of the systems.

Finally, we conclude with a few thoughts on the sense in which our four landscape applications can be considered successful. Our first three examples were primarily developed as proofs of concept in research and development. From an internal perspective, we consider these applications successful at providing an interpretation and synthesis of large volumes of information that we think usefully encapsulated scientific understanding of large, complex, and abstract problems. Of course the "acid test" for decision-support applications is that managers find them useful, understand them, and actually put them into service addressing real-world management problems more efficiently and effectively than before. Our final example of project-level planning was highly successful in these terms.

This landscape evaluation tool is now being implemented on all seven Districts of the Okanogan and Wenatchee National Forests, on an area of more than 1.6 million ha, prior to implementing any landscape restoration project under its Strategy. Moreover, between the draft and final stages of this chapter, the US Fish and Wildlife Service in their Revised Recovery Plan and Critical Habitat Rule (CHR) for the northern spotted owl recommended that methods such as ours can serve as an example of how to assess and restore ecological patterns and processes to eastern Washington and Oregon forest landscapes (USFWS 2011, 2012).

Acknowledgments We gratefully acknowledge the Naches Ranger District interdisciplinary team (IDT) for all of their fine work in data development, analysis, and landscape planning, and especially for allowing us to use them as guinea pigs when prototyping these new methods. Special thanks go to Jim Bailey—IDT Leader/Fire & Fuels/Air Quality analysis, Jacquie Beidl—Archaeologist/Heritage Resources analysis, Barry Collins—Engineering/Road Management analysis, Bill Garrigues—Hydrology/Soils analysis, Jodi Leingang—Invasive Species/Understory Vegetation/Range analysis, Dave Lucas and Miles Porter—Timber Sale Layout and Design, Phil Monsanto and Chris Ownby—Silviculture/Landscape Evaluation/GIS Analysis, Matt Dahlgreen—Silviculture, field validation, and landscape prescription development, Sue Ranger—Recreation/Visual Resources analysis, Joan St. Hilaire—Wildlife analysis, Gary Torretta—Fisheries analysis, and Irene Davidson—District Ranger. We are grateful for the reviews of three anonymous referees whose comments substantially improved the manuscript.

References

Ager AA, Finney MA, Kerns BK, Maffei H (2007) Modeling wildfire risk to northern spotted owl (*Strix occidentalis caurina*) habitat in central Oregon, USA. For Ecol Manage 246:45–56

Alatalo RV (1981) Problems in the measurement of evenness in ecology. Oikos 37:199–204

Allen CD, Macalady AK, Chenchouni H, Bachelet D, McDowell N, Vennetier M, Kitzberger T, Rigling A, Breshears DD, Hogg EH, Gonzalez P, Fensham R, Zhang Z, Castro J, Demidova N, Lim JH, Allard G, Running SW, Semerci A, Cobb N (2010) A global overview of drought and heat-induced tree mortality reveals emerging climate change risks for forests. For Ecol Manage 259:660–684

Barrett TM (2001) Models of vegetation change for landscape planning: a comparison of FETM, LANDSUM, SIMPPLLE, and VDDT. General Technical Report RMRS-GTR-76. Ogden, UT: U.S. Department of Agriculture, Forest Service, Rocky Mountain Research Station, p 14

Benda L, Miller D, Andras K, Bigelow P, Reeves G, Michael D (2007) NetMap: A new tool in support of watershed science and resource management. For Sci 53:206–219

Cary GJ, Keane RE, Gardner RH, Lavorel S, Flannigan MD, Davies ID, Li C, Lenihan JM, Rupp TS, Mouillot F (2006) Comparison of the sensitivity of landscape-fire-succession models to variation in terrain, fuel pattern, climate and weather. Landscape Ecol 21(1):121–137

Eyre FH (ed) (1980) Forest cover types of the United States and Canada. Society of American Foresters, Washington

Finney MA, Seli RC, McHugh CW, Ager AA, Bahro B, Agee JK (2007) Simulation of long-term landscape-level fuel treatment effects on large wildfires. Int J Wildland Fire 16:712–727

Forthofer J, Shannon K, Butler B (2009) Simulating diurnally driven slope winds with *WindNinja*. In: Proceedings of 8th symposium on fire and forest meteorological society. Kalispell, MT

Gaines WL; Singleton PH; Ross RC (2003) Assessing the cumulative effects of linear recreation routes on wildlife habitats on the Okanogan and Wenatchee National Forests. USDA Forest Service, General Technical Report PNW-GTR-586

Gaines WL, Harrod RJ, Dickinson JD, Lyons AL, Halupka K (2010) Integration of northern spotted owl habitat and fuels treatments in the eastern Cascades, Washington, USA. For Ecol Manage 260:2045–2052

Gärtner S, Reynolds KM, Hessburg PF, Hummel SS, Twery M (2008) Decision support for evaluating landscape departure and prioritizing forest management activities in a changing environment. For Ecol Manage 256:1666–1676

Gedalof Z, Peterson DL, Mantua NJ (2005) Atmospheric, climatic, and ecological controls on extreme wildfire years in the northwestern United States. Ecol Appl 15:154–174

Girvetz E, Shilling F (2003) Decision support for road system analysis and modification on the Tahoe National Forest. Environ Manage 32:218–233

Hessburg PF, Agee JK (2003) An environmental narrative of Inland Northwest US forests 1800–2000. For Ecol Manage 178:23–59
Hessburg PF, Mitchell RG, Filip GM (1994) Historical and current roles of insects and pathogens in eastern Oregon and Washington forested landscapes. General Technical Report PNW-GTR-327. U.S. Department of Agriculture, Forest Service, Pacific Northwest Research Station, Portland, OR
Hessburg PF, Smith BG, Salter RB (1999a) Detecting change in forest spatial patterns. Ecol Appl 9(4):1232–1252
Hessburg PF, Smith BG, Kreiter SD, Miller CA, Salter RB, McNicholl CH, Hann WJ (1999b) Historical and current forest and range landscapes in the interior Columbia River Basin and portions of the Klamath and Great Basins: Part I: Linking vegetation patterns and landscape vulnerability to potential insect and pathogen disturbances. General Technical Report PNW-GTR-458. Portland, OR: U.S. Department of Agriculture, Forest Service, Pacific Northwest Research Station. p 357
Hessburg PF, Smith BG, Salter RB (1999c) Using natural variation estimates to detect ecologically important change in forest spatial patterns: a case study of the eastern Washington Cascades. PNW Res. Pap. PNW-RP-514, p 65
Hessburg PH, Smith BG, Miller CA, Kreiter SD, Salter RB (1999d) Modeling change in potential landscape vulnerability to forest insect and pathogen disturbances: Methods for forested subwatersheds sampled in the midscale interior Columbia River Basin assessment. USDA Forest Service, Pacific Northwest Research Station, General Technical Report PNW-GTR-454. Portland, OR, p 56
Hessburg PF, Smith BG, Salter RB, Ottmar RD, Alvarado E (2000a) Recent changes (1930s–1990s) in spatial patterns of interior northwest forests, USA. For Ecol Manage 136:53–83
Hessburg PF, Salter RB, Richmond MB, Smith BG (2000b) Ecological subregions of the Interior Columbia Basin, USA. Appl Veg Sci 3(2):163–180
Hessburg PF, Reynolds KM, Salter RB, Richmond MB (2004) Using a decision support system to estimate departures of present forest landscape patterns from historical conditions: an example from the Inland Northwest Region of the United States. In: Perera AH, Buse LJ, Weber MG (eds) Emulating natural forest landscape disturbances: concepts and applications. Columbia University Press, New York
Hessburg PF, Agee JK, Franklin JF (2005) Dry forests and wildland fires of the inland Northwest USA: contrasting the landscape ecology of the pre-settlement and modern eras. For Ecol Manage 211:117–139
Hessburg PF, Reynolds KM, Keane RE, James KM, Salter RB (2007a) Evaluating severe wildfire danger and prioritizing vegetation and fuels treatments. For Ecol Manage 247:1–17
Hessburg PF, James KM, Salter RB (2007b) Re-examining fire severity relations in pre-management era mixed conifer forests: inferences from landscape patterns of forest structure. Landscape Ecol 22(1):5–24
Hessburg PF, Reynolds KM, Salter RB, Dickinson JD, Gaines WL, Harrod RJ (2013) Landscape evaluation for restoration planning on the Okanogan-Wenatchee National Forest. USA. Sustainability 5(3):805–840
Hill MO (1973) Diversity and evenness: a unifying notation and its consequences. Ecology 54:427–431
Huff MH, Ottmar RD, Alvarado E, Vihnanek RE, Lehmkuhl JF, Hessburg PF, Everett RL (1995) Historical and current forest landscapes of eastern Oregon and Washington. Part II: Linking vegetation characteristics to potential fire behavior and related smoke production. General Technical Report PNW-GTR-355. U.S. Department of Agriculture, Forest Service, Pacific Northwest Research Station, Portland, OR, p 43
Hummel S, Agee JK (2003) Western spruce budworm defoliation effects on forest structure and potential fire behavior. Northwest Sci 77:159–169
Hummel S, Calkin DE (2005) Costs of landscape silviculture for fire and habitat management. For Ecol Manage 207:385–404

Hummel S, Barbour RJ, Hessburg PF, Lehmkuhl JF, (2001) Ecological and financial assessment of late-successional reserve management. PNW-RN-531. USDA Forest Service Pacific Northwest Research Station, Portland, OR

Hunn ES (1990) Nch'i-Wana, ''The Big River'': Mid-Columbia Indians and Their Land. University of Washington Press, Seattle

Janssen R, Goosen H, Verhoeven ML, Verhoeven JTA, Omtzigt AQA, Maltby E (2005) Decision support for integrated wetland management. Environ Model Softw 30:215–229

Keane RE, Parsons R, Hessburg PF (2002) Estimating historical range and variation of landscape patch dynamics: limitations of the simulation approach. Ecol Model 151(1):29–49

Keane RR, Hessburg PF, Landres P, Swanson FJ (2009) The use of historical range and variability (HRV) in landscape management. For Ecol Manage 258:1025–1037

Landres P, Morgan P, Swanson F (1999) Overview of the use of natural variability in managing ecological systems. Ecol Appl 9:1279–1288

Langston N (1995) Forest dreams, forest nightmares: the paradox of old growth in the Inland Northwest. University of Washington Press, Seattle, p 368

Lehmkuhl JF, Raphael MG (1993) Habitat patterns around northern spotted owl locations on the Olympic Peninsula. Wash J Wildl Manage 57:302–315

Littell JS, McKenzie D, Peterson DL, Westerling AL (2009) Climate and wildfire area burned in western US ecoprovinces, 1916–2003. Ecol Appl 19:1003–1021

McGarigal K, Marks BJ (1995) FRAGSTATS: spatial pattern analysis program for quantifying landscape structure. General Technical Report PNW-GTR-351. U.S. Department of Agriculture, Forest Service, Pacific Northwest Research Station, Portland, OR, p 122

McKenzie D, Gedalof Z, Peterson DL, Mote P (2004) Climatic change, wildfire, and conservation. Conserv Biol 18:890–902

McNulty SG, Aber JD (2001) U.S. national climate change assessment on forest ecosystems: an introduction. Bioscience 51(9):720–722

Millar CI, Stephenson NL, Stephens SL (2007) Climate change and forests of the future: managing in the face of uncertainty. Ecol Appl 17:2145–2151

Miller BJ, Saunders MC (2002) The NetWeaver reference manual. Pennsylvania State University, College Park

Moeur M, Stage AR (1995) Most similar neighbor: an improved sampling inference procedure for natural resources planning. Forest Sci 41(2):337–359

Moritz MA, Hessburg PF, Povak NA (2011) Native fire regimes and landscape resilience.In: McKenzie D, Miller C, Falk DA (eds) The landscape ecology of fire. Springer (in press)

O'Hara KL, Latham PA, Hessburg PF, Smith BG (1996) A structural classification of Inland Northwest forest vegetation. West J Appl For 11(3):97–102

Ottmar RD, Vihnanek RE, Wright CS (1998) Stereo photo series for quantifying natural fuels. In: Mixed-conifer with Mortality, Western Juniper, Sagebrush, and Grassland Types in the Interior Pacific Northwest, vol I. PMS 830, National Wildfire Coordinating Group, National Interagency Fire Center, Boise, ID

Pyne SJ (1982) Fire in America: a cultural history of wildland and rural fire. University of Washington Press, Seattle, p 680

Reeves GH, Benda LE, Burnett KM, Bisson PA, Sedell JR (1995) A disturbance-based ecosystem approach to maintaining and restoring freshwater habitats of evolutionarily significant units of anadromous salmonids in the Pacific Northwest. Proc Am Fish Soc Symp 17:334–349

Reynolds KM (1999a) EMDS users guide (version 2.0): knowledge-based decision support for ecological assessment. General Technical Report PNW-GTR-470. U.S. Department of Agriculture, Forest Service, Pacific Northwest Research Station, Portland, OR, p 63

Reynolds KM (1999b) NetWeaver for EMDS user guide (version 1.1): a knowledge base development system. General Technical Report PNW-GTR-471. U.S. Department of Agriculture, Forest Service, Pacific Northwest Research Station, Portland, OR, p 75

Reynolds KM (2001a) Using a logic framework to assess forest ecosystem sustainability. J Forest 99:26–30

Reynolds KM (2001b) Fuzzy logic knowledge bases in integrated landscape assessment: examples and possibilities. General Technical Report PNW-GTR-521. U.S. Department of Agriculture, Forest Service, Pacific Northwest Research Station, Portland, OR, p 24

Reynolds KM, Hessburg PF (2005) Decision support for integrated landscape evaluation and restoration planning. For Ecol Manage 207:263–278

Robbins WG (1994) Colony and empire: the capitalist transformation of the American West. University Press of Kansas, Lawrence

Robbins WG (1997) Landscapes of promise: the Oregon story, 1800–1940. University of Washington Press, Seattle, p 392

Robbins WG (1999) Landscape and environment: ecological change in the Intermontane Northwest. In: Boyd R (ed) Indians, fire, and the land in the Pacific Northwest. Oregon State University Press, Corvallis, pp 219–237

Saaty TL (1992) Multicriteria decision making: the analytic hierarchy process. RWS Publications, Pittsburgh

Saaty TL (1994) Fundamentals of decision making and priority theory with the analytic hierarchy process. RWS Publications, Pittsburgh

Sauer CO (1971) Sixteenth century North America: the land and the people as seen by the Europeans. University of California Press, Berkeley, p 19

Schmidt KM, Menakis JP, Hardy CC, Hann WJ, Bunnell DL (2002) Development of coarse-scale spatial data for wildland fire and fuel management. Gen. Tech. Rep. RMRS-GTR-87. U.S. Department of Agriculture, Forest Service, Rocky Mountain Research Station, Fort Collins, CO

Seaber PR, Kapinos PF, Knapp GL (1987) Hydrologic Unit Maps. Water-Supply Paper 2294. U.S. Geological Survey, Washington, DC

Simpson EH (1949) Measurement of diversity. Nature 163:688

Spies TA (1998) Forest structure: a key to the ecosystem. Northwest Sci 72(2):34–39

Spittlehouse DL, Stewart RB (2003) Adapting to climate change in forest management. J Ecosyst Manage 4(1):7–17

Staus NL, Strittholt JR, Dellasala DA (2010) Evaluating areas of high conservation value in Western Oregon with a decision-support model. Conserv Biol 24:711–720

Stephens SL, Millar CI, Collins BM (2010) Operational approaches to managing forests of the future in Mediterranean regions within a context of changing climates. Environ Res Lett 5:024003 (9p)

Stolle L, Lingnau C, Arce JE (2007) Mapeamento da fragilidade ambiental em áreas de plantios florestais, pp 1871–1873 in Anais XIII Simpósio Brasileiro de Sensoriamento Remoto, Florianópolis, Brasil, 21–26 abril 2007, INPE

Twery MJ, Knopp PD, Thomasma SA, Rauscher HM, Nute DE, Potter WD, Maier F, Wang J, Dass M, Uchiyama H, Glende A, Hoffman RE (2005) NED-2: a decision support system for integrated forest ecosystem management. Comput Electron Agric 49:24–43

USDA (1994) Record of decision for the amendments to Forest Service and BLM planning documents within the range of the Northern Spotted Owl and standards and guidelines for management of habitat for late-successional and old-growth forest within the range of the Northern Spotted Owl. USDA Forest Service, Portland, Oregon, and BLM, Moscow, Idaho

USDA (2006) Terrestrial species assessments: region 6 forest plan revisions. USDA Forest Service, Pacific Northwest Region. Portland, OR

USDA Forest Service (USDA-FS) (2010) The Okanogan-Wenatchee National Forest Restoration Strategy: adaptive ecosystem management to restore landscape resiliency. (http://www.fs.fed.us/r6/wenatchee/forest-restoration/). Last accessed Jan 12, 2011

USDA-NRCS (U.S. Department of Agriculture-Natural Resources Conservation Service) (2005) Soil Survey Geographic (SSURGO) Database. http://www.ncgc.nrcs.usda.gov/products/datasets/ssurgo/. Last accessed on Jan 12, 2011

USDA-NRCS (U.S. Department of Agriculture-Natural Resources Conservation Service) (2006) National geospatial datasets. http://soildatamart.nrcs.usda.gov/. Last accessed on Jan 12, 2011

USDI (1992) Final draft recovery plan for the Northern Spotted Owl. Report of the USDI, Washington, D.C.
USFWS (2011) Revised recovery plan for the Northern Spotted Owl (Strix occidentalis caurina); U.S. Fish and Wildlife Service, Portland, OR, USA, p 277
USFWS (2012) Endangered and threatened wildlife and plants; designation of revised critical habitat for the Northern Spotted Owl. Final Rule; 50 CFR, Part 17, FWS–R1–ES–2011–0112; 4500030114, RIN 1018–AX69, Federal Register 77, 233, Tuesday, Dec 4, 2012, Rules and Regulations; US Government Printing Office, Washington, DC, USA; pp 71876–72068
van Wagtendonk JW (2007) The history and evolution of wildland fire use. Fire Ecology 3(2):3–17
Wang J, Chen J, Ju W, Li M (2010) IA-SDSS: a GIS-based land use decision support system with consideration of carbon sequestration. Environ Model Softw 25:539–553
Westerling AL, Swetnam TW (2003) Interannual and decadal drought and wildfire in the western United States. EOS Trans AGU 84(545):554–555
Westerling AL, Hidalgo HG, Cayan DR, Swetnam TW (2006) Warming and earlier spring increase western U.S. forest wildfire activity. Science 313:940–943
White R (1991) It's your misfortune and none of my own: a new history of the American West. University of Oklahoma Press, Norman 644 p
White R (1992) Discovering nature in North America. J Am Hist 79:874–891
White R (1999) Indian land use and environmental change. In: Boyd R (ed) Indians, fire, and the land in the Pacific Northwest. Oregon State University Press, Corvallis, pp 36–49
White MD, Heilman GE Jr, Stallcup JE (2005) Science assessment for the Sierra Checkerboard Initiative. Conservation Biology Institute, Encinitas
Wu J, Loucks OL (1995) From balance of nature to hierarchical patch dynamics: a paradigm shift in ecology. Q Rev Biol 70:439–466

Ecological Research Reserve Planning

David M. Stoms

Abstract Guidelines for assessing sites as potential reserves for scientific study tend to be very general. Translating these imprecise qualities into measurable ranking criteria in a large landscape can be particularly challenging. Here, we evaluate the usefulness of the Ecosystem Management Decision Support (EMDS) system to assess suitability of sites to serve as a potential new reserve to support the research and teaching mission of the University of California Merced campus. The assessment was performed iteratively at three geographic scales, narrowing the spatial scope but increasing the detail of the criteria at each subsequent stage. The products of the assessment were the identification of a small number of high suitability parcels to be field inspected, and a flexible, transparent framework for assessing new areas in the future. Although the criteria might vary among reserve systems, our basic approach and use of EMDS could be readily adapted to other reserve systems in the U.S. and abroad.

Keywords Ecological reserve · Reserve planning · Spatial decision support · EMDS

1 Introduction

Many programs establish protected areas for a variety of purposes such as conservation of biodiversity and natural ecosystems. One specialized form of this is the protection of areas dedicated to scientific research and monitoring of environmental conditions and processes. In the United States, programs that designate scientific

D. M. Stoms (✉)
Bren School of Environmental Science & Management, University of California, Santa Barbara, Santa Barbara, CA 93106-5131, USA
e-mail: stoms.david@gmail.com

California Energy Commission, Sacramento, CA 95814, USA

K. M. Reynolds et al. (eds.), *Making Transparent Environmental Management Decisions*, Environmental Science and Engineering,
DOI: 10.1007/978-3-642-32000-2_8,

reserves (hereafter, ecological reserves) include the U.S. Forest Service (research natural areas, Lugo et al. 2006), the National Oceanic and Atmospheric Administration (National Estuarine Research Reserves, http://www.nerrs.noaa.gov/), and teaching and research reserves operated by academic institutions (Ford and Norris 1988). Such sites are known by many names: e.g., research natural areas, research reserves, natural reserves, experimental forests, biological stations, and field stations. I will refer to sites dedicated to protection for scientific study as *ecological research reserves*. In addition to supporting the research needs of the managing institution, ecological research reserves are often aggregated into networks of sites to facilitate comparative studies between ecoregions or to monitor phenomena across a greater range of conditions than is contained in any one institution's system. Examples include the Long-Term Ecological Research network funded by the National Science Foundation (Franklin et al. 1990; Hobbie et al. 2003) and the National Ecological Observatory Network (Keller et al. 2008). Similar programs and networks can be found throughout the world, and indeed there are proposals for international networks of research reserves (Peters et al. 2008).

Ecological research reserves provide many valuable services. They provide irreplaceable training opportunities for the next generation of scientists. Greater understanding of the natural world and its response to management practices is created, both directly from experimental studies but also from analysis of data archives collected at reserves. As Michener et al. (2009) observed, many of these discoveries would not be possible from isolated field studies, but only through the existence over time of a dedicated site with infrastructure, protection from incompatible uses, and a critical mass of creative minds. Further, findings from ecological research reserves may be incorporated into management policy for other lands in similar landscapes (Burke and Lauenroth 1993).

A thoughtful selection of sites is required to achieve these purposes. Land uses on the site and its neighborhood should be compatible with the mission of the reserve. If the mission is to isolate effects of management treatments, the ideal site would be relatively undisturbed. The ecosystems at the site should be important to study, either due to their rarity, their sensitivity to management practices or regional/global change, or the degree to which they represent biophysical environments that lack coverage by another reserve. In the context of significant climatic change, biophysical environment settings move across the landscape and are redefined. In this context, having a broad range of environmental settings represented within research reserve networks becomes especially important to detecting and understanding a variety of ecological and climatic changes.

The costs of initial acquisition and ongoing maintenance and management must also be considered. Setting aside land for a reserve may be politically controversial, either with neighboring landowners, or with other constituencies who had different interests in the site. Although research reserve networks generally have qualitative criteria for evaluating the suitability of individual sites, they usually lack an explicit, operational procedure for comparing and ranking candidate sites (Stoms et al. 1998). Use of a formal, explicit, transparent, and repeatable approach for reserve site selection is especially important where decisions are closely scrutinized.

Approaches used for selecting new reserve sites may be usefully divided into "bottom-up" versus "top-down" types. In the more common bottom-up approach, a specific site is nominated for consideration and then evaluated against the criteria. One could say that a site is suitable if it meets the minimal criteria, but one cannot say, without additional evidence, that it is the "best" site for addition to the network. A top-down approach, in contrast, rates all potential sites in an assessment region and identifies those that best meet the criteria. Obviously, the top-down process needs information for the criteria for all eligible sites. If the region of interest is large, the information available for all sites will be less detailed than for a single site in a bottom-up process. For example, the ecological condition of a single parcel could be assessed with a field survey, whereas the assessment of a large region may require broad-scale mapped data from remotely sensed or other sources.

The University of California Natural Reserve System (UC-NRS) (Norris 1968; Ford and Norris 1988) is the world's most extensive example of an academic ecological reserve program, with 34 natural reserves affiliated with eight of its ten campuses. In the early 21st century, the University of California (UC) was planning for the development of a tenth campus in the Central Valley near the city of Merced (Fig. 1). Anticipating that UC would establish one or more new NRS research and teaching reserves, a generic top-down decision-support framework was developed for selecting new site(s) based on existing UC guidelines (University of California 1984, published in the Appendix of this chapter). The framework was then adapted for three geographically nested stages (Fig. 1), where the criteria were tailored to the available spatial data appropriate to each scale.

Stage 1 assessed general suitability of the neighboring Sierra Nevada or San Joaquin Valley. Stages 2 and 3 further refined the criteria to assess site suitability for an NRS reserve in vernal pool-grassland habitat, in more detail, and over progressively smaller areas. This chapter describes how EMDS was used to frame the decision using a knowledgebase applied to (1) assessing suitability of the existing NRS reserves and (2) identifying specific land parcels of high suitability for a candidate reserve for the new Merced campus. Readers who wish to learn more about the details of the analyses that populated the data links of the knowledgebase are referred to Stoms et al. (2000, 2002).

2 Case Study

2.1 *Assessment of Existing University of California Reserves*

The UC-NRS employs a set of guidelines for evaluating the suitability of sites as new reserves (University of California 1984, published in the Appendix of this chapter). These guidelines are organized hierarchically. The topmost level is organized in three categories of suitability—scientific, academic, and administrative.

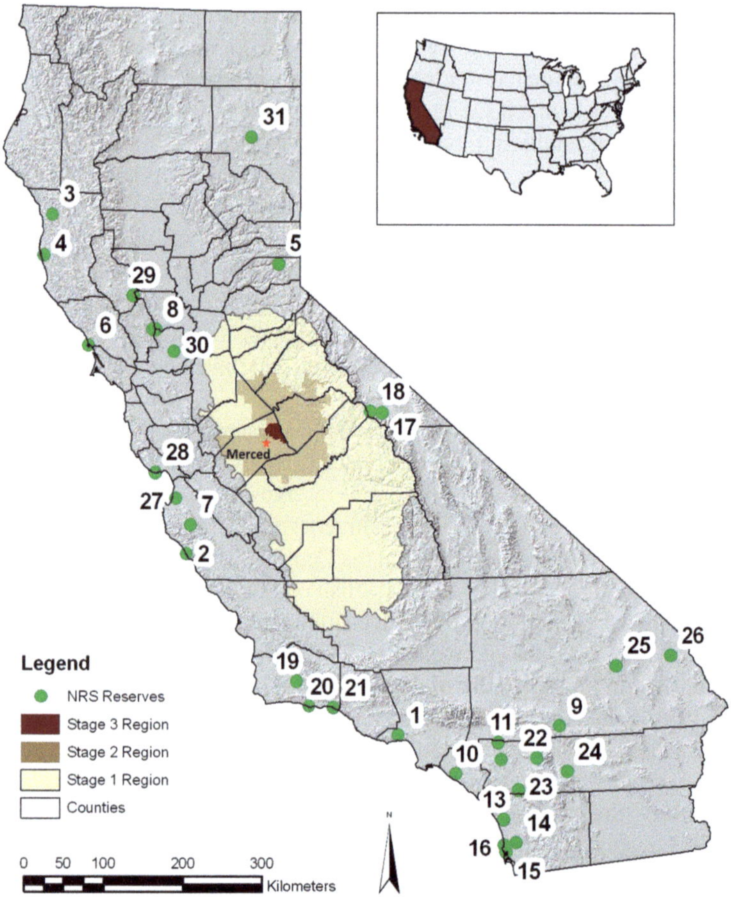

Fig. 1 Map of California showing three hierarchically organized assessment regions. To support a logical and systematic search for a new ecological research reserve with an under-represented ecosystem for the UC Merced campus, three stages of assessment (corresponding with the three mapped regions) were implemented in an hierarchically structured decision support model (see Stages 1–3 in the text). *Numbered points* identify existing NRS reserves at the time of the analysis (see Table 2 for the reserve names)

Scientific guidelines refer to the biological significance of the site as well as the integrity ("viability") of its ecosystems. Academic guidelines include the number of disciplines that could use the site for teaching or research, and the accessibility from the managing campus for those purposes. The administrative guidelines deal with filling gaps in representation of California's natural ecosystems, and the costs and manageability of the site. All these are only general guidelines, however, and do not specify how they should be measured, weighted relative to each other (Malczewski 1999; Jiang and Eastman 2000), or combined into an overall suitability rating (Hopkins 1977). Each assessment committee retains some discretion to determine how the guidelines will be interpreted, whether with precisely measured variables, or with qualitative estimates of suitability.

The guidelines have several characteristics worth noting. First, they are organized hierarchically. The overall measure of the suitability of a site as a new ecological research reserve is based on three logical antecedents (i.e., the scientific, academic, and administrative criteria). Each of these guidelines is similarly predicated on additional antecedents specific to each criterion. Second, many guidelines are semantically and quantitatively imprecise, such as "close to a campus" and "include typical samples of widely distributed habitat types" Such guidelines are poorly represented by crisp threshold values. For example, it would be unreasonable to consider sites up to 25 km from campus as suitable but totally unsuitable if they are one meter beyond that threshold.

On balance, these characteristics of extant guidelines suggest the use of a fuzzy, knowledge-based approach as highly suited to the problem of reserve selection, wherein decision rules would be formulated as a series of equally applied propositions (Malczewski 2002, Chaps. 1 and 2). The propositions would be evaluated not as "true" or "false", in a strict Boolean sense (e.g., distance from campus $\leq$25 km from campus), but as continuous truth values, in which distance from campus, for example, is mapped into a membership function, in the set "close to campus." Formulating the problem in a knowledgebase both formalizes the relationship of the guidelines and the linkage to actual data, and provides insights into which factors are critical in determining the most suitable sites. The knowledgebase also provides a flexible decision-support environment in which the analyst can manipulate the criteria and their weightings, in a gaming sense, to determine the relative influence of various criteria.

The NRS guidelines were interpreted into a logic network using EMDS for ArcView®. As described earlier, the overall proposition that a site is suitable was decomposed according to the three primary criteria of scientific, academic, and administrative suitability (Fig. 2). A poor rating for any of those criteria should yield a relatively poor overall rating, and an excellent site should rate highly in all three criteria. The top network therefore uses an AND node to combine criteria. An AND node functions similarly to a MINIMUM operator, but in EMDS is based on a complex formula in which the minimum value strongly influences the result. The scientific criterion was represented by two subcriteria—viability of ecosystems and the significance of the habitat in a site. Likewise, academic suitability was represented by two subcriteria—the level of potential academic use and proximity to a campus. Administrative suitability was a product of the three subcriteria—the ability of an assessment unit to fill representation gaps, to add geographic balance to the NRS network, and to have favorable acquisition terms.

To make the logic network operational, EMDS requires that each terminal node of the network connect to an elementary topic at which data are evaluated. However, it is occasionally impractical that all criteria and subcriteria be mapped over large spatial domains. For instance, it is not feasible to assess potential acquisition terms for all properties of an assessment region that is thousands of

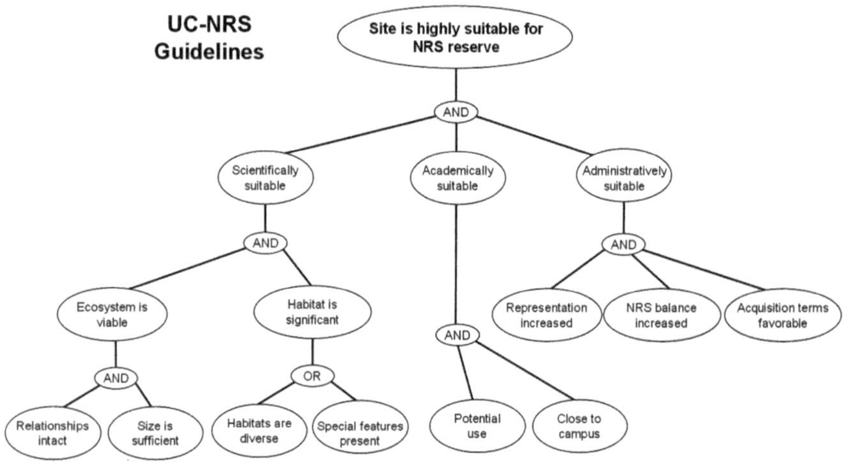

Fig. 2 The UC-NRS guidelines represented as a conceptual network of propositions. The overall proposition asserts that the "site is highly suitable for an NRS reserve." It consists of three sub-networks, joined by an "AND" node, which evaluates the degree to which the assessment unit is scientifically, academically, AND administratively suitable

square kilometers. Consequently, the logic network was adapted (Table 1, Fig. 3). The UC guidelines were not specific about how to measure ecosystem integrity or habitat significance. In view of the types of data available for the entire state of California, two sources were selected for each criterion. Integrity was characterized by a measure of the area affected by roads (Stoms 2000) and the area of undeveloped habitat in an assessment unit. Significance was assessed as the combination—number of plant community types and number of rare elements. Academic suitability was limited to a single criterion based on travel time from the site of the managing campus, because the level of current and potential use for teaching and research could not be determined for all sites. Administrative suitability was a product of the ability of an assessment unit to fill representation gaps and to add balance to the NRS network. The representation subtopic was defined as either being distinct from the environments represented by the other NRS reserves or filling gaps from other programs or agencies (in this case from the California Gap Analysis Project, Davis et al. 1998). To ensure accurate comparison of existing reserve sites with sites to be considered in Stage 1 for the proposed new campus, the spatial units for the assessment had to be comparable. Thus, 6×6 mile square township areas (approximately 9,400 ha each) from the Public Land Survey were used as assessment units. All spatial information was aggregated to townships. These assessment units are larger than the actual NRS reserves.

The existing system of reserves was then assessed by the knowledgebase to determine (1) how well existing reserves met the selection criteria, as assessed by this knowledgebase and the available data, and (2) to establish a baseline for

Table 1 Data links in the Stage 1 knowledgebase and how they were derived

Data link to proposition (VARIABLE NAME)	Assumption or explanation	Data source	GIS processing steps
Road-effect zone is small (ROAD)	Areas with greater area affected by roads exhibit lower ecological integrity	California Gap Analysis Project (Davis et al. 1998) using TIGER roads layer (US Census)	Apply a buffer on roads, buffer width is related to road class. Summarize percentage of area in assessment units within the "road effect zone"
Native habitat is sufficiently large (SIZE)	Size should be sufficient so that the natural balance of the community may be maintained with the survival of the plant and animal elements assured	California Gap Analysis Project shapefile (Davis et al. 1998)	Sum the total area with native vegetation in each assessment unit
Number of California Natural Diversity Database communities (DIVERSE)	The more diversity, the better the site is	California Gap Analysis Project (Davis et al. 1998)	Sum the number of unique plant community types in each assessment unit
Rare elements are present (RARENUM)	Site has added value if it also possesses special features such as rare or endangered species/ habitats	California Natural Diversity Data Base, California Fish and Game	Sum the number of unique elements in each assessment unit from 1992 version of California Natural Diversity Data Base
Research accessible (TRAVEL)	Maximum travel time allowed is 1 h at 50 km/h for teaching, but is a continuous decreasing function for research needs	TIGER roads (US Census)	Travel time estimated by road segment, then a 'cost' distance is estimated. Total time is minimum time to each assessment unit
Fills NRS PCA gaps (NRSMINPCA)	Greater environmental distance from a reserve means site is less-well represented	Biophysical factors, various sources	Principal components analysis of nine biophysical factors, compute multivariate Euclidean distance from each NRS reserve. Find minimum value for each pixel, average values for each assessment unit

(continued)

Table 1 (continued)

Data link to proposition (VARIABLE NAME)	Assumption or explanation	Data source	GIS processing steps
Fills other gaps (VULNCOM)	Site gets added value if it also contributes to representation goals of other institutions	California Gap Analysis Project (Davis et al. 1998)	Score of assessment unit as weighted by vulnerability and area of all plant community types
Far from other NRS reserves (NRSDIST)	Maximizes geographic distribution of reserves	Locations of NRS reserves	Calculate Euclidean distance from all reserves; then determine minimum distance to each assessment unit

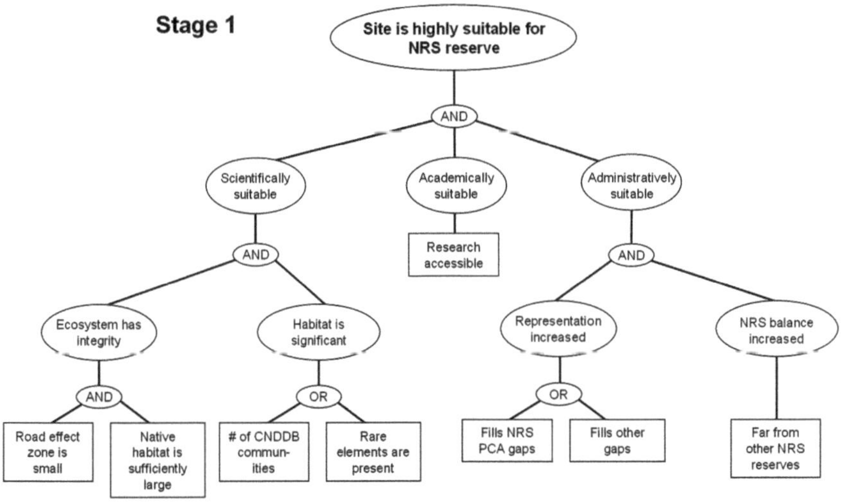

Fig. 3 The knowledgebase as implemented in Netweaver for the top-level proposition that the "site is highly suitable for an NRS reserve." Networks are shown as *ovals* and data links as *rectangles*. This network was used both in the assessment of existing reserves and in screening for Stage 1

comparing potential reserve sites with those previously identified by UC as highly suitable. At the time of this analysis, there were 34 reserves in the NRS; two of these are island reserves (Santa Cruz and Año Nuevo). Because of their uniqueness relative to onshore reserves, they were excluded from this assessment.

One of the overarching aims of the NRS program is to provide representation of California's environmental diversity. Representativeness is important for providing comprehensive opportunities for teaching and research. In addition, it is more efficient to design a network of reserves that is representative so that research results are relevant over the greatest spatial extent (Burke and Lauenroth 1993). Because the concept of representativeness is so fundamental yet imprecise, this section focuses in greater detail on how this criterion of administrative suitability was measured.

Measuring representativeness of a set of sites is not a trivial problem, however, and many different methods have been proposed. Some researchers have arbitrarily divided the primary environmental gradients into segments and classified the combinations of factors (Engelking et al. 1994; Pressey et al. 1996). Others have used continuous data sets but clustered them into classes using statistical techniques (Mackey et al. 1988; Kirkpatrick and Brown 1994; Belbin 1995; Hargrove and Hoffman 2004). Nominally, the NRS uses an unpublished habitat classification and hopes to have 70 % of the extant types represented (Ford and Norris 1988). The reserve descriptions on the NRS web site (http://nrs.ucop.edu/) and published brochures often (but not always) list the habitat types or plant communities that occur, but the classification of habitats and communities are not consistent between reserves or with recognized classification systems.

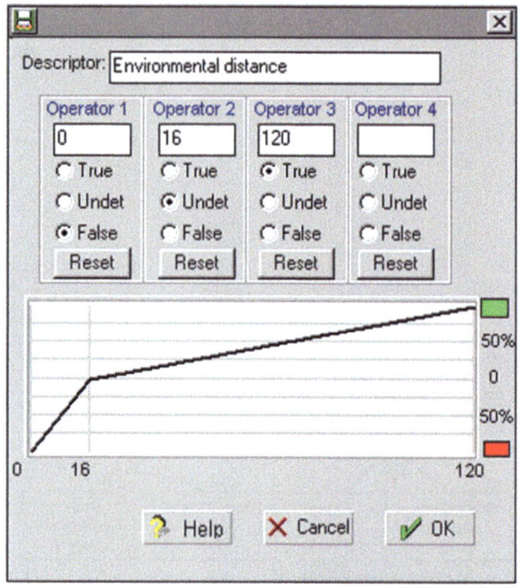

Fig. 4 A fuzzy membership function for the environmental distance of a site from its most similar NRS site

To maintain a continuous data space with measured environmental distance from existing reserves (Faith and Walker 1996), environmental variability of California was characterized via principal components analysis (PCA) of biophysical data sets covering climatic, topographic, and soil productivity factors. The PCA transformed data [on mean annual precipitation, January and July mean temperature and seasonal difference, solar irradiance, degree-day heat and cold sums, equivalent elevation (elevation adjusted for latitude), and soil productivity (Stoms and Hargrove 2000)] into a reduced data space capturing 91 % of the multi-variance. Finding the Euclidean distance of each 1-km^2 grid cell to the most similar research site represented by the NRS provided a quantitative measure of poorly represented environments (Michener et al. 2009). This data value was then transformed into a truth value through a fuzzy membership function defined in NetWeaver (Fig. 4). The resulting evaluation assessed the proposition that each 1-km^2 grid cell was adequately represented by an existing research reserve.

From this representativeness assessment, some regions of the state, particularly the south and central coast (where most of the UC campuses are located), appeared relatively well-represented. In fact, some habitats such as coastal wetlands were over-represented, and these reserves scored relatively low under this criterion. Regions that were less well-represented by the NRS included subalpine environments, some deserts, the mid-elevation conifer zone, and the Central Valley. The UC-Merced campus occurs in the Central Valley, close to mid-elevation Sierra Nevada conifer forests. New reserves for this campus would aid the goal of better representing California's landscapes if placed in Sierra Nevada mixed conifer forest.

To fill this gap, the assessment region for Stage 1 (Fig. 1) included most of the San Joaquin Valley and the west slope of the southern Sierra Nevada. Truth values for the proposition that existing NRS reserves were *highly suitable* for that purpose ranged from −1.000 (totally false) to 0.519 (moderately true) with a mean value of −0.087. Within the second level of networks, the maximum values for assessment units were higher. The range for *scientifically suitable* was −1.000 to 0.974 (mean 0.223); *academically suitable* was −0.119 to 1.000 (mean 0.547); and *administratively suitable* was −1.000 to 1.000 (mean −0.109). Thus, at least some assessment units scored very high for individual sets of criteria, compared to the lower maximum truth value for the overall ranking. In fact, even the lowest scoring assessment units tended to score very high in one or two of the three main criteria, but they all had one or two criteria on which they scored very low (Table 2). These reserves would appear to have been selected because they excelled at meeting some criteria but displayed limitations in other respects.

On the other hand, the highest scoring reserves did not exhibit very high scores for every suitability criterion; they were instead characterized by the absence of extremely low scores. In addition, sites with near-average scores for all three suitability criteria tended to score higher overall than sites with extreme values. Some degree of trade-off between criteria was always present. For example, compare the truth values for the highest rated (Jepson Prairie) and the second lowest rated (Kendall-Frost Mission Bay) NRS reserves (Table 2). Jepson Prairie rated as partially true in all three subcriteria, which generated a high *overall suitability*. Kendall-Frost Mission Bay, in contrast, was outstanding in *academic suitability*, being close to the UC San Diego campus, but rated poorly in the other criteria.

The interpretation of these results must be used with caution because the logic and data used in our assessment were not identical to those used in selecting these reserves initially. The truth values describe the assessment unit containing the reserves, not the specific parcel, which may have special properties not represented in our regional-scale database. That is, this assessment is based on what information would be available to a planner looking broadly across the state for new reserves, whereas an evaluation team for a specific parcel of land would typically be privy to more specialized information. The most common criterion that caused the low scores was the fact that reserves were close to one another and therefore did not contribute highly to representing the ecological and geographic diversity of the state. They also tended to be located in urban areas (with low ecological integrity as measured by area affected by roads and lack of native habitats) and as a result had average to poor *scientific suitability* values. Most assessment units scored moderately high for the *academic suitability* criterion, which was based solely on travel time from the sponsoring campus. The key point is that the lowest scoring reserves were not poor in all aspects, but instead were excellent in some and poor in others.

Table 2 Truth value of the assessment units containing NRS reserves for top-level criteria in ascending numerical order

Reserve assessment unit	Overall suitability	Scientifically suitable	Academically suitable	Administratively suitable
15–Kendall-Frost Mission Bay Marsh	−1.000	−1.000	1.000	−0.985
16–Scripps Coastal	−1.000	−1.000	1.000	−1.000
24–Boyd Deep Canyon Desert Center	−0.701	0.582	0.000	−0.780
32–Quail Ridge	−0.615	0.416	1.000	−0.741
8–Stebbins Cold Canyon	−0.604	0.564	1.000	−0.737
21–Carpinteria Salt Marsh	−0.597	−0.349	1.000	−0.700
25–Sweeney Granite Mountains Desert Research Center	−0.570	0.650	0.000	−0.678
2–Landels-Hill Big Creek	−0.475	0.922	0.000	−0.613
20–Coal Oil Point	−0.437	−0.424	1.000	−0.561
9–Burns Pinon Ridge	−0.222	0.056	0.000	−0.298
13–Dawson Los Manos Canyon	−0.100	−0.114	1.000	−0.273
17–Sierra Nevada Aquatic Research Laboratory	−0.092	0.546	−0.085	−0.207
18–Valentine Camp	−0.076	0.199	−0.119	−0.124
22–James San Jacinto Mountain	0.013	0.543	0.755	−0.214
23–Emerson Oaks	0.015	0.071	0.737	−0.141
12–Motte Rimrock	0.015	−0.160	1.000	−0.069
26–Sacramento Mountains	0.081	0.198	0.000	0.290
4–Hans Jenny Pygmy Forest Reserve	0.091	0.176	0.000	0.372
7–Hastings Natural History Reservation	0.109	0.421	0.000	0.233
14–Elliot Chaparral	0.121	0.255	1.000	−0.097
3–Angelo Coast Range	0.125	0.466	0.000	0.284
11–Box Springs	0.135	0.083	1.000	−0.053
10–San Joaquin Freshwater Marsh	0.153	−0.050	1.000	0.183
6–Bodega Marine	0.169	0.168	0.000	0.845
28–Younger Lagoon	0.179	0.292	1.000	−0.041
5–Chickering American River	0.189	0.974	0.000	0.157
31–Eagle Lake Field Station	0.247	0.482	0.000	1.000
19–Sedgwick	0.295	0.401	0.966	0.078
27–Fort Ord	0.339	0.193	0.886	0.231
29–McLaughlin	0.430	0.650	0.353	0.398
1–Stunt Ranch Santa Monica Mountains	0.466	0.268	1.000	0.475
30–Jepson Prairie	0.519	0.655	1.000	0.287
Average	**−0.087**	**0.223**	**0.547**	**−0.109**
Minimum	**−1.000**	**−1.000**	**−0.119**	**−1.000**
Maximum	**0.519**	**0.974**	**1.000**	**1.000**

Numbers preceding the reserve name correspond to the site numbers in Fig. 1. Note that the assessment units are larger than the actual reserves

2.2 Stage 1

With the knowledgebase of ecological reserve suitability guidelines and the assessment of existing reserves in hand, the next task was to perform a three-stage suitability screening assessment specific to identifying potential reserve sites for the proposed UC Merced campus. The Stage 1 region for assessing suitability was derived from our representativeness assessment of the existing NRS network. It included the west side of the central and southern Sierra Nevada, plus a portion of the San Joaquin Valley surrounding the proposed site for the UC Merced campus. This region contains many environments not well-represented by the existing NRS network, does not contain any reserves or overlap with any other UC campus. The combined area totaled over 63,000 km^2 in 15 counties (Fig. 1).

As Stage 1 screened a large geographical region, the analysis needed to be at a coarse scale. For this stage therefore, small "planning watersheds" (mean size of approximately 3,300 ha, Menning et al. 1997) were used as the assessment units in the Sierra Nevada portion of the region. For the San Joaquin Valley, where planning watersheds have not been delineated, 6 $\times$ 6 mile townships (approximately 9,400 ha each) from the Public Land Survey were used. All spatial information was aggregated to planning watersheds or townships. These assessment units were larger than a typical UC-NRS site, but compatible with the resolution of the regional data on biological, environmental, and administrative factors (Stoms et al. 1998).

A GIS database was compiled for the data links needed by the knowledgebase (Table 1, Fig. 3) for each of the approximately 1,400 assessment units. Most of the information to calculate data links for the knowledge-based network came from the California Gap Analysis Project (Davis et al. 1998) that mapped land cover and land ownership for the entire state. These and additional data were interpreted by various GIS analyses, as described in Table 1 and Stoms et al. (2000), to generate the data evaluated in the elementary topics (the lowest level of the logic network).

Applying the knowledge-based logic network to the data links from the GIS database generated truth values for every assessment unit in the Stage 1 region (Fig. 5). No units absolutely met the suitability proposition (i.e., truth value = +1.000). One unit had a value of 0.815, but the next highest assessment units displayed values below 0.7. *Travel time* from campus was a very strong constraint on rankings, such that the highest ranked units fell within a 2-h driving distance. Within that radius, there were units that scored above 0.6. Beyond that time-limited distance from campus, most assessment units in the Sierra Nevada and at the southwestern corner of the study area showed low positive values. Most units in the agricultural Central Valley displayed negative scores as a result of combined low *ecological integrity* and *representativeness* values. Assessment units in the highest elevation zone of the Sierra Nevada scored very high on *ecosystem integrity* but low for *travel time* and *representativeness* (some units were inaccessible by road because they reside in designated wilderness). Thus many sites displayed high truth values for some criteria but low values with at least one other.

The range of *suitability* truth values for the UC Merced assessment region was similar in comparison with the range for the existing reserves (Table 2). That is, existing NRS reserves were on average slightly more suitable than the average assessment unit in the Stage 1 assessment region. Existing reserves fared slightly better on *scientific criteria*, on average, than the assessment region, but relatively poorer for *administrative criteria*. The *scientific criteria* results could be explained by the large number of assessment units that were highly impacted by urban and agricultural land uses. This result appears to be related to the distinctiveness of the study area, in both ecological and geographical distance from existing reserves. By these measures of similarity, the existing reserves tend to be relatively redundant while the Stage 1 assessment region tends to be dissimilar with the existing network. By the *academic suitability* criterion, the existing reserves scored higher on average than the Stage 1 assessment region, because they were generally closer to their sponsoring campus. Most of the assessment units for the Stage 1 region lie beyond a 2-h travel time and therefore had low truth values for *academic suitability*. These results indicate that at least some assessment units in the Stage 1 region were comparably suitable to existing NRS reserves, based on the NRS guidelines.

2.3 Stage 2

The Stage 1 suitability assessment identified several areas having a roughly circular outline (strongly influenced by *travel time*), where *suitability* truth values of assessment units were consistently greater than 0.2, and mostly greater than 0.4 (Fig. 5). This circular area became the Stage 2 assessment subregion for more detailed application of the logic network. The Stage 2 assessment subregion included portions of five counties (Tuolumne, Stanislaus, Merced, Mariposa, and Madera). Within this area, there are several sites currently managed for conservation or research purposes that could be considered for NRS use or to complement an existing NRS reserve without additional university management. This observation created a potential opportunity to establish a series of ecological research reserves along an ecological gradient over several thousand meters of elevation range, which could be especially valuable to support global change studies (Zhang et al. 1997).

The goal of the Stage 2 assessment was to further narrow the pool of potential sites for NRS reserves and more specifically to assess suitability to represent the vernal pool/grassland habitat type. Vernal pool and grassland ecosystems were targeted because of their regional ecological significance and their close association with the proposed location for the new campus. The vicinity of the proposed Merced campus is considered the largest region of dense vernal pool habitat in California (Holland 2000). Large, dense vernal pool complexes are more likely to contain a diversity of pool sizes, depths, and duration of inundation, and therefore

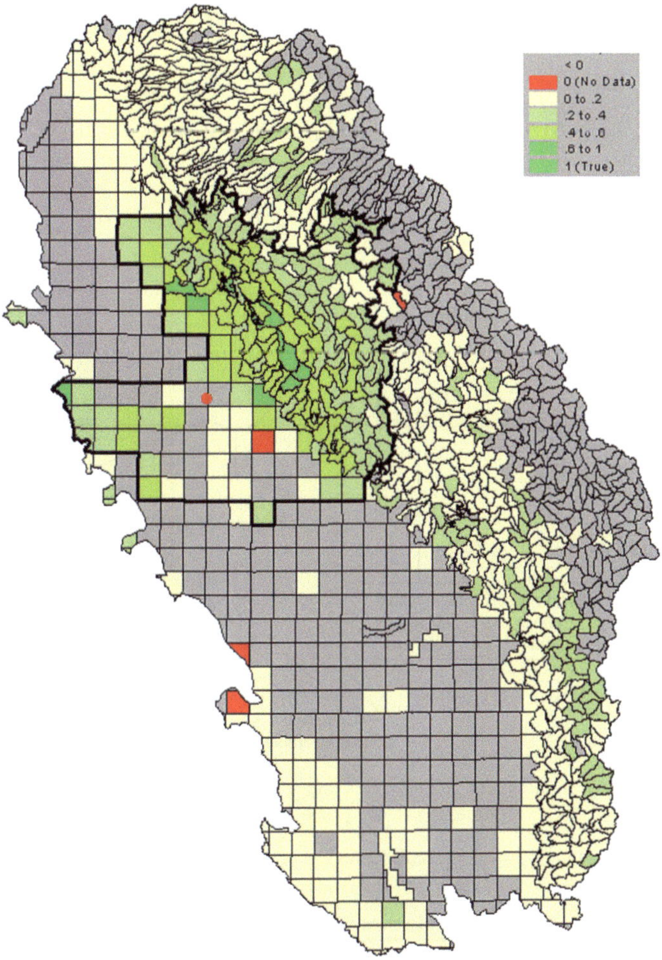

Fig. 5 Map of suitability in the Stage 1 assessment region with the outline (*bold*) of the proposed Stage 2 region. The city of Merced is shown as a *red dot*

support more species than sites with small or less dense complexes (Mead 1996). Vernal pools are considered one of the most threatened ecosystems in California, with a significant proportion of their distribution lost to cultivation or urbanization (Jones and Stokes Associates 1987). These seasonal pools form during winter rains in small depressions above an impermeable layer and then dry up in the long summer drought. Vernal pools are associated with many rare and endangered species that have evolved on the unusual soil chemistry and highly fluctuating hydrology (Mead 1996; Holland 2000). The vernal pool/grassland habitat near Merced has also been identified as critical to the recovery of several endangered species (U. S. Fish and Wildlife Service 1998).

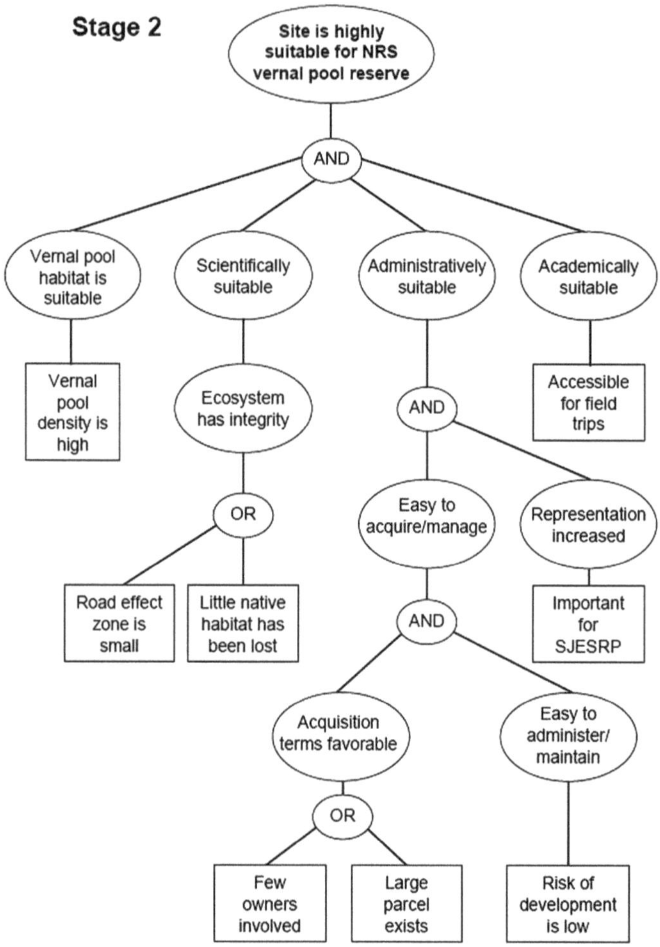

Fig. 6 The network for the Stage 2 proposition that the "site is highly suitable for an NRS vernal pool reserve." Networks are shown as *ovals* and data links as *rectangles*. With kind permission from Springer Science+Business Media (Stoms et al. 2002, p 549, Figure 2)

In Stage 2 the NRS guidelines were interpreted into a logic network that was similar to Stage 1, starting with the three primary criteria of *scientific*, *academic*, and *administrative suitability* (Fig. 6). There were three primary differences distinguishing Stage 1 from Stage 2. First, because the Stage 2 assessment was focused on a specific habitat type, a fourth network was added to specifically test the assertion that "vernal pool/grassland habitat is suitable," based on vernal pool quality and density (Holland 1998). In a sense, the "*habitat significance*" criterion was detached from the *scientific suitability* network and promoted to a top-level network. Second, the smaller size of the Stage 2 assessment subregion permitted more detailed information to be included in the logic networks. The third change was that Stage 1 criteria were not carried over to Stage 2 if they no longer helped distinguish relative suitability. For instance, in Stage 2, all assessment units adequately filled gaps in the NRS, so the criterion was no longer useful.

The nature of the existing vernal pool density map (Holland 1998) necessitated the use of a different multi-criteria analysis function in EMDS. This map contained area polygons classified as having high, medium, or low densities of vernal pools. Any given assessment unit therefore might have some combination of these three, as well as land with no vernal pools. Because these class data could not be converted to continuous values, pool density was calculated as a weighted sum of the proportions of the assessment unit in each density class. This calculated value could then be translated into a truth value that "vernal pool density is high."

Scientific suitability in Stage 2 was characterized by the integrity of the ecosystem in terms of area affected by roads (Stoms 2000) and land use conversion from photo-interpreted maps of farmland use (Fig. 6). *Academic suitability* was defined solely by travel time from the proposed campus site as modeled over the road network using previously stated speed and distance assumptions. Because of the large size of the assessment subregion, data on individual parcels were not available. Instead, the potential ease of acquisition was based on the number of landowners and size of largest parcel in a unit. The assumption behind these criteria was that it would be less desirable to assemble a reserve from many small parcels with multiple owners rather than from a few larger parcels. The risk of development as it may impact compatible uses in neighboring units and therefore the ease of management was based on a simple model of future urban growth (see Stoms 2000 for details). The statewide data from the California Gap Analysis Project that were used in Stage 1 were generally replaced with more detailed maps of land use/land cover from the California Farmland Mapping and Monitoring Program (http://conservation.ca.gov/dlrp/fmmp/Pages/Index.aspx). Site importance was also based on data from the San Joaquin Endangered Species Recovery Plan (U. S. Fish and Wildlife Service 1998).

The Stage 2 assessment subregion encompassed over 12,628 km^2 (Fig. 7) or 20 % of the Stage 1 assessment region (Fig. 1). To allow finer resolution of the Stage 2 assessment, the assessment units were redefined so that they would not be bisected by major roads. Thus, most assessment units were delineated as blocks of unroaded area bounded by roads. Where the size of unroaded units was excessively large, they were further subdivided by watershed or township boundaries. This process delineated 623 assessment units, ranging in size from 136 to 12,285 ha, with a mean size of 2,027 ha (slightly less than half the size of Stage 1 assessment units). These assessment units are still larger than many UC-NRS sites, but they are compatible with the resolution of the regional data on biological, environmental, and administrative factors (Stoms et al. 1998).

The overall suitability, therefore, gave the highest truth values (0.601–0.924) to a small set of contiguous assessment units surrounding, and including, the proposed campus site (Fig. 7). A few additional assessment units had moderately high scores just north or south of the most highly rated units. The areas with the highest density of vernal pools occurred along the grassy base of the Sierra Nevada, with a

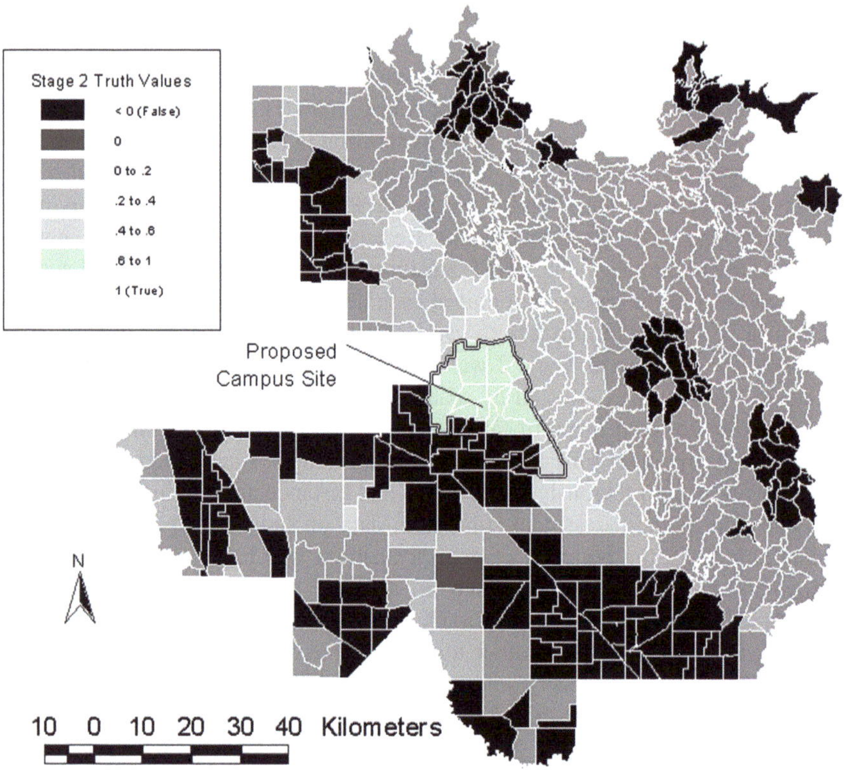

Fig. 7 Map of truth values for vernal pool site suitability for sites within the stage 2 assessment region. The *bold black outline* with *white inner line* shows the Stage 3 planning boundary. With kind permission from Springer Science+Business Media (Stoms et al. 2002, p 551, Figure 3)

secondary zone in the wetlands near the various wildlife refuges west of Merced. The large extent of dense vernal pool complexes in these assessment units would most likely contain a broader diversity in pool size, depth, duration of inundation, and number of species than sites with small or less dense complexes (Mead 1996). These same locations were also important for the San Joaquin Endangered Species Recovery Plan (U. S. Fish and Wildlife Service 1998). The travel time from campus criterion favored assessment units closer to the proposed campus site, which also contain some of the highest suitability vernal pool sites. The criteria relating to ease of acquisition and management, where such information was available, rated the ranchlands in the vernal pool zone among the highest suitability sites. Assessment units in the Sierra Nevada had high ecological integrity but rated very low to low positive values because they contained few or no vernal pools. Otherwise, assessment units tended to have negative truth values, and were unsuitable for a new NRS vernal pool habitat reserve.

The area containing and surrounding the proposed UC Merced campus contained a very dense complex of vernal pools, among the best remaining examples in the Central Valley (Holland 2000). These assessment units also achieved a high level of concurrence with the UC-NRS guidelines for their *scientific*, *academic*, and *administrative suitability*, were relatively intact ecologically, contained few roads or converted lands, consisted of larger ranches rather than small farms or rural residential lots; and were within an easy commuting distance for class field trips. There were other vernal pool complexes within the Stage 2 assessment subregion that perhaps rivaled those near the campus in size and density; however, the sites containing them did not meet the university guidelines as well as those closer to the campus. Only the highly rated assessment units in the vicinity of the proposed campus (Fig. 7) were further evaluated in Stage 3. Other vernal pool sites were not evaluated further unless no available parcels could be found among the Stage 3 candidates.

2.4 Stage 3

In Stage 3, the NRS guidelines were coded into a logic network that was similar to Stages 1 and 2, starting with the same three primary criteria (*scientific, administrative, and academic suitability*); however, *academic suitability* based on travel time from campus was considered uniform across all parcels (after Stage 2 analysis) and dropped from the Stage 3 model). The scientific and administrative suitability of assessment units was then assessed only for vernal pool/grassland habitat in Stage 3 as it was in Stage 2. The logic was similar to that of previous studies that incorporated vernal pool diversity and density, potential threat of development, parcel size, and condition and defensibility of the site (Mead 1996, Reiner and Swenson 2000). In particular, suitability was based on representing the diversity of pool communities, which differ significantly among landforms and parent soil materials (Smith and Verrill 1996, Holland 2000, Reiner and Swenson 2000). In the absence of biological inventory data, soil mapping units (Arkley 1954) were used to infer the pedological diversity and any associated biological diversity. Second, the smaller size of the assessment region permitted detailed parcel-level information to be included in the logic networks. For instance, information on actual and potential land use from assessor's records was used to estimate the existing capital investment and land value that influences the degree of difficulty in acquiring parcels for a new reserve. The full logic network and terminal data nodes are shown graphically in Fig. 8.

Stage 3 analysis narrowed the domain to 430 km^2 (Figs. 1, 9), or 3 % of the Stage 2 assessment area (Figs. 1, 5), and less than 1 % of Stage 1 area (Fig. 1). To allow finer resolution of the Stage 3 assessment, the assessment units were redefined as 298 assessor's parcels from Merced County. The parcel coverage, purchased from the county, contained attribute data about current use, zoning and general plan designation, and the owner. Several large ranches encompass multiple parcels so the number of landowners is smaller than the number of parcels.

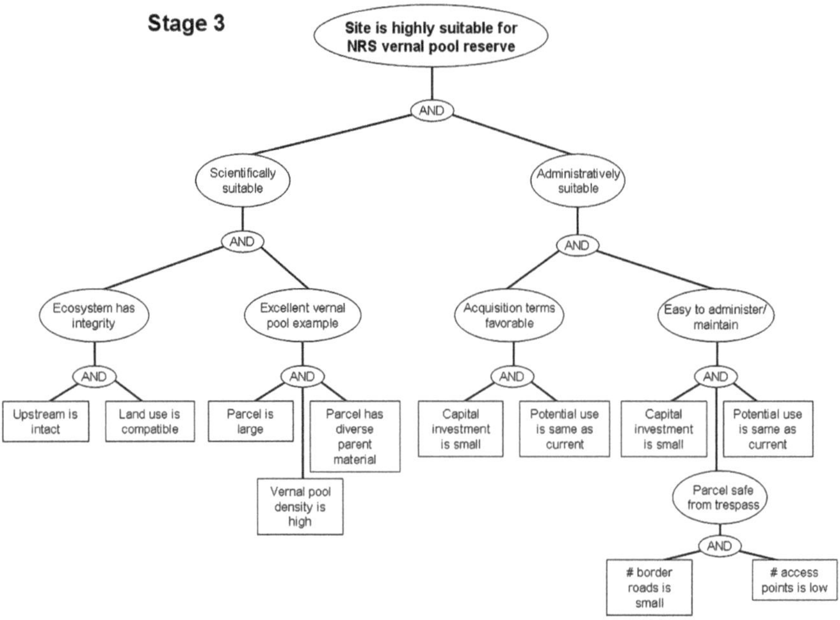

Fig. 8 The network for the Stage 3 proposition that the "site is highly suitable for an NRS vernal pool reserve." Networks are shown as *ovals* and data links as *rectangles*. With kind permission from Springer Science+Business Media (Stoms et al. 2002, p 552, Figure 4)

Three clusters of parcels labeled A, B, and C (Fig. 9) had the highest overall suitability (greater than 0.9). Most parcels had relatively high suitability for most factors, except for those on the edges of the Stage 3 area that are currently agricultural, or are zoned for development, and those that are influenced by irrigation canals or paved roads. The two criteria that had the most influence on the ratings were vernal pool ratings and trespass factors. The majority of the Stage 3 area had only a single soil parent material type in each parcel and therefore relatively low diversity. A few parcels at lower elevations tended to have two or three soil types and presumably greater biological diversity. Also, the density of pools was greatest across the middle of the study area. Generally, the parcels in the north of the Stage 3 region tended to have lower suitability as a prime example of vernal pool complexes because of lower pool density and soil diversity. The areas that rated highest for ease of administration and maintenance were those with fewer roads. The most highly rated parcels had truth values too close to confidently select one or more as the appropriate site for a reserve, given the nature of the methods and quality of the data. What this suggests is that there are many locations that would potentially make excellent reserves. Thus, there is a good deal of flexibility to negotiate with landowners to identify lands within this set of suitable parcels that could be made available to the university by acquisition or management agreement.

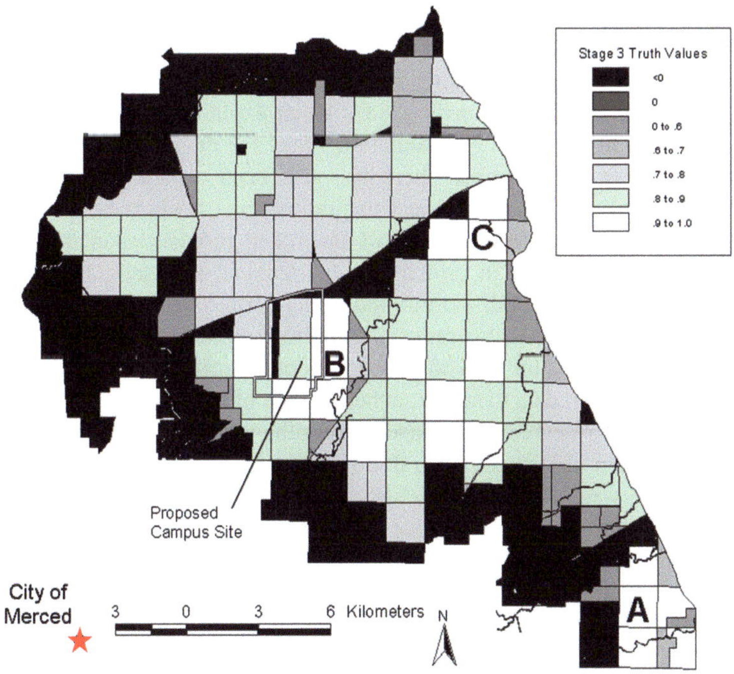

Fig. 9 Map of truth values for vernal pool site suitability for sites within the Stage 3 assessment region. The *bold line* indicates the boundary of the proposed UC Merced campus site. With kind permission from Springer Science+Business Media (Stoms et al. 2002, p 553, Figure 5)

3 Conclusions and Future Directions

Selection of ecological research reserves tends to be opportunistic, where one or more known sites are compared against formal or informal criteria. The UC-NRS guidelines, for example, define a general set of qualities UC reserves should possess, but provide little specific guidance for a quantitative, systematic, and repeatable process for selecting sites for the NRS network. This is not uncommon among organizations that designate lands for research reserves. The Forest Service, for instance, has similar guidelines for its network of research natural areas (Stoms et al. 1998). The informal process for assessing a single candidate site is analogous to how most people purchase a house. They may have a particular neighborhood in mind that constrains the search area. Then they look at a few houses. Once they find one that has the features they must have (e.g., close enough to work) and many they want (e.g., nice style), they make a decision to make an offer. But selecting a reserve is essentially a permanent commitment on behalf of the institution with important implications for the scientific validity of management recommendations, political controversy over land uses, and costs to the institution, taxpayers, or donors. In high profile cases, the screening process must be explicit and transparent.

The original impetus for creating the UC-NRS was the actual loss of a wild, natural study site (Ford and Norris 1988), which explains the criterion favoring undisturbed condition. Although maintaining sites for monitoring baseline reference conditions is still valuable, there has also been a paradigm shift toward scientific study of coupled human and natural systems (Collins et al. 2007; Liu et al. 2007). Study sites for this type of integrated research span the land use spectrum (Liu et al. 2007). In principle, a hierarchical knowledgebase could be developed that considers both natural and human criteria for assessing potential research site suitability, but this remains a future research need.

EMDS provides a solid decision-support tool for organizing criteria, expert knowledge, and spatial data to structure the problem of screening sites for potential ecological research reserves. NetWeaver allows the analyst to structure objectives hierarchically from the most general down to the specific. This hierarchical structure superbly matched the UC-NRS guidelines. The fuzzy logic approach facilitates the interpretation of vague or imprecise criteria such as "close to campus" into measurable continuous values. Using the fuzzy membership function standardizes all criteria, both qualitative and quantitative, to a common measurement scale with a common interpretation of the truth of the proposition associated with the criterion. Fuzzy logic is based upon a rigorous mathematical foundation with formal operators, avoiding the ad hoc nature of many multicriteria analyses. On the other hand, NetWeaver also supports the use of many mathematical operations, for example, we used multiplication within NetWeaver to calculate the weighted sum of vernal pool density classes, as described above.

One particularly attractive feature of EMDS is how it tracks the truth value of all propositions at all levels of the knowledgebase hierarchy. This transparency allows a decision maker or stakeholder to explore precisely how assessment units were rated as they were. Not only does this increase understanding of the complex information and hopefully greater acceptance of the results, but it can also reveal errors in logic and where a criterion may be improperly specified. This wealth of information throughout the knowledgebase hierarchy can be rather daunting, however. We found it particularly useful to lay out the whole hierarchy in poster form to absorb the big picture. The poster contained the full logic network, with each proposition and data link portrayed with both a map of its values (Figs. 5, 7, and 9) and the fuzzy membership function (Fig. 4). Plotting this in poster format facilitated reviews of the knowledgebase and its implications for the suitability assessment. We note that other studies found utility in presenting the map hierarchy along with the logic representations (e.g., see Hessburg et al. 2007; Reynolds et al. 2009; Staus et al. 2010).

EMDS can be readily adapted to various geographic scales of assessment. In this case study, assessments for the UC-NRS guidelines were performed at three scales. The criteria at top levels remained the same, but they were modified at the lower levels (stages) in response to the availability and appropriate usage of higher resolution data. For example, in Stage 1, ecological condition was defined by a measure of road density and area of native habitat from a low resolution (1:100,000) land cover map. In Stage 2, percent conversion was calculated from a higher resolution (1:24,000) land use map. In Stage 3, condition was based on the intactness of the catchment upstream of the parcel and the compatibility of the specific land use with management of a reserve. Thus the finest resolution data were only required for a relatively small area. While this does not guarantee that excellent parcels were not overlooked at the coarser scales, it expedited analysis and because of its explicitness can be subjected to review by regional experts.

The vernal pool reserve case study described in this chapter represents just one type of application related to siting and designing reserves that a fuzzy logic assessment of suitability could support. In this example, the decision maker's concern was principally with the suitability of each assessment unit to support the mission of the reserve system. Where setting aside land for scientific or conservation purposes potentially conflicts with other interests, the knowledgebase may be expanded to incorporate the objectives of other stakeholders. Wood and Dragicevic (2007) illustrate this type of expanded scope using fuzzy logic for assessing potential marine protected areas. They assessed two conflicting objectives—biodiversity conservation and fisheries profit-maximization—as separate primary topics of their knowledgebase. Interestingly, some criteria were (appropriately) applied to both objectives (e.g., species persistence).

The assessment units used to identify a candidate reserve do not necessarily make good boundaries for its final design. Once a general location has been selected for consideration as a reserve, multi-criteria analysis could also be used to design its final configuration. Bojorquez-Tapia et al. (2003, 2004) demonstrated the use of multi-criteria suitability analysis with optimization modeling to reach consensus on the location of boundaries for a park (2004) and a biosphere reserve (2003). The design for the park addressed maximizing conservation value within the boundary and minimizing conflicts at the boundaries. The biosphere reserve design balanced winter habitat for Monarch butterflies with timber production from the local forest. The use of a fuzzy logic knowledgebase to design reserve boundaries has not been reported in the scientific literature to date; however, in principle, the approach might be the reverse of that used in the present study. In the reserve boundary design case, the assessment units would be smaller than a reserve and the assessment would involve assembling a set of units that make a logical whole. With respect to the case study in this chapter, a final reserve might be assembled from a set of contiguous assessor's parcels similar to the clusters labeled in Fig. 9. In a related example, Staus et al. (2010) used EMDS to map high conservation value lands in western Oregon to identify contiguous blocks of land worth investigating further based on their value and condition for both terrestrial and aquatic biodiversity.

Designing networks or systems of conservation areas is frequently done by selecting a set of sites that optimally satisfies specific conservation targets (e.g., a percentage of the area of each habitat type, Margules and Pressey 2000). Optimization is used to minimize the cost (or total area) of the network. Economic cost is often difficult to estimate in these planning projects, so planners sometimes use the size of the assessment or planning units as a surrogate for cost, i.e., meet conservation targets in the least total area. The quality or integrity of the unit can also be considered a type of cost. Davis et al. (1996, 1999) used a multi-criteria measure of "unsuitability" to standardize quality in cost terms, combined with area, to select units that minimized total area and unsuitability. Humphries et al. (2008) adapted this framework but used a knowledgebase approach to assess unsuitability for conservation. The biological significance criteria in conservation area network selection is detached from suitability and treated as a minimal constraint in the reserve system optimization problem. Similarly, cost is detached from the suitability proposition and becomes the objective function to be minimized by the network.

Stoms et al. (2005) proposed using a logic network to assess the suitability of terrestrial sites based on (1) the risk to coastal ecosystem features where the land was not in protected status, and (2) the importance of coastal resources at risk. This idea extended the traditional measure of site suitability based on intrinsic factors to include its geographic neighborhood. The land-sea linkage proposed by Stoms et al. (2005) would also incorporate ecological linkages between sites and ecosystems.

The theme of this chapter has been the application of fuzzy logic networks in planning for ecological research reserves. It should be noted that the basic framework is equally appropriate for screening for other types of protected areas, such as national wildlife refuges, nature preserves, regional parks, and urban open space. More generally, EMDS can support suitability assessment for any form of land use where the criteria are either precise or imprecise.

Acknowledgments The financial support for this study by the University of California, Office of the President is gratefully acknowledged. Frank Davis conceived of the idea of applying EMDS to this application and guided the development of the knowledgebase. Alex Glazer of the UC-NRS provided valuable comments. Keith Reynolds and Noah Goldstein gave technical advice on applying the EMDS software. Jennifer McDonald assisted with GIS analysis, and the Merced County Association of Governments and the Map and Imagery Laboratory at UCSB helped provide data.

A.1 4 University of California Natural Reserve System–NRS Acquisition Guidelines June 1984

A.1.1 Scientific Criteria

General. The objective of the Natural Reserve System (NRS) is to develop and maintain, for educational and scientific study, a system of natural reserves broadly representing California's diversity of natural environment. A site with many habitat types will make a bigger contribution to the NRS than one with only a single habitat type. However, there may be occasions when a feature of special interest will override the usually important requirement for diversity. Ecosystems totally free of human influence are no longer to be found, and in reality, units of a system of natural reserves will fall within a spectrum with undisturbed ecosystems on the one hand and ecosystems heavily influenced by humans on the other hand. With care and good judgment, the reserves will be bunched as closely as possible to the undisturbed end of the spectrum with samples of selected ecosystems of significant merit elsewhere along the spectrum.

Criteria. (1) Viable ecosystem: Ecosystem viability is a prime requisite in establishing a natural reserve. The natural relationships should be essentially intact (i.e., an ecosystem operating as much as possible under its own influences), and the reserves should be of sufficient size so that the natural balance of the community may be maintained with the survival of the plant and animal elements assured. Boundary configuration is an important contributor to viability. The boundaries must be located so as to encompass the critical landscape features necessary to maintain the ecosystem. An ideal reserve will be buffered from the detrimental impact of adjacent land uses. In some instances, a disturbed ecosystem will revert to its formerly undisturbed condition and may be considered as a candidate natural reserve. In other instances, a candidate natural reserve will be a remnant ecosystem not meeting the test of viability, but with value for study during whatever time is left before the natural reserve value is lost.

(2) Habitat significance: Reserves should possess exceptional value in illustrating, interpreting, and protecting examples of the major habitat types of California. The most desirable situation is a reserve with a large diversity of habitats. This maximizes the academic yield for its acquisition cost by providing a large variety of things to see and do on a given field trip as well as maximizing the variety of research possibilities at a given location. It is easy to become enamored with the unusual and overlook the common. Therefore, it is important that the NRS guard against unbalancing its system in favor of unusual values and take care to include typical samples of widely distributed habitat types. However, a reserve has added value if it also possesses special features, such as:

- Important variations of the common habitat types, such as different successional stages (including important human-induced successional stages) or variations in soil parent material.
- Significant gene pools, such as isolated populations or populations at extreme limits of the range of a species or habitat type.
- "Type localities," for example, the location where a species, soil type, geological type, etc., are first described.
- Transition zones (ecotones) and interfaces between adjacent habitat types.
- The presence of a rare or an endangered habitat type or the presence of a rare or endangered species.
- The presence of a feature of geological, archaeological, or paleontological importance.

In some cases, unusual features will be deliberately acquired because they are judged to have special value to the NRS.

A.1.2 Academic Criteria

General. There is an increasing awareness of the need for establishing natural reserves. Federal, state, and private agencies involved are stepping up their levels of participation allowing the NRS to concentrate on its special ability to serve the needs of higher education. Worthy sites lacking a high degree of academic usefulness can be left to the other agencies to protect.
Criteria. Of particular importance is acquisition of sites enjoying current academic use, but not yet in the system. Some sites are not presently being used because of budget stringencies or other reasons which, if eliminated, would result in future academic use. This potential for future use is an important criterion. The larger the variety of disciplines that can be accommodated, the more useful the reserve will be. This is somewhat a matter of degree, since most reserves will be useful for more than the one biological science, but only in special cases will a reserve also be useful for such other disciplines as geology, paleontology, and archaeology. Extended field trips and studies in remote locations play an important role in field biology and these needs should be met by the NRS, but the backbone of undergraduate education is the normal 3-h laboratory period. Sites close to a campus will naturally receive more use and make a correspondingly high contribution to the NRS.

A.1.3 Administrative Criteria

General. Once the scientific and academic value of a candidate reserve is established, there are a number of administrative criteria that help to establish acquisition priorities.
Criteria. Since it is an NRS objective to have samples of as many habitat types as possible, there is importance in filling NRS habitat voids. There is special importance if a potential acquisition will also fill a habitat void in natural reserves programs administered by other agencies. This is not to imply that the opposite situation—protection "in depth"—is to be avoided. On the contrary, there are advantages to be gained in this. An additional criterion is the balanced growth of the NRS. It is important that the NRS be distributed geographically around the state as well as among the various campuses of the university. Favorability of the terms of acquisition is, of course, an important criterion. Responsiveness to this criterion affects the ability to build the best system with the resources available. Similarly, the ease in administering a site (trespass, maintenance of facilities, etc.) and the availability of maintenance funds will influence its relative priority.

References

Arkley RJ (1954) Soils of eastern Merced County, California. University of California, College of Agriculture. Agricultural Experiment Station, Berkeley

Belbin L (1995) A multivariate approach to the selection of biological reserves. Biodivers Conserv 4:951–963

Bojorquez-Tapia LA, Brower LP, Castilleja G, Sanchez-Colon S, Hernandez M, Calvert W, Diaz S, Gomez-Priego P, Alcantar G, Melgarejo ED, Solares MJ, Gutierrez L, Juarez MD (2003) Mapping expert knowledge: redesigning the monarch butterfly biosphere reserve. Conserv Biol 17:367–379

Bojorquez-Tapia LA, de la Cueva H, Diaz S, Melgarejo D, Alcantar G, Jose Solares M, Grobet G, Cruz-Bello G (2004) Environmental conflicts and nature reserves: redesigning Sierra San Pedro Martir National Park, Mexico. Biol Conserv 117:111–126

Burke I, Lauenroth WK (1993) What do LTER results mean? extrapolating from site to region and decade to century. Ecol Model 67:19–35

Collins SL, Swinton SM, Anderson CW, Benson BJ, Brunt J, Gragson T, Grimm NB, Grove M, Henshaw D, Knapp AK, Kofinas G, Magnuson JJ, McDowell W, Melack J, Moore JC, Ogden L, Porter JH, Reichman OJ, Robertson GP, Smith MD, Vande Castle J, Whitmer AC (2007) Integrative science for society and the environment: a strategic research initiative. Publication #23 of the US LTER Network. LTER Network Office, Albuquerque, NM

Davis FW, Stoms DM, Andelman S (1999) Systematic reserve selection in the USA: an example from the Columbia Plateau ecoregion. Parks 9:31–41

Davis FW, Stoms DM, Church R.L, Okin WJ, Johnson KN (1996) Selecting biodiversity management areas. In: Sierra Nevada Ecosystem Project: final report to Congress, vol. II. Assessments and scientific basis for management options. University of California, Centers for Water and Wildlands Resources, Davis, pp 1503–1528

Davis FW, Stoms DM, Hollander AD, Thomas KA, P. A. Stine, Odion D, Borchert MI, Thorne JH, Gray MV, Walker RE, Warner K, Graae J (1998) The California Gap Analysis Project-final report. University of California, Santa Barbara. Available at: http://www.biogeog.ucsb.edu/projects/gap/gap_rep.html

Engelking LD, Humphries HC, Reid MS, DeVelice RL, Muldavin EH, Bourgeron PS (1994) Regional conservation strategies: assessing the value of conservation areas at regional scales. In: Jensen ME, Bourgeron PS (eds) Eastside forest ecosystem health assessment. Forest service general technical report PNW, Portland, US, pp 208–223

Faith DP, Walker PA (1996) Environmental diversity: on the best-possible use of surrogate data for assessing the relative biodiversity of sets of areas. Biodivers Conserv 5:399–415

Ford LD, Norris KS (1988) The University of California natural reserve system. Bioscience 38:463–470

Franklin JF, Bledsoe CS, Callahan JT (1990) Contributions of the long term ecological research program: an expanded network of scientists, sites, and programs can provide crucial comparative analyses. Bioscience 40:509–523

Hargrove WW, Hoffman FM (2004) A flux atlas for representativeness and statistical extrapolation of the AmeriFlux Network. Oak Ridge National Laboratory Technical Memorandum ORNL/TM–2004/112. Available at http://www.geobabble.ornl.gov/flux-ecoregions

Hessburg PF, Reynolds KM, Keane RE, James KM, Salter RB (2007) Evaluating wildland fire danger and prioritizing vegetation and fuels treatments. For Ecol Manag 247:1–17

Hobbie JE, Carpenter SR, Grimm NB, Gosz JR, Seastedt TR (2003) The US Long Term Ecological Research program. Bioscience 53:21–32

Holland RF (1998) Great Valley vernal pool distribution, photorevised 1996. In: Witham CW, Bauder ET, Belk D, Ferren WR, Ornduff R (eds) Ecology, conservation, and management of vernal pool ecosystems. California Native Plant Society, Sacramento, pp 71–75

Holland RF (2000) The Flying M Ranch: soils, plants, and vernal pools in eastern Merced County. Fremontia 27(28):28–32

Hopkins L (1977) Methods for generating land suitability maps: a comparative evaluation. J Am Inst Plan 43:386–400
Humphries HC, Bourgeron PS, Reynolds KM (2008) Suitability for conservation as a criterion in regional conservation network selection. Biodivers Conserv 17:467–492
Jiang H, Eastman JR (2000) Application of fuzzy measures in multi-criteria evaluation in GIS. Int J Geogr Inf Sci 14:173–184
Jones and Stokes Associates (1987) Sliding towards extinction: the state of California's natural heritage. California Nature Conservancy, San Francisco
Keller M, Schimel DS, Hargrove WW, Hoffman FM (2008) A continental strategy for the National Ecological Observatory Network. Front Ecol Environ 6:282–284
Kirkpatrick JB, Brown MJ (1994) A comparison of direct and environmental domain approaches to planning reservation of forest higher plant communities and species in Tasmania. Conserv Biol 8:217–224
Liu J, Dietz T, Carpenter SR, Alberti M, Folke C, Moran E, Pell AN, Deadman P, Kratz T, Lubchenco J, Ostrom E, Ouyang Z, Provencher W, Redman CL, Schneider SH, Taylor WW (2007) Complexity of coupled human and natural systems. Science 317:1513–1516
Lugo AE, Swanson FJ, González OR, Adams MB, Palik B, Thill RE, Brockway DG, Kern C, Woodsmith R, Musselman R. (2006) Long-term research at the USDA Forest Service's experimental forests and ranges. Bioscience 56: 39–48
Mackey BG, Nix HA, Hutchinson MF, MacMahon JP, Fleming PM (1988) Assessing representativeness of places for conservation reservation and heritage listing. Environ Manag 12:501–514
Malczewski J (1999) GIS and multicriteria decision analysis. Wiley, New York, p 392
Malczewski J (2002) Fuzzy screening for land suitability analysis. Geogr Environ Model 6:27–39
Margules CR, Pressey RL (2000) Systematic conservation planning. Nature 405:243–253
Mead DL (1996) Determination of available credits and service areas for ESA vernal pool preservation banks. In: Witham CW, Bauder ET, Belk D, Ferren JWR, Ornduff R (eds.) Ecology, conservation, and management of vernal pool ecosystems. pp 274–281, California Native Plant Society, Sacramento, California
Menning KM, Erman DC, Johnson KN, Sessions J (1997) Modeling aquatic and riparian systems, assessing cumulative watershed effects, and limiting watershed disturbance. Sierra Nevada ecosystem project. Final report to Congress University of California, Centers for Water and Wildlands Resources, Davis, California, pp 33–51
Michener WK, Bildstein KL, McKee A, Parmenter RR, Hargrove WW, McClearn D, Stromberg M (2009) Biological field stations: research legacies and sites for serendipity. Bioscience 59:300–310
Norris KS (1968) California's Natural Land and Water Reserve System. Bioscience 18:415–417
Peters DP, Groffman PM, Nadelhoffer KJ, Grimm NB, Collins SL, Michener WK, Huston MA (2008) Living in an increasingly connected world: a framework for continental-scale environmental science. Front Ecol Environ 6:229–237
Pressey RL, Ferrier S, Hager TC, Woods CA, Tully SL, Weinman KM (1996) How well protected are the forests of north-eastern New South Wales? Analyses of forest environments in relation to formal protection measures, land tenure, and vulnerability to clearing. For Ecol Manage 85:311–333
Reiner R, Swenson R (2000) Saving vernal pools in the Cosumnes River watershed. Fremontia 27(28):33–37
Reynolds KM, Hessburg PF, Keane RE, Menakis JP (2009) National fuel-treatment budgeting in US federal agencies: capturing opportunities for transparent decision-making. For Ecol Manag 258:2373–2381
Smith DW, Verrill WL (1996) Vernal pool-soil-landform relationships in the Central Valley, California. In: Witham CW, Bauder ET, Belk D, Ferren JWR, Ornduff R (eds) Ecology, conservation, and management of vernal pool ecosystems. pp 15–23, California Native Plant Society, Sacramento, California

Staus NL, Strittholt JR, DellaSala DA (2010) Evaluating areas of high conservation value in western Oregon with a decision-support model. Conserv Biol 24:711–720

Stoms DM (2000) GAP management status and regional indicators of threats to biodiversity. Landscape Ecol 15:21–33

Stoms DM, Borchert MI, Moritz MA, Davis FW, Church RL (1998) A systematic process for selecting representative Research Natural Areas. Nat Area J 18:338–349

Stoms DM, Davis FW, Andelman SJ, Carr MH, Gaines SD, Halpern BS, Hoenicke R, Leibowitz SG, Leydecker A, Madin EMP, Tallis H, Warner RR (2005) Integrated coastal reserve planning: making the land-sea connection. Front Ecol Environ 3:429–436

Stoms DM, Hargrove WW (2000) Potential NDVI as a baseline for monitoring ecosystem functioning. Int J Remote Sens 21:401–407

Stoms DM, McDonald JM, Davis FW (2000) Knowledge-based site suitability assessment for new NRS reserves for the proposed UC Merced campus. University of California, Santa Barbara. Online at http://www.biogeog.ucsb.edu/projects/smer/nrs_report.pdf

Stoms DM, McDonald JM, Davis FW (2002) Fuzzy assessment of land suitability for scientific research reserves. Environ Manag 29:545–558

University of California (1984) NRS acquisition guidelines. Oakland, California (Reprinted in the appendix of this chapter)

U. S. Fish and Wildlife Service (1998) Recovery plan for upland species of the San Joaquin Valley. California, Region 1, Portland, Oregon

Wood LJ, Dragicevic S (2007) GIS-based multicriteria evaluation and fuzzy sets to identify priority sites for marine protection. Biodivers Conserv 16:2539–2558

Zhang X-S, Gao Q, Yang D-A, Zhou G-S, Ni J, Wang Q (1997) A gradient analysis and prediction on the northeast China transect (NECT) for global change study. Acta Botanica Sinica 39:785–799

Forest Conservation Planning

Michael D. White and James R. Strittholt

Abstract This chapter reviews the application of Ecosystem Management Decision Support (EMDS) to conservation planning in a large forest landscape. A significant challenge faced by conservation planners is explicitly defining and mapping values of interest. EMDS is a powerful decision support application development tool that can be used to facilitate explicit and consistent definition of subjective conservation values, which can then be used in a variety of spatial assessments. We describe a case study for the Sierra Checkerboard Initiative where we used EMDS to assess and map three conservation values across a largely forested 4,856 km^2 landscape in the northern Sierra Nevada of California: biodiversity, mature forest connectivity, and passive recreation.

Keywords Forest conservation · Conservation planning · Biodiversity · Forest connectivity · Recreation · Spatial decision support · EMDS

Abbreviations

GIS	Geographic Information Systems
EMDS	Ecosystem Management Decision Support System
HCVF	High Conservation Value Forests
FSC	Forest Stewardship Council
USFS	U.S. Forest Service

M. D. White (✉)
Tejon Ranch Conservancy, PO Box 216, Frazier Park, CA 93225, USA
e-mail: mwhite@tejonconservancy.org

J. R. Strittholt
Conservation Biology Institute, 136 SW Washington Avenue, Suite 202,
Corvallis, OR 97333, USA
e-mail: stritt@consbio.org

K. M. Reynolds et al. (eds.), *Making Transparent Environmental Management Decisions*, Environmental Science and Engineering,
DOI: 10.1007/978-3-642-32000-2_9,

1 Introduction

Conservationists are concerned with the protection of natural resources in the face of human-induced changes. Natural resources targeted for protection are varied and often closely tied to subjective value systems. For example, a focus of the early conservation movement in the United States was preserving clean water, productive soils, and sustainable yields of fish, wildlife, and forest products for human uses (Pinchot 1947). However, at the same time, early preservationists were making non-utilitarian arguments for protection of wild nature for wilderness and spiritual values (Callicott 1990). In recent decades, the targets of conservation have largely shifted to the preservation of biodiversity, ecosystem services, and associated aesthetic values (Chan et al. 2006; Daily et al. 1997; Groves 2003; Margules and Pressey 2000; Swart et al. 2001). Thus, conservationists may be concerned with protection of specific values such as rare species or habitats, or more subjective values such as wilderness experiences or ecological health. Not only are many of these values ill-defined, often no data exist to directly describe them in a spatially-explicit fashion, which is essential for effective conservation planning. In spite of these limitations, conservation planners must integrate available data, often imperfect for the intended application, to describe the distribution of and relationships between conservation targets and values of interest. Furthermore, conservation planning results have to be conveyed in a transparent and understandable fashion to on-the-ground conservation practitioners, funders, and public policy decision-makers, who frequently have little or no technical background. This chapter discusses our experience in applying EMDS as a conservation planning tool. In particular, we discuss its utility in explicitly defining and mapping subjective conservation values for a large, mostly forested region in the northern Sierra Nevada, California.

2 Background

Systematic conservation planning has been practiced since the 1970s (Groves 2003), and has become more sophisticated with the development of new theoretical models for conservation reserve planning, the evolution of geographic information systems (GIS), better spatially-explicit data describing natural resources, and increasingly powerful computing systems. To be effective, conservation planning requires clear definition of targeted conservation values and the metrics used to describe them (Margules and Pressey 2000). However, given the complexity and variability of ecosystems and the subjective human perception of value, it is often difficult to develop broadly-applicable quantitative models to describe many conservation values. Knowledge-based reasoning is well suited for conservation planning because it allows representation of the conceptual factors that contribute to conservation values and the logical relationships between them

(Reynolds 2001, 2003). In knowledge-based models, data used to evaluate the conceptual factors can be specified, and the manner in which available data are used to arrive at a conclusion defined. Incorporating fuzzy logic into knowledge-based models allows imprecise information typical of natural resources science to be used in modeling (Reynolds et al. 2000). Knowledge-based reasoning can be a powerful approach for (1) describing conservation values, in terms of landscape characteristics or conditions (e.g., acres of forest, numbers of special status species, levels of habitat fragmentation), (2) defining the specific data that will be used to evaluate these conditions, and (3) documenting the logical relations between the factors used to describe conservation values and the data used to evaluate them.

EMDS is a knowledge-based decision support application development tool that uses spatial data and maps, and applies hierarchically organized fuzzy logic networks to these data, in a GIS environment. Gordon et al. (2004) found it to be one of the most versatile and comprehensive of over 30 software packages tested for forest assessment and planning. Although EMDS provides a potentially useful tool for conservationists, it has been used relatively infrequently in conservation planning. White et al. (2005a) used EMDS to quantify the distribution of conservation values in the northern Sierra Nevada, California (described in a case study in this chapter). However, we are aware of relatively few conservation planning applications of EMDS (Strittholt et al. 2006, 2007; Staus et al. 2010; Manzuli 2005; Stoms et al. 2000, 2002, Chap. 8, this volume; Humphries et al. 2008).

EMDS has been successfully used to map High Conservation Value Forests (HCVF, Strittholt et al. 2007; Staus et al. 2010). The HCVF concept, which is sometimes referred to as Principal 9 of the Forest Stewardship Council (FSC) forest stewardship guidelines, refers to forests that contain one or more high conservation values (Jennings et al. 2003). High conservation value is defined by criteria indicative of forests with significant biological, environmental, and social values (for examples, see Strittholt et al. 2007; Staus et al. 2010). HVCF is an important concept gaining global acceptance within forest ecosystems not only within the FSC context, but also outside of this specific certification system.

Strittholt et al. (2007) used EMDS to map endangered forests in the Alberta Foothills Ecoregion of Canada, a subset of HCVF that some forest conservation groups consider inappropriate for industrial timber harvest. Areas of high conservation value in the Alberta Foothills were identified as those with a combination of high landscape values and high biodiversity values. High landscape values were defined using degree of landscape integrity and amount of historical impacts from oil and gas development and logging. High biodiversity values were defined using distinct bird, fish, caribou, grizzly bear, forest, and natural heritage conservation values, i.e., concentrations of rare plants and animals. The analysis identified portions of the Alberta Foothills that merited greater protection than received, and allowed an explicit evaluation of how special land management provisions may affect high conservation value forests.

Staus et al. (2010) used EMDS to identify HCVF in western Oregon as part of evaluating proposed changes to the Bureau of Land Management (BLM) portions of the Northwest Forest Plan. Using EMDS, model outputs highlighted areas in western Oregon, including significant BLM lands that were important to protect to maintain stated forest conservation values.

Manzuli (2005) used EMDS within a monitoring and evaluation program for a gas pipeline project to assess the conservation status of the Chiquitano Dry Forest in Bolivia. Manzuli defined conservation opportunities within the study area as a function of four conditions (amount of forest clearings, fuel-wood collection pressure, browsing cattle pressure, and logging pressure), and used various landcover, demographic, and social data sets to evaluate contributions of each condition within a fuzzy logic network. This analysis allowed evaluation of the contributions of both direct and indirect factors on the conservation status of the Chiquitano Dry Forest. It also established a baseline description of conservation status against which future assessments could be compared.

Stoms et al. (2002, Chap. 8) evaluated the suitability of sites for a scientific research reserve using EMDS. Research site suitability is a function of how well a candidate site meets particular scientific, academic, and administrative criteria. However, these criteria can be imprecise (e.g., site is "close" to campus) and the manner in which criteria should be combined and weighted is undefined. The fuzzy knowledge-based approach of EMDS is well-suited to such a problem, and allowed the investigators to explicitly define and evaluate a knowledge-based system that incorporated standard, but imprecise site suitability criteria.

The role of socioeconomic suitability factors when targeting land for conservation purposes was considered by Humphries et al. (2008) using EMDS. In this analysis, EMDS was used to develop suitability rating scenarios for analysis units based on ownership (public vs. private), number of land cover types, and area. These suitability ratings were then converted into "cost" scenarios, and then incorporated into a conservation reserve-selection algorithm.

Strittholt et al. (2006) employed EMDS to carry out a global conservation assessment that emphasized biological (e.g., high wildness and biodiversity values), socioeconomic (e.g., high levels of aboriginal involvement and a low cost to value ratio), and political (e.g., stable government) considerations. The outcome of this EMDS model helped direct conservation philanthropy in several locations in the world that best met the specific desired conditions set forth by the anonymous foundation that commissioned the study.

3 Case Study: Sierra Checkerboard Initiative

We often conduct conservation planning work with partners across large landscapes, much of it in North American forests. These forests provide numerous important ecosystem services that are increasingly threatened by the synergistic effects of timber harvest, mining, rural residential development, road building, altered fire regimes, forest diseases, alien species invasions, pollution, and global climate change (Sierra Nevada Ecosystem Project SNEP 1996; Lindenmayer and Franklin 2002; The Millennium Assessment 2005).

The California Sierra Nevada supports a diverse array of complex and interrelated conservation values. It lies within one of the world's biodiversity hotspots (Mittermeier et al. 1999; Conservation International 2005) and is one of the most floristically diverse regions in North America (Shevock 1996). Nearly 30 % of California's total runoff originates in Sierran mountains (Kattelmann 1996). Its mountain forests are an important source of wood products and store hundreds of millions of metric tons of carbon (Zhu and Reed 2012). Valuable rangelands are located in nearby foothills and intermountain valleys. In addition, the Sierra Nevada provides highly sought-after recreational and scenic resources for millions of people.

Land in the northern Sierra Nevada is characterized by a checkerboard ownership pattern, a legacy of the United States government's granting of alternate square miles to the Central Pacific Railroad during construction of the transcontinental railroad in the 1860s (Duane 1999). Many individuals and private corporations now hold these land grants, which are interspersed with public lands administered by the U.S. Forest Service. This checkerboard ownership pattern, where private industrial timberland is interspersed with public lands important for biodiversity protection and recreational uses, creates great challenges for regional conservation and land management.

Recognizing the threats to conservation values posed by this ownership pattern, the Trust for Public Land (TPL) initiated the Sierra Checkerboard Initiative (White et al. 2005a) to work towards a more sustainable land management and ownership landscape in the region. To prioritize their conservation work in the region, TPL commissioned the Conservation Biology Institute to better quantify and understand the distribution of conservation values across a 4,856 km^2 northern Sierran landscape (Fig. 1).

Working with TPL and other conservation partners in the region, we identified three primary conservation values as the focus of our work–biodiversity, mature forest connectivity, and passive recreation. Our challenge was twofold: to quantify these values in a spatially-explicit fashion using available data, and to spatially integrate the data across the landscape in a way that could be clearly articulated to conservation partners and the general public. We found EMDS ideally suited for this analysis. EMDS allowed us to clearly model the magnitude and distribution of these three conservation values in the Sierra Checkerboard Initiative area. The EMDS logic networks also allowed us to integrate numerous disparate data sets and to easily update the analysis as data sets were refined. Our assumptions and approach were transparent, and the logic networks could be easily modified to evaluate the influence of individual model components.

In our model, an evaluation unit (the most basic landscape unit) supported high conservation values if it supported high biodiversity value, high mature forest connectivity value, *or* high passive recreation value (Fig. 2, Table 1). In NetWeaver, we constructed a hierarchical fuzzy logic network that defined these relations. The three component logic networks under the premise of high

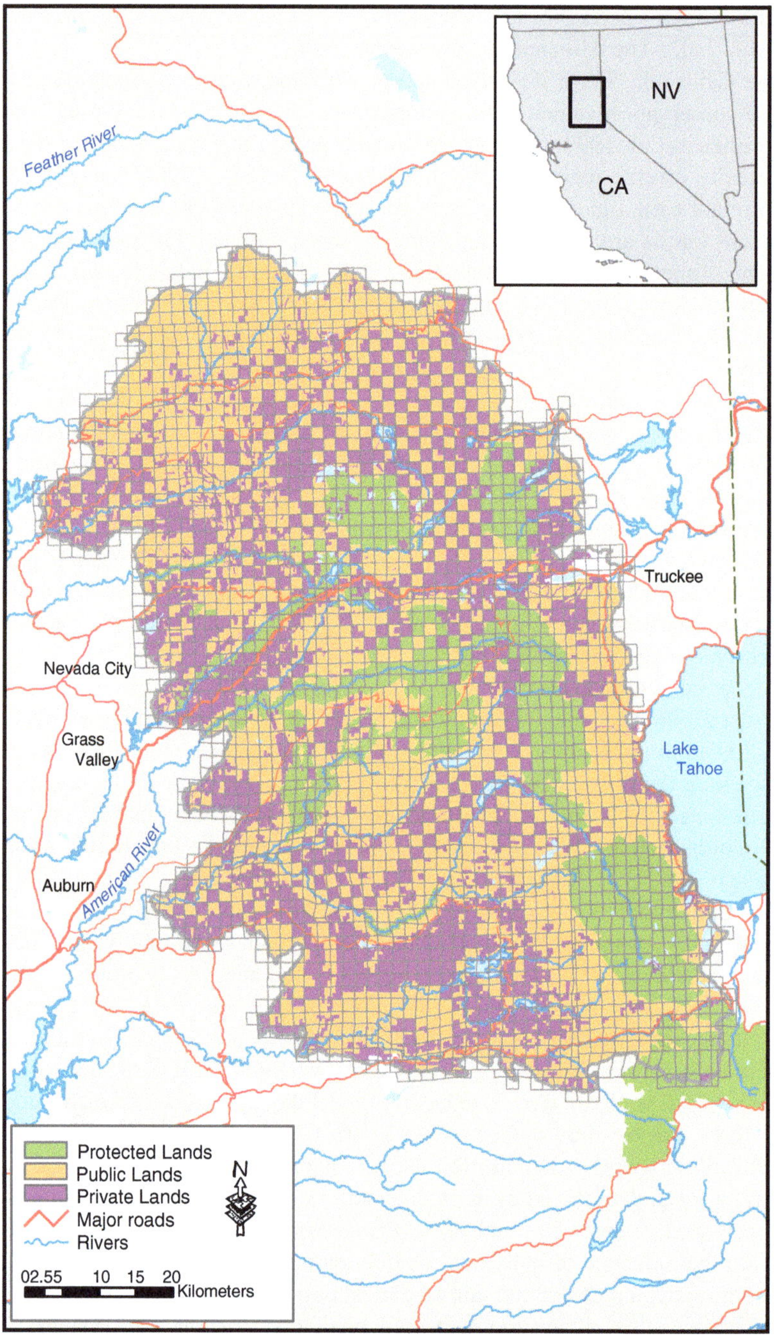

Fig. 1 Location of the Sierra Checkerboard Initiative study area showing the grid of EMDS evaluation units. Protected areas were defined as Wilderness, Wild and Scenic Rivers, and Inventoried Roadless Areas

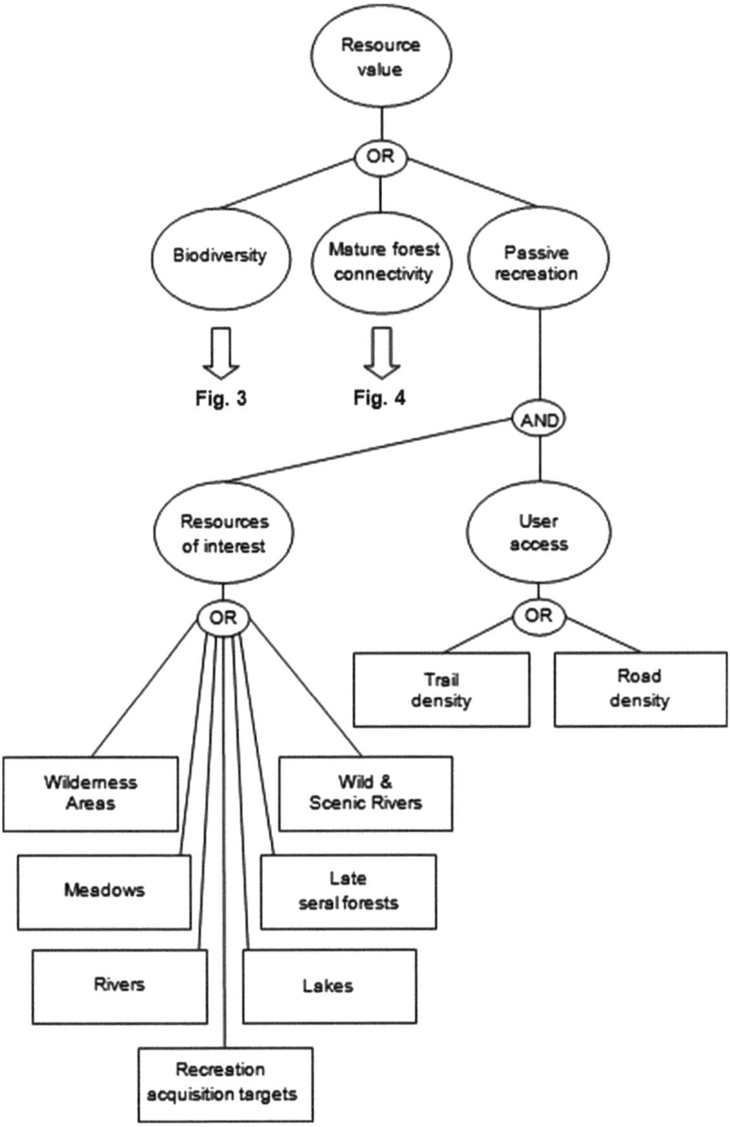

Fig. 2 The resource value network showing the logic for evaluating the passive recreation conservation value of an evaluation unit

conservation value are discussed in the following sections. In all logic networks, circles represent presumed conditions or states, small ovals are logic operators, and boxes represent input data. Evaluation units were Sections of land within the U.S. Public Land Surveying System, generally 2.59 km^2 (or ~ 1 mi^2).

Table 1 Outline of propositions and subpropositions used in the Sierra Checkerboard Initiative model to map resource values in the study area, and the elementary topics used to assess the evidence in support of them

Propositions and subpropositions	Elementary topics
1. The site supports high **biodiversity**	–
A. The site supports high **existing biodiversity**	–
i. The site supports high **terrestrial biodiversity**	–
a. The site has good **landscape condition**	Road density, Roadless Area, Human impact
b. The site supports **under-represented habitats**	Priority vegetation types
c. The site supports **terrestrial special elements**	Sensitive terrestrial species, Late seral forests, Rare edaphic features
ii. The site supports high **aquatic biodiversity**	–
a. The site has high **watershed condition**	Reservoir volume, Roads near streams, Roadless area of watershed, Roads on steep slopes
b. The site supports **aquatic special elements**	Sensitive aquatic habitats, Sensitive aquatic species
B. The site supports high **future biodiversity**	Biodiversity in neighborhood
i. The site supports **potential biodiversity**	Rare edaphic features
a. The site has low **development density**	Relative development
b. The site has high **site productivity**	Site class
2. The site supports high **mature forest connectivity**	–
A. The site supports high existing mature forest connectivity	–
i. The site has low **mature forest fragmentation**	Mean patch size, Number of patches, Mean distance to nearest neighbor, Proportion mature forest, Total core area index
ii. The site has low **mature forest fragmentation in the neighborhood**	Mean patch size in neighborhood, Number of patches in neighborhood, Mean distance to nearest neighbor in neighborhood, Proportion mature forest in neighborhood, Total core area index in neighborhood
B. The site supports high **future mature forest connectivity**	–
i. The site has high **mature forest recovery potential**	–
a. The site has low **development density**	Relative development
b. The site has high **site productivity**	Forest site class
3. The site supports passive recreation	–
A. The site supports recreational **resources of interest**	Wilderness areas, Meadows, Rivers, Lakes, Wild & Scenic Rivers, Late seral forest, Recreation acquisition targets
B. The site has good **user access**	Trail density, Road density

Bolded words correspond to the labels in the logic network figures

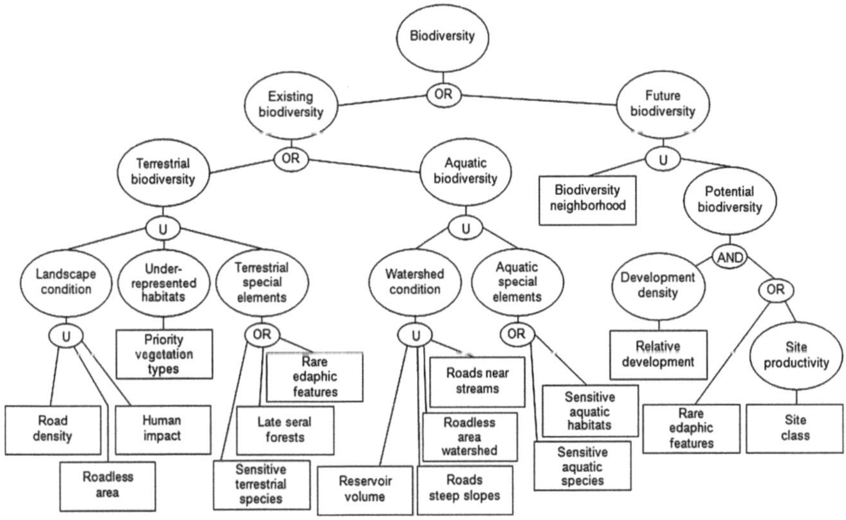

Fig. 3 Biodiversity conservation value network

3.1 Biodiversity Value

The objective of the biodiversity value assessment was to identify areas supporting the rich diversity of flora and fauna in the northern Sierra Nevada. Since comprehensive, fine-scaled data describing biodiversity patterns did not exist for the study area, we used a number of indirect measures that we assumed could serve as surrogates for biodiversity. The interplay of the diverse geology and terrain, range of elevations, climate variation, and variety of aquatic habitats is largely responsible for the characteristic biodiversity of the Sierra Nevada region. Our conceptual model (Fig. 3) recognized that terrestrial and aquatic systems make important biodiversity contributions, and that areas with low existing biodiversity values can recover to support high biodiversity values in the future.

Terrestrial biodiversity value was assessed via three factors: landscape condition, presence of under-represented habitats, and presence of special elements. Areas of high landscape condition (i.e., areas that are relatively unaltered by human modifications such as road building and residential development) often support intact ecosystem processes, allow for species movements through the landscape, and support species that are sensitive to human modifications of the landscape, such as mammalian carnivores (Saunders et al. 1991; Trombulak and Frissell 2000; Brooks et al. 2002; Crooks 2002), and were thus considered to be good predictors of biodiversity. Landscape condition was evaluated using a UNION operator, which averaged the truth values of the data inputs of percent roadless area, road density, and area of human footprint. Under-represented habitats were those not well represented in protected areas in the region (Scott et al. 2001), and were a priority for protection in the northern Sierra. A representation analysis of vegetation communities within protected areas in the region was used to identify under-represented habitats, with high-priority habitats (i.e., those with less than 20 % of their extent in protected areas) or medium-priority habitats (i.e., those with 20–35 % of their extent in protected areas) selected as important contributors to biodiversity value. Special elements were defined as rare species or habitats (e.g., mature forests) or rare edaphic features (e.g., serpentine geology) that contribute to regional biodiversity. The logic network for terrestrial biodiversity also used a UNION operator to evaluate the collective contributions of the three terrestrial biodiversity factors.

Aquatic biodiversity was treated similarly in our logic network but used only two factors, watershed condition and presence of special aquatic elements (aquatic habitat types were not mapped at an adequate resolution to perform a representation analysis). As with terrestrial landscapes, fragmentation and human modifications, such as dams and diversions, can alter aquatic ecosystem processes, degrade habitat quality, and facilitate invasions of non-native species (Poff et al. 1997; Kattelmann 1996). We used data describing road density on steep slopes, road density near streams, percent of watersheds that were roadless, and size (volume) of reservoirs to assess good watershed condition via a UNION operator. Aquatic special elements were defined as species and habitats with regionally restricted distributions that contributed to the unique biodiversity of the region, such as wetlands, fens, and springs. Watershed condition and presence of special aquatic elements were evaluated via a UNION operator.

Although much of the northern Sierra Nevada ecosystem has been historically managed for targets other than biodiversity (U.S. Forest Service 2001), the ecosystem is dynamic, and with appropriate management actions, is capable of recovering conditions that support specific biodiversity values. We wanted to ensure that areas currently supporting low biodiversity values, but that have a high potential for recovery, were not disregarded. Our biodiversity logic network (Fig. 3) therefore included an element for high future biodiversity values, assessed as the UNION of high potential biodiversity values (i.e., biodiversity values resulting from ecosystem recovery) and existing biodiversity values in the surrounding neighborhood (defined as a 5 km^2 area in this analysis). Thus, to support high future biodiversity values, an area must be capable of recovery and have high existing biodiversity values in its immediate vicinity that can serve as a recovery source. Existing biodiversity values were assessed via the logic network described above, and averaged within a 5 km^2 neighborhood around each evaluation unit. The logic network for potential biodiversity assumed that the evaluation unit must have both low development density, and either support rare edaphic features or have high site productivity, as measured by site class. We assumed vegetation growth and recovery rates following disturbance would be positively related to site productivity, and thus site class would be indicative of potential biodiversity.

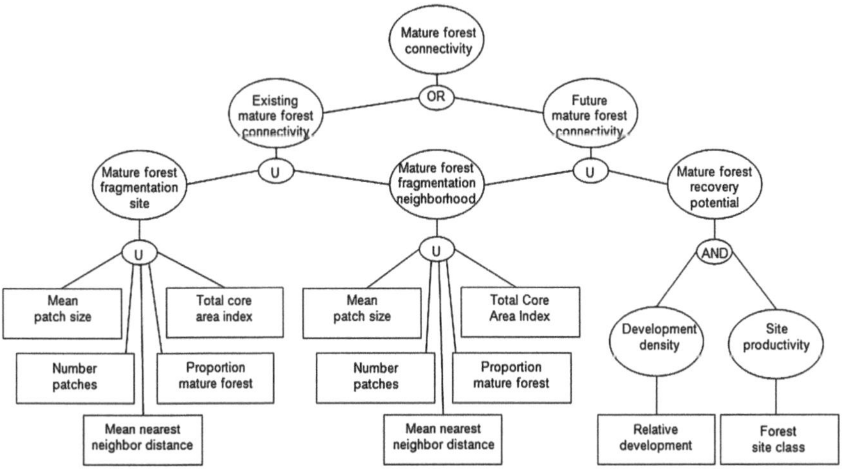

Fig. 4 The mature forest connectivity network showing the logic for evaluating the mature forest connectivity value of an evaluation unit

3.2 Mature Forest Connectivity Value

Mature forests (most trees in a forest plot >28 cm diameter at breast height) and especially the late-successional component (forests with many of its trees >50 cm diameter) provide many important ecosystem functions, wildlife habitat, and benefits to human society (Franklin and Fites-Kaufmann 1996; Graber 1996; Marcot 1997). However, commercially important forest types in the Sierra Nevada, such as west-side mixed-conifer forests, are deficient in mature forest characteristics relative to their pre-settlement conditions (Franklin and Fites-Kaufmann 1996). The loss of mature forests in the northern Sierra Nevada has eliminated habitat and decreased habitat connectivity for associated wildlife species throughout the Sierra Checkerboard Initiative study area. Thus, mature forest connectivity was identified as a conservation value in our assessment. For the purposes of our assessment, we defined mature forests as forest stands with >40 % canopy cover and 28 cm diameter or larger trees.

As with biodiversity value, our NetWeaver model for mature forest connectivity recognized that conservation value can be associated with existing or future mature forest connectivity (Fig. 4). The supposition of high existing mature forest connectivity was assessed as the UNION of the degree of fragmentation of mature forests in the evaluation unit and fragmentation within a 5 km^2 neighborhood around the unit, i.e., the collective contribution of mature forest fragmentation at local and neighborhood scales. At both scales, fragmentation was assessed as the UNION of five fragmentation metrics (proportion of mature forest, mean nearest neighbor distance, total core area index, mean patch size, and number of patches). These fragmentation metrics were calculated by running FRAGSTATS

(McGarigal et al. 2002) on the mature forest data layer. Our model for future mature forest connectivity assumed that an evaluation unit must have both high potential to recover and have mature forests with high connectivity in the immediate vicinity. Thus, future mature forest connectivity was assessed by evaluating the UNION of existing mature forest fragmentation within the 5 km^2 neighborhood and the potential for mature forest recovery within an evaluation unit. Mature forest recovery potential was high where development density (number of dwelling units per unit area) was low and productivity (site class) was high.

3.3 Passive Recreation Value

The northern Sierra Nevada provides a wide variety of recreational opportunities, and demand for these resources is high and increasing (Duane 1996). The U.S. Forest Service (USFS) collects recreation visitor information in eight categories: developed recreation, motorized and mechanized travel, dispersed recreation, winter sports, resorts and cabins, hunting, fishing, and nature study (U.S. Forest Service 2001). The Sierra Checkerboard Initiative focused on passive recreation values, which would exclude motorized and mechanized travel, some winter sports (e.g., downhill skiing), and resorts and cabins.

In our model, for a site to support high passive recreation values it must display both good user access AND support resources of recreation interest (Fig. 2). User access was considered good if there was either good trail OR road access, as measured by trail or road density. We defined resources of high recreation interest as rivers, lakes, Wild and Scenic Rivers, late seral forests, meadows, Wilderness Areas, and areas identified by public land managers in the region as land acquisition targets. The proposition that a site supported resources of high recreation interest was evaluated by considering the presence of any of the seven resources types via an OR operator.

3.4 Membership Functions and Data

Logic networks terminate in links to elementary topics where data are evaluated. We used fuzzy membership functions to assess *degrees of truth* or evidence (Reynolds 2003) in support of each proposition and subproposition (Table 1). For the majority of variables in our logic networks, we established linear relationships between the maximum and minimum values in the given data set and the full support (+1) and no support (−1) truth values (Fig. 5). For example, the presence of a Wilderness Area designation fully supported the proposition–*supports resources of recreational interest*–in the passive recreation value network (Fig. 2). The corresponding no support value in the membership function was set at the minimum value (no percent Wilderness Area) and the full support value was set to

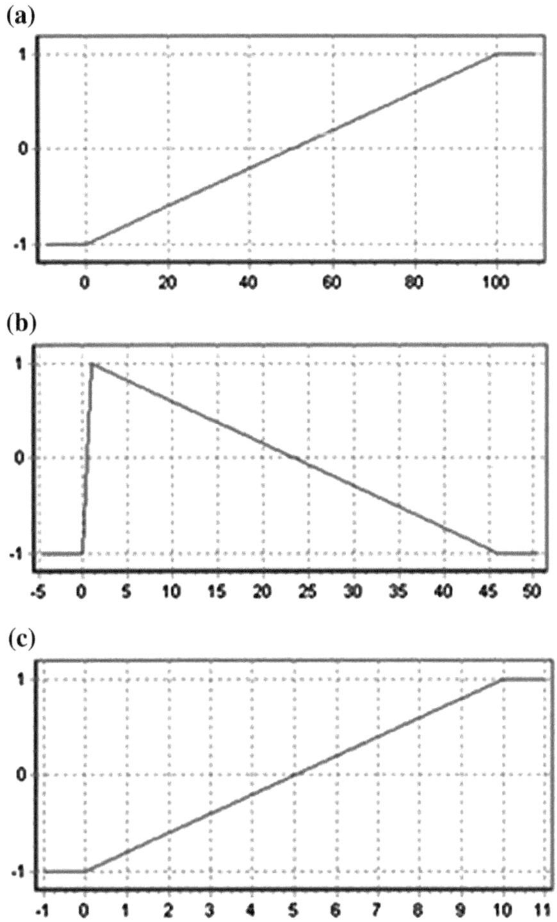

Fig. 5 Examples of membership functions used in the Sierra Checkerboard Initiative: **a** percentage of evaluation unit in a Wilderness Area, **b** number of mature forest patches in an evaluation unit, and **c** percentage of evaluation unit area with lakes. The ordinate in all figures is the level of support for the proposition under evaluation

the maximum value (100 % Wilderness Area, Fig. 5a). Conversely, the number of patches of mature forest was inversely related to the proposition–*low mature forest fragmentation of site*–in the mature forest connectivity value network (Fig. 4). Corresponding end members of the membership function were no support, at the minimum value of 0–no patches of mature forest–and full support at a value of 1 (one patch of mature forest), and again no support at the maximum value in the number of mature forest patches (mature forest is highly fragmented, Fig. 5b).

Where appropriate we also modified the relationship between the range of values in the datasets and their degree of support for each proposition. For example, the presence of lakes supported the proposition–*supports resources of recreational interest*–in the passive recreation value network (Fig. 2). Because lakes are such an attractive recreational resource, we assumed that relatively low lake area in an evaluation unit would provide as much recreational interest as high lake area. Thus, we arbitrarily set the value connoting full support for the proposition at an arbitrary threshold of ≥ 10 % lake area within the evaluation unit (Fig. 5c), and the value for no support at the minimum value of 0 [no percentage area of lake(s)].

Metadata for data used in the analysis were described in White et al. (2005b). All data sets were summarized for each evaluation unit in the study area prior to running EMDS.

3.5 Sierra Checkerboard Initiative Results

While the main focus of this chapter was to describe how we used EMDS to construct models of conservation values for a forest conservation planning application, it is illustrative to review the output maps for the respective conservation values in light of the formulation of their models.

Figure 6 shows the results of the Biodiversity Value component of our model (i.e., the results of the logic network in Fig. 3). The Biodiversity Value logic network was driven by landscape condition (such as presence of roadless areas), known occurrences of special habitats and species for both terrestrial and aquatic environments, presence of under-represented terrestrial habitats (such as mixed conifer forests), and high forest productivity and recovery potential. These features occurred within mid-montane forests on the west slope of the Sierra Nevada along the western half of the study area (Fig. 6). The eastern half of the study area, largely supporting upper montane and alpine habitats, had fewer of these features, and thus lower biodiversity value, as defined in our model.

The Mature Forest Connectivity Value component of our model (i.e., logic network in Fig. 4) displayed a pattern of higher values in the western half of the study area and lower values in the eastern half (Fig. 7). The western half of the study area supports a greater proportion of mature forest, configured in large patches, and with greater core area, than does the eastern half of the study area, and these forests exhibit relatively high productivity and recovery potential. The upper montane forests in the eastern half of the study area tended to be more fragmented than in the western half as a result of a naturally fragmented condition along the crest of the Sierra Nevada, where significant amounts of exposed granite occur, and where urban and rural development near Truckee and the Tahoe basin were evident.

Passive Recreation Value (i.e., the logic network in Fig. 2) exhibited a different pattern than either Biodiversity or Mature Forest Connectivity Values. Higher Passive Recreation Values occurred in the eastern half of the study area (Fig. 8). This result was largely driven by the greater availability of recreation resources (e.g., rivers, lakes, meadows, Wilderness Areas) along the crest of the Sierra Nevada.

The result for Resource Value, which combined Biodiversity Value, Mature Forest Connectivity Value, and Passive Recreation Value (Fig. 2), showed high

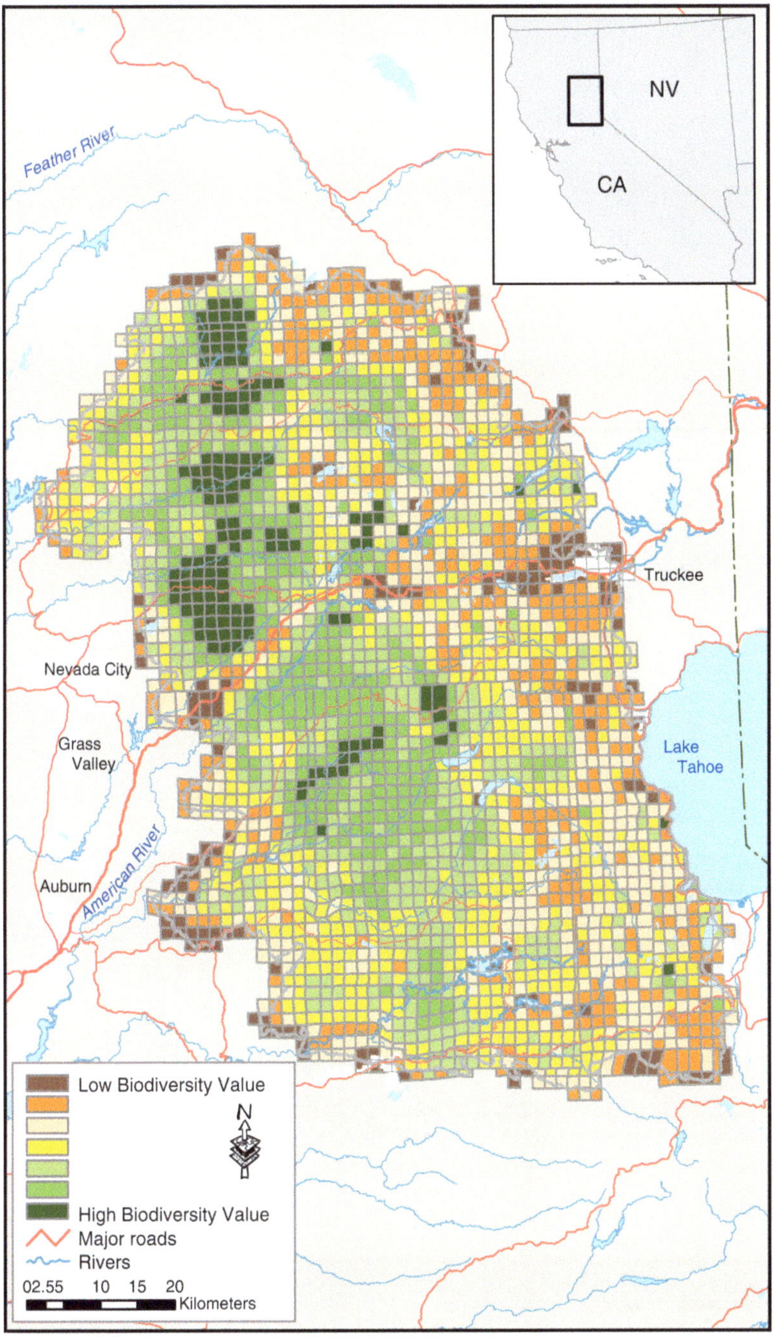

Fig. 6 EMDS results for the biodiversity value

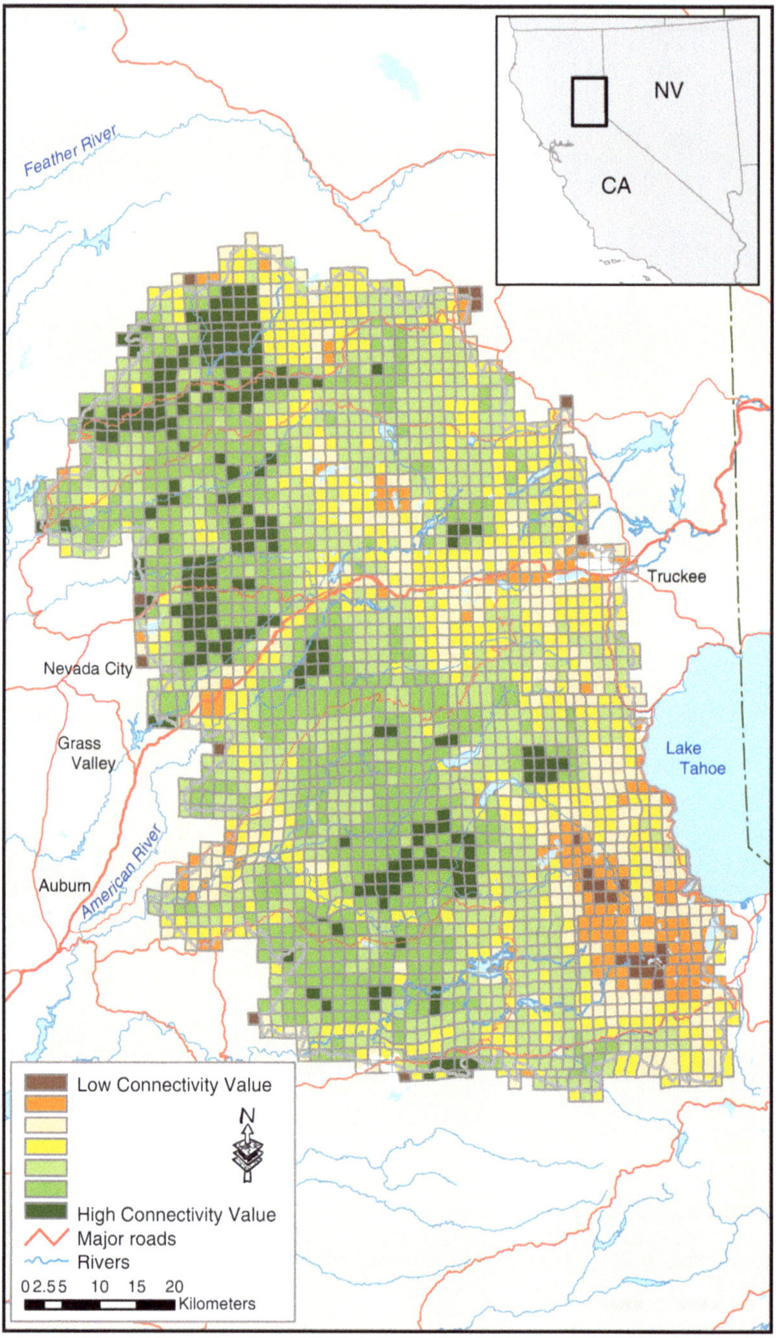

Fig. 7 EMDS results for mature forest connectivity value

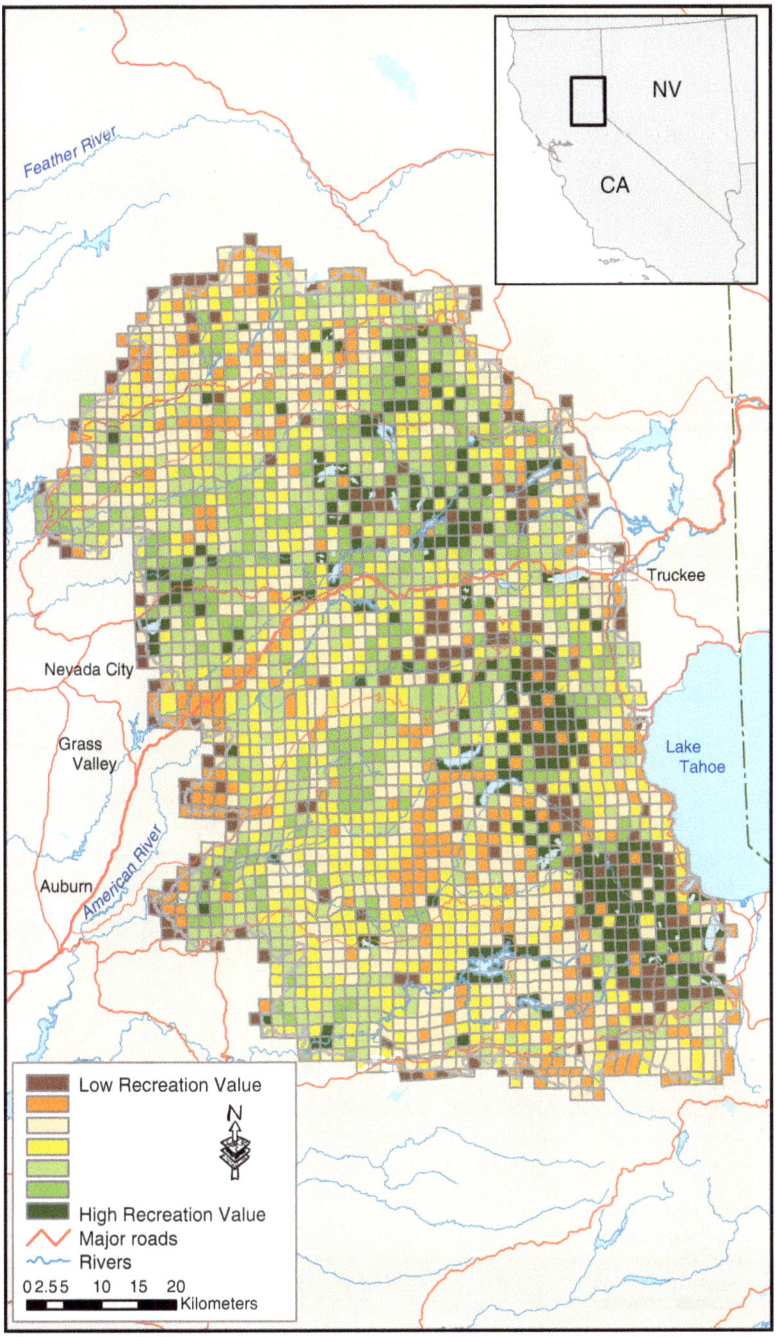

Fig. 8 EMDS results for passive recreation value

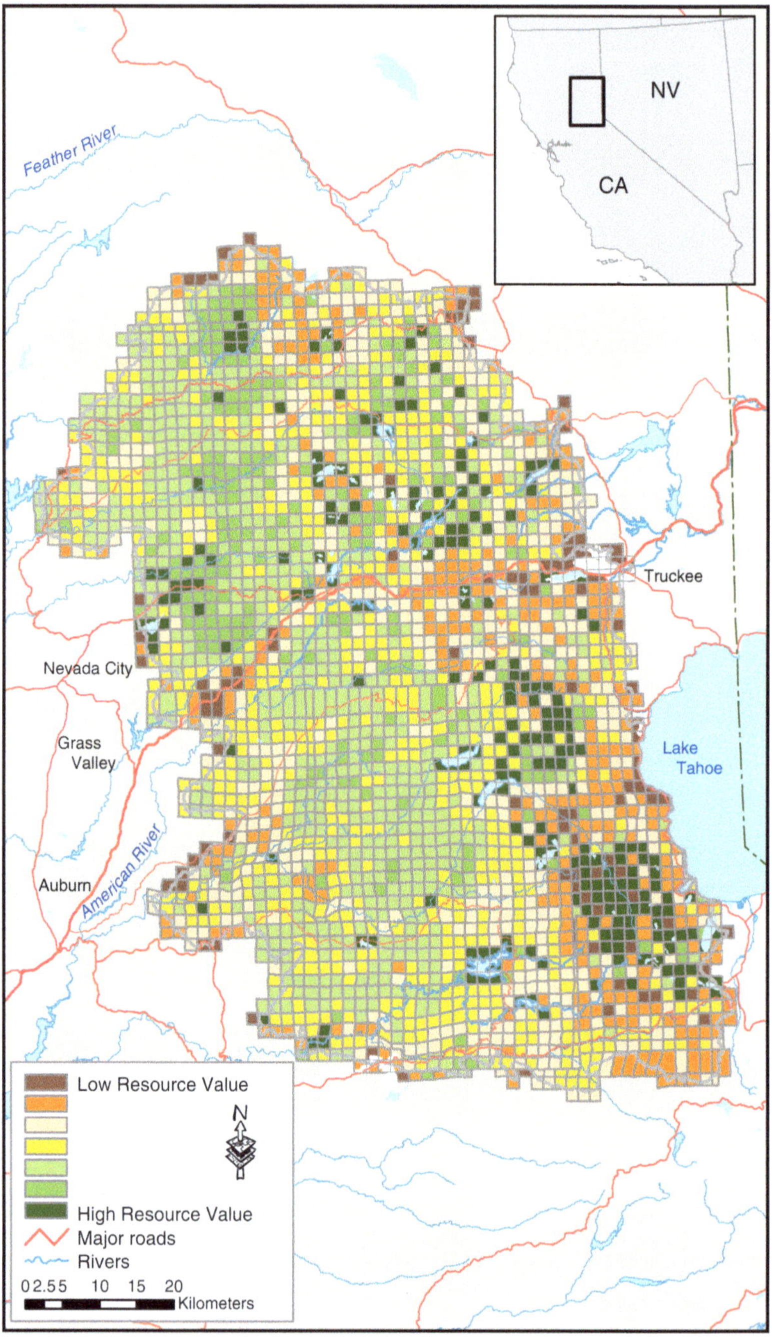

Fig. 9 EMDS results for resource value

values in the eastern (driven by Passive Recreation Value) and western halves (driven by Biodiversity and Mature Forest Connectivity Values) of the study area (Fig. 9). These results were used by TPL to identify and prioritize geographic areas supporting individual conservation values, or their specific elements. Conservation strategies were developed for the Sierra Checkerboard Initiative (White et al. 2008) to address specific values in priority areas. For example, privately-owned landscape areas or watersheds that were currently in good condition could be prioritized for fee simple purchase or could be considered for conservation easements. In other areas, for example, on actively managed forest lands, management agreements could be negotiated with owners who possess lands that currently display a low to moderate conservation value, but with good chance of recovery.

In developing conservation strategies, we used the results of our EMDS models to identify portions of the study area that supported specific conservation values, or key elements of conservation values (e.g., high integrity watershed basins). For these lands, we developed desired future conditions to maintain or enhance these values, identified potential conservation implementation strategies, and potential funding mechanisms to achieve these outcomes. The results of the EMDS analysis were used by TPL and their partners to clearly explain the rationale for their decisions and to garner support for the Sierra Checkerboard Initiative.

4 Conclusions

EMDS is a useful tool for conservation planners, allowing them to explicitly design logic models to define conservation values, the interrelations between factors contributing to conservation value, and the influence of specific data sets in determining the results. We have successfully applied EMDS in several conservation planning applications that demonstrate its utility. In the Sierra Checkerboard Initiative, we used EMDS to define and integrate three distinct conservation values: biodiversity, mature forest connectivity, and passive recreation. This approach had a number of important advantages: (1) we were able to apply our conceptual model regarding conservation values to a knowledgebase; (2) the knowledgebase linked our concept with spatial data representing conservation values in a transparent manner; (3) we produced spatially explicit maps of the results with intuitive results; (4) our logic models were defined such that our conservation partners and end-users could understand our assumptions, model structure, and results; (5) the influence and interrelationships of specific variables within the knowledgebase could be understood by all observers; and (6) the use of EMDS allowed us to easily integrate disparate data sets across a large and complex landscape.

Acknowledgments Funding for the Sierra Checkerboard Initiative was provided by grants from the Richard and Rhoda Goldman Fund, the Bella Vista Foundation, and The Sierra Fund to the Trust for Public Land. Keith Reynolds provided valuable guidance with EMDS and NetWeaver software. Members of the Sierra Checkerboard Initiative Science Advisory Panel—Michael Barbour, Frank Davis, Don Erman, David Graber, Bob Heald, and Bill Zielinski provided valuable input and advice. We thank Sue Britting for her contributions to the project. Wayne Spencer provided useful comments on this manuscript. We thank the Trust for Public Land's Sierra Checkerboard Initiative team—Dave Sutton, Elise Holland, Robin Park, Carl Somers, Trish Strickland, and Dan Martin for their support.

References

Brooks TM, Mittermeier RA, Mittermeier CG, da Fonseca GAB, Rylands AB, Konstant WR, Flick P, Pilgrim J, Oldfield S, Magain G, Hilton-Taylor C (2002) Habitat loss and extinction in the hotspots of biodiversity. Conserv Biol 6:91–100

Callicott JB (1990) Whither conservation ethics? Conserv Biol 4:15–20

Chan KMA, Shaw MR, Cameron DR, Underwood EC, Daily GC (2006) Conservation planning for ecosystem services. PLoS Biol 4(11):e379. doi:10.1371/journal.pbio.0040379

Conservation International (2005) Biodiversity hotspots revisited. Center for Applied Biodiversity Science at Conservation International. http://www.biodiversityhotspots.org/xp/hotspots

Crooks KR (2002) Relative sensitivities of mammalian carnivores to habitat fragmentation. Conserv Biol 16:488–502

Daily GC, Alexander S, Ehrlich PR, Goulder L, Lubchenco J, Matson PA, Mooney HA, Postel S, Schneider SH, Tilman D, Woodwell GM (1997) Ecosystem services: benefits supplied to human societies by natural ecosystems. Issues Ecol 2:1–18

Duane T (1996) Recreation in the Sierra. In: Sierra Nevada ecosystem project, final report to Congress, vol II, Assessments and scientific basis for management options. University of California at Davis, Centers for Water and Wildland Resources. http://ceres.ca.gov/snep/pubs/

Duane T (1999) Shaping the Sierra: nature, culture, and conflict in the changing West. University of California Press, Berkeley

Franklin JF, Fites-Kaufmann JA (1996) Assessment of late successional forests in the Sierra Nevada. In: Sierra Nevada ecosystem project, final report to Congress, vol II, Assessments and scientific basis for management options. University of California at Davis, Centers for Water and Wildland Resources. http://ceres.ca.gov/snep/pubs/

Gordon SN, Johnson KN, Reynolds KM, Crist P, Brown N (2004) Decision support systems for forest biodiversity: evaluation of current systems and future needs. Final Report—Project A10 National Commission on Science and Sustainable Forestry. www.ncssf.org

Graber DM (1996) Status of terrestrial invertebrates. In: Sierra Nevada ecosystem project, final report to Congress, vol II, Assessments and scientific basis for management options. University of California at Davis, Centers for Water and Wildland Resources. http://ceres.ca.gov/snep/pubs/

Groves CR (2003) Drafting a Conservation Blueprint: A Practitioner's Guide to Planning for Biodiversity. Island Press, Washington, DC

Humphries HC, Bourgeron PS, Reynolds KM (2008) Suitability for conservation as a criterion in regional conservation network selection. Biodivers Conserv 17:467–492

Jennings S, Nussbaum R, Judd N, Evans T (2003) The high conservation value forest toolkit, 1st edn. Proforest. http://hcvnetwork.org/

Kattelmann R (1996) Hydrology and water resources. In: Sierra Nevada ecosystem project, final report to Congress, vol II, Assessments and scientific basis for management options. University of California at Davis, Centers for Water and Wildland Resources. http://ceres.ca.gov/snep/pubs/

Lindenmayer DB, Franklin JF (2002) Conserving forest biodiversity: a comprehensive multiscaled approach. Island Press, Washington, DC

Manzuli AG (2005) Knowledge-based monitoring and evaluation system of land use: assessing the ecosystem conservation status in the influence area of a gas pipeline in Bolivia. Doctoral dissertation Mathematisch-Naturwissenschaftlichen Fakultäten der Georg-August-Universität Göttingen, Germany

Marcot BG (1997) Biodiversity of old forests of the West: a lesson from our elders. In: Kohm KA, Franklin JF (eds) Creating a forestry for the 21st century: the science of ecosystem management. Island Press, Washington, DC

Margules CR, Pressey RL (2000) Systematic conservation planning. Nature 405:243–253

McGarigal K, Cushman SA, Neel MC, Ene E (2002) FRAGSTATS: spatial pattern analysis program for categorical maps. Computer software program produced by the authors at the University of Massachusetts, Amherst. Available at www.umass.edu/landeco/research/fragstats/fragstats.html

Millennium Assessment (The) (2005) Millennium ecosystem assessment synthesis report. www.millenniumassessment.org/en/index.aspx

Mittermeier RA, Meyers N, Mittermeier CG (1999) Hotspots: earth's biologically richest and most endangered terrestrial ecoregions. CEMEX, Mexico City

Pinchot G (1947) Breaking new ground. Brace and Company, Harcourt, New York

Poff NL, Allan JD, Bain MB, Karr JR, Prestegaard KL, Richter BD, Sparks RE, Stromberg JC (1997) The natural flow regime: a paradigm for river conservation and restoration. Bioscience 47:769–784

Reynolds KM (2001) Fuzzy logic knowledge bases in integrated landscape assessment: examples and possibilities. General technical report PNW-GTR-521. U.S. Department of Agriculture, Forest Service, Pacific Northwest Research Station. Portland, OR

Reynolds K (2003) Ecosystem management decision support extension. U.S. Department of Agriculture, Forest Service, Pacific Northwest Research Station. Portland, OR

Reynolds KM, Jensen M, Andreasen J, Goodman I (2000) Knowledge-based assessment of watershed condition. Comput Electron Agric 27:315–333

Saunders DN, Hobbs RJ, Margules CR (1991) Biological consequences of ecosystem fragmentation: a review. Conserv Biol 5:18–32

Scott JMFW, Davis RG, McGhie RG, Wright RG, Groves C, Estes J (2001) Nature reserves: do they capture the full range of America's biological diversity? Ecol Appl 11:999–1007

Shevock JR (1996) Status of rare and endemic plants. In: Sierra Nevada ecosystem project, final report to Congress, vol II, Assessments and scientific basis for management options. University of California at Davis, Centers for Water and Wildland Resources. http://ceres.ca.gov/snep/pubs/

Sierra Nevada Ecosystem Project (SNEP) (1996) Sierra Nevada ecosystem project, final report to Congress, vol I, Assessment summaries and management strategies. University of California at Davis, Centers for Water and Wildland Resources

Staus N, Strittholt JR, DellaSala DA (2010) Evaluating high conservation value in western Oregon, USA, with a decision support model. Conserv Biol 24:711–720

Stoms DM, McDonald JM, Davis FW (2000) Knowledge-based site suitability assessment for new NRS reserves for the proposed UC Merced campus. University of California, Santa Barbara. http://www.biogeog.ucsb.edu/projects/snner/nrs_report.pdf

Stoms DM, McDonald JM, Davis FW (2002) Fuzzy assessment of land suitability for scientific research reserves. Environ Manage 29:545–558

Strittholt JR, Frost PA, Staus N, Rustigian H, Heilman G Jr (2006) A global conservation prioritization assessment using EMDS. A special report for the Full Circle Foundation, Hong Kong

Strittholt JR, Staus N, Heilman G Jr, Bergquist J (2007) Mapping high conservation value and endangered forests in the Alberta Foothills using spatially explicit decision support tools. Special report for Limited Brands. http://consbio.org/what-we-do/mapping-high-conservation-value-and-endangered-forests-in-the-alberta-foothills-using-spatially-explicit-decision-support-tools?searchterm=alber

Swart JAA, van der Windt HJ, Keulartz J (2001) Valuation of nature in conservation and restoration. Restor Ecol 9(2):230–238
Trombulak SC, Frissell CA (2000) Review of the ecological effects of roads and terrestrial and aquatic communities. Conserv Biol 14:18–30
U.S. Forest Service (USFS) (2001) Sierra Nevada forest plan amendment, final environmental impact statement. United States Department of Agriculture, Forest Service, Pacific Southwest Region, Jan
White MD, Heilman GE, Stallcup JA (2005a) Science assessment for the Sierra checkerboard initiative. Prepared for the trust for public land, July. http://consbio.org/files/sierra-assessment-p1-report.pdf
White MD, Heilman GE, Stallcup JA (2005b) Online technical appendix to the science assessment for the Sierra checkerboard initiative. Prepared for the trust for public land, July. http://consbio.org/what-we-do/science-assessment-for-the-sierra-checkerboard-2/full-metadata-index/science-assessment-for-the-sierra-checkerboard
White MD, Heilman GE, Budge NA (2008) Conservation stratgey for implementing the Sierra checkerboard initiative. Prepared for the trust for public land, Jan. http://consbio.org/files/Final%20Sierra%20Checkerboard%20Conservation%20Strategy.pdf
Zhu Z, Reed BC (eds) (2012) Baseline and projected future carbon storage and greenhouse-gas fluxes in ecosystems of the Western United States. U.S. Geological Survey Professional Paper 1797, p 192. http://pubs.usgs.gov/pp/1797/

Wildlife Habitat Management

Sean N. Gordon, Heather McPherson, Lowell Dickson, Joshua Halofsky, Chris Snyder and Angus W. Brodie

Abstract Wildlife biologists have been designing habitat models for over 50 years; however, the use of the Ecosystem Management Decision Support system (EMDS) in association with modeling is a relatively recent addition to the field. EMDS has proven its usefulness to habitat modeling and evaluation through successful application to a number of large landscape (>5000 km^2) studies. While EMDS cannot be used to model wildlife populations directly, past efforts have incorporated population data, along with a variety of other indicators. Here, we provide an overview of previous modeling efforts, with an emphasis on the northern spotted owl (*Strix occidentalis caurina*). We then review applications of EMDS to wildlife modeling, before delving into a case study. In that study, the Washington State Department of Natural Resources (DNR) used EMDS to assess the impacts of alternative forest management strategies on dispersal habitat for spotted owls. We discuss how expert workshops were used to define three separate EMDS models to assess and score patches of foraging, roosting, and movement habitat across a large landscape. We used the habitat scores to develop a dispersal habitat model outside of EMDS, which incorporated graph theory concepts and a variable resistance landscape surface to assess the connectivity of owl dispersal habitat.

Keywords Wildlife habitat · Habitat management · Northern spotted owl · Dispersal habitat · Spatial decision support · EMDS

S. N. Gordon (✉)
Institute for Sustainable Solutions, Portland State University, P.O. Box 751, SUST, Portland, OR 97207-0751, USA
e-mail: sean.gordon@pdx.edu

H. McPherson · L. Dickson · J. Halofsky · C. Snyder · A. W. Brodie
Washington Department of Natural Resources, 1111 Washington Street SE, Olympia, WA 98504, USA
e-mail: angus.brodie@wadnr.gov

K. M. Reynolds et al. (eds.), *Making Transparent Environmental Management Decisions*, Environmental Science and Engineering,
DOI: 10.1007/978-3-642-32000-2_10,

1 Historical Background

Responsibility for wildlife management in the United States is shared between states and the federal government. The federal Fish and Wildlife Service (USFWS) has the regulatory responsibility for protecting terrestrial and aquatic species, as mandated in the Endangered Species Act (ESA 1973) and other laws. State agencies work within federal guidelines (and supplementary state regulations) to do most of the day-to-day wildlife management. It is within this context that, in 1997, the Washington Department of Natural Resources (DNR) signed a habitat conservation plan (HCP) with the USFWS. The HCP is a relatively recent legal mechanism, requiring land managers to commit to a conservation plan in exchange for certainty that no further requirements will be imposed by the USFWS for the duration of the agreement (70 years in the case of the DNR). The DNR HCP addresses two federally protected terrestrial species, the northern spotted owl (NSO, *Strix occidentalis caurina*) and the marbled murrelet (*Brachyramphus marmoratus*), salmonid habitats and those of a number of other late-successional and old growth associated species of concern. It covers approximately 650,000 hectares of state forest lands in western Washington within the range of the NSO. The DNR's conservation objective is to provide habitat that makes a significant contribution to demographic support, maintenance of species distribution, and facilitates dispersal for NSOs.

By state law, the DNR is required to set a sustainable harvest level for state-owned forest trust lands, which they completed in 2004 using a coarse statewide analysis (WADNR 2004). The state's trust responsibilities support public education in Washington. They are also required to develop detailed tactical forest plans for each HCP planning unit. The DNR began this process with their South Puget Planning Unit, where they manage approximately 145,000 acres within a broader context of private and federal lands (Fig. 1). As specified in the HCP, management on DNR lands should be coordinated with management of federal species conservation areas. Areas within two miles of federal late-successional reserves (LSRs) were reviewed for designation as NSO nesting, roosting, and foraging habitat (NRF), while lands falling between large federal reserves were given special consideration for dispersal habitat. Two areas within the HCP South Puget Planning Unit, the Elbe-Tahoma and Black Diamond, contain approximately 80 % (66,000 acres) of the designated dispersal lands defined in the HCP.

Management efforts preceding the development of the South Puget Forest Land Plan Environmental Impact Statement had revealed a number of limitations to the original HCP dispersal habitat definition. First, it used a simple threshold approach, so that any stand evaluated either meets or does not meet the standard. This approach did not recognize stands meeting several but not all criteria. Second, initial dispersal definitions were considered minimal criteria for meeting the needs of dispersing owls, and more recent thinking questions whether they provide adequate structural complexity, especially to support foraging during dispersal (Buchanan 2004). Finally, the definition was weak on landscape configuration. It required that

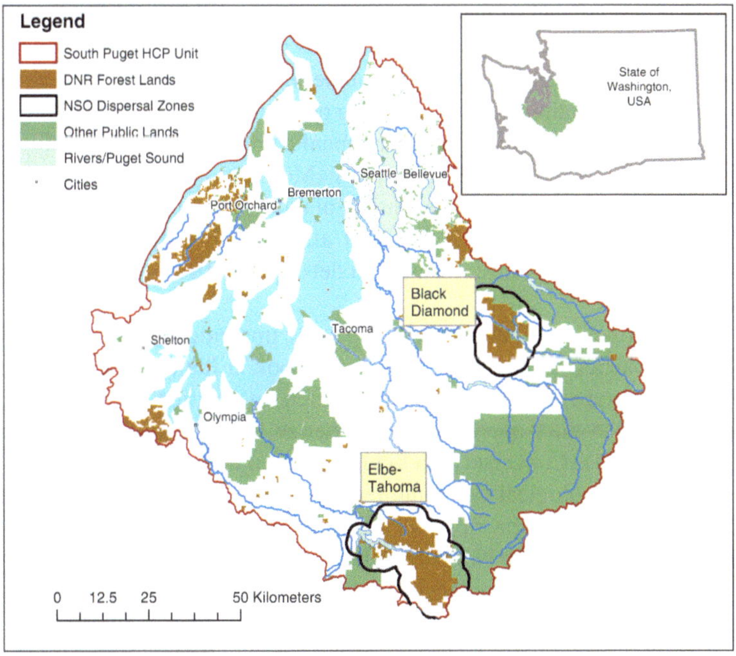

Fig. 1 Northern spotted owl dispersal areas analyzed within the South Puget HCP planning unit

50 % of each watershed unit achieve dispersal standards, but it was not clear how well such a requirement would actually support dispersal. These weaknesses, as well as a mandate for investigating ways to improve owl management in the HCP, caused the DNR to search for new methods to assess habitat for the NSO, especially dispersal habitat. Above all, the DNR needed a standardized tool that could be used to compare results of habitat assessments under multiple management scenarios over time.

2 Modeling Applications in This Problem Domain

2.1 Wildlife Modeling

Wildlife management has been defined as the "manipulation or protection of a population to achieve a goal" (Caughley and Sinclair 1994 p. 1). Predicting how populations might respond to different management options is often complex, because it is dependent on a variety of demographic and habitat relationships. This complexity has led to the development of a wide variety of computational models. In this section, we review some of the basic elements of wildlife modeling and summarize efforts related to the NSO, to place our case study within this broader context and discuss the utility of EMDS in this domain.

Models for wildlife management, as in other disciplines, can fulfill a number of purposes, including defining and formalizing problems, clarifying ideas, organizing concepts, identifying knowledge gaps, communicating information, developing and testing hypotheses, and making predictions (Morrison et al. 2006). Researchers and managers have developed a wide variety of modeling approaches and structures to meet this range of objectives. Roloff et al. (2001) broadly classified wildlife models into three categories: habitat models, population models, and spatial population models.

A further distinction in habitat modeling is the implementation of coarse versus fine filter approaches. Coarse filter approaches look at the distribution of ecological community types over a landscape, often in relation to some historical baseline, while fine filter approaches characterize habitat availability for specific species (Noss 1987). Conversely, individual species are sometimes used as indicators of broader ecological communities or habitat types, as has been the case with the NSO, often thought of as an indicator species for the condition of old-growth forests in the Pacific Northwest (Dawson et al. 1987; Lee 1985; WADNR 2010b). The indicator species concept has been recently criticized, and the validity of various refinements is hotly debated (e.g., see Cushman et al. 2010; Lindenmayer et al. 2002; MacNally and Fleishman 2004; Simberloff 1998; Wiens et al. 2008).

Probably the earliest type of wildlife model developed was the qualitative habitat matrix, which simply relates certain habitat types to the presence/absence of certain species. Such habitat "crosswalks" typically use either a binary classification (habitat/non-habitat) or a few classes (high/medium/low). Although simplistic, this approach is widely used because assessments attempt to include several species across large landscapes (Hulse et al. 2002; Johnson and O'Neil 2001).

Another early modeling approach focuses on population numbers rather than habitat. Such numerical population models predict future population numbers based on current population and simple reproductive and survival functions but without reference to available habitat. The Leslie matrix is one of the best known examples (Leslie 1945). One of the earliest management plans for the NSO was based on an aspatial population model (U.S. Department of Agriculture 1984). While more attention is now given to spatial habitat models, a few important population modeling tools continue to be available (e.g., see Possingham and Davies 1995; Lacy 2000; Applied Biomathematics 2011).

A second generation of NSO modeling incorporated meta-population theory by combining simple habitat assumptions with stochastic models of demography and dispersal. These models assumed a certain percentage of the landscape to be suitable habitat, which then influenced the success and survivorship of juvenile dispersing owls. Using a sensitivity analysis, Lande (1988) found that the probability of successful dispersal was the second in importance of six basic life history parameters, and integrated this factor into his model using unique terms for the proportion of the region in suitable habitat, and the number of territories a dispersing juvenile might search. Doak (1989) further separated dispersal into within- and between-habitat cluster dispersal.

This increasing spatial awareness was integrated into a second major conservation strategy for the NSO, which called for larger habitat blocks (each large enough for 20 territories), spaced over the landscape, so that they were no more than 12 miles apart (Thomas et al. 1990). Little was known about dispersal needs, but the associated "50-11-40 rule" assumed that maintaining at least 50 % of the landscape between habitat blocks in trees $\geq 11''$ in diameter with ≥ 40 % canopy cover would be sufficient. The models stemming from this plan were the first to incorporate a specific spatial structure for the landscape into a dispersal model, although in an idealized fashion where the size and spacing of habitat clusters for each model run were uniform (Lamberson et al. 1994). A few environmental groups successfully blocked this "Thomas Strategy" in court, due largely to arguments over estimates of population demographics, but also involving the unrealistic modeling assumptions of a uniform landscape and maximally efficient dispersal patterns (Harrison et al. 1993).

Schumaker (1996) took another step towards realism by beginning with a raster map of binary predicted habitat (suitable/unsuitable) over the Northwest Forest Plan area (ROD 1994). He was also the first to test the applicability of more abstract landscape metrics by comparing them to results from a dispersal simulation. Schumaker's individual-based dispersal model has continued to evolve and be applied to a wide range of species. The first model version, PATCH, evolved into the current version called HexSim (Schumaker et al. 2004).

Akçakaya and Raphael (1998) also began with the Northwest Forest Plan habitat map, which they aggregated to 5.7 km^2 mapping units and identified areas of sufficient habitat to support one or more NSO home ranges. They applied the commercialized RAMAS GIS spatial meta-population software to NSO population modeling and modeled dispersal as a function of population size, distance, and patch size. Their research focus was on the effect of uncertainty in life history parameters, and they ran the RAMAS model using a range of life history inputs. In contrast to Lande (1988), they found that neither total NSO abundance nor population viability was sensitive to dispersal rate; however, survival was not integrated with dispersal rate.

While some NSO studies focused on population numbers, others have relied on habitat as a surrogate for population numbers. A further step in modeling complexity from simple habitat matrices includes additional detail on habitat features and more quantitative assessment. This sort of modeling, now widely used in wildlife habitat regulation and management, is referred to as habitat suitability or effectiveness modeling (USFWS 1981). Habitat suitability indices (HSI) score specific habitat features (e.g., number of trees $>20''$ diameter) on a common scale and combine scores into an overall suitability index. Individual indicators are typically combined via one of three types of functions: evaluating indicators in terms of a most limiting factor (minimum), as equal and compensating factors (average), or as cumulative factors (sum). Often indicators are grouped by the species life history need they meet (e.g., nesting, roosting, or foraging habitat indicators). Habitat effectiveness models (or habitat evaluation procedures, as they are sometimes called) generally incorporate HSI-like habitat quality measures, along with habitat quantity measures to derive estimates or indices of population carrying capacity (Roloff et al. 2001).

HSIs can be developed from expert knowledge or derived empirically. An empirical approach, which has been recently applied to the NSO, is referred to as occupancy modeling; i.e., taking known nesting sites and determining which habitat factors can best predict them. Zabel et al. (2003) worked with a biological team to create six different habitat maps in northern California and 27 statistical model variations. They found that the official habitat map performed poorly, while the best model performance occurred at the core home range level (200 ha), combining different relationships between nesting/roosting and foraging habitat quantities. For the first major monitoring report on the Northwest Forest Plan (ROD 1994), Lint (2005) examined both NSO population trends and changes in habitat area. Nesting habitat suitability was predicted based on occupancy data using the Biomapper software (Hirzel et al. 2004). In addition to reporting habitat quantities by administrative units, they also calculated a number of landscape metrics using the Subdivision Analysis extension for ArcView (Lang 2004) and FRAGSTATS (McGarigal and Marks 1995) as part of the Patch Analyst extension for ArcView GIS (Rempel and Carr 2003). Available data were insufficient to model dispersal habitat suitability, so they used Thomas Strategy assumptions of 11-inch tree diameter and 40 % canopy cover.

As part of the Oregon Coastal Landscape Analysis and Modeling study, McComb et al. (2002) were first to explore possible effects of different future forest management options on NSO habitat in a spatially-realistic manner. Their HSI, constrained by variables in a forest projection model, consisted of tree densities in a variety of size classes. Their 25 m^2 mapping unit, considerably finer-grained than past efforts, employed a moving window approach, and tested seven possible models at four different scales (from 0.56 to 1810 ha).

Finally, in the most comprehensive spotted owl modeling effort to date, Sutherland et al. (2007) combined landscape modeling with multiple scales of habitat assessment and a population model for a study designed to support the Canadian Spotted Owl Recovery Team in British Columbia. They used the cell-based SELES landscape dynamics system (Fall and Fall 2001) to project alternative management scenarios, which were evaluated for NSO habitat using a multi-stage process. The habitat process began with a site-level HSI, which was then evaluated at a broader scale for viable owl territories. These territories were then analyzed for spatial connectivity using concepts from graph theory applied to the landscape of habitat suitability ratings (Urban and Keitt 2001). A meta-population model was then applied based on territories and connectivity, and outputs were combined within a Bayesian belief network to rank potential habitats.

2.2 EMDS Applications in This Problem Domain

Although this chapter describes the first application of EMDS (Reynolds et al. 2002) specifically for the NSO, it has been used in a variety of wildlife habitat-related projects. Table 1 summarizes a number of such modeling applications, which we discuss in more detail below.

One of the first applications to use EMDS assessed watershed condition under a newly established regional plan for monitoring federal forestlands in the Pacific Northwest, USA (Reeves et al. 2004). The monitoring program has used EMDS since 2001, which included a 10-year assessment of Northwest Forest Plan effectiveness (Gallo et al. 2005). These EMDS models have focused on aquatic habitat conditions for anadromous fish species (salmonids) because of their threatened status under the federal Endangered Species Act; however, models are also designed to evaluate aquatic and riparian conditions more generally. Because of this broad assessment objective and the inherent difficulties of developing statistical models for anadromous fish habitat, the program has relied on an expert workshop approach for model development and parameterization (Gordon and Gallo 2011). EMDS models have also been developed for other watershed assessment work, but from a regulatory rather than management perspective, including for the US Environmental Protection Agency (Reynolds et al. 2000), and for the state of California's North Coast Watershed Assessment Program (Dai 2004; Walker 2007). Extending assessments to prioritize watersheds for restoration, by incorporating feasibility and efficacy considerations, has been demonstrated as well (Reynolds and Peets 2001).

EMDS has also been used for fish and wildlife habitat evaluation as part of broader assessments. For example, Humphries et al. (2008) used EMDS to score the suitability of land for conservation reserves in the Interior Columbia Basin USA. Given the large planning area (58 million ha), they used a very coarse metric of habitat based on 35 vegetated cover types. Instead of evaluating specific habitat attributes (e.g., canopy cover), they considered disturbances, including road density, vegetation change, and disturbance regime changes. They also incorporated a basic spatial component by evaluating the size of each polygon, and where small, they integrated the scores of its neighbors. To derive optimized reserve networks, a measure of uncertainty was introduced by combining the EMDS polygon suitability scores with 15 potential acquisition cost scenarios using site-selection software.

White et al. (2005, and Chapter 9) used EMDS to identify conservation priorities in the 620,000 ha Sierra Nevada Checkerboard Initiative landscape (mixed public and private lands) west of Lake Tahoe, California, USA. For each 2.56 km^2 (1 mi^2) analysis unit, they assessed a large number of attributes ($\sim$50) pertaining to biodiversity, mature forest connectivity, recreation, and threats. For biodiversity, they combined coarse-filter (vegetation communities) and fine-filter (numbers of sensitive species) indicators for both aquatic and terrestrial habitats. Mature forest connectivity was evaluated using a variety of metrics calculated for each

Table 1 Key aspects of EMDS applications related to wildlife management

Application	Extent	Analysis units	Types of indicators	Criteria source(s)	Spatial metrics	Sensitivity/ Uncertainty
NWFP watershed condition (Gallo et al. 2005)	NW Forest Plan area (100,000 km^2)	Subwatersheds (40–120 km^2)	- watershed-level vegetation and roads - stream reach level physical and biological factors	Expert knowledge and interpretation of the literature	None	None
Desert tortoise (Heaton et al. 2008)	Army training center and surroundings (20,000 km^2)	Public land survey sections (2.59 km^2)	- habitat - population - threats	(Not specified)	Various proximity metrics	6 model structure alternatives
US Forest Indicators (Reynolds et al. 2008)	United States	Forest Service regions Ecoregions Counties States	Coarse filter changes in forest types to fine filter metrics of endangered species listings and population counts	- Mostly fairly recent history - A few historical reference conditions - Focus on trends	None	None
R1 National Forest Planning (Jensen et al. 2009)	3 Forest Service planning zones (580,000 km^2)	Ecological subsections (100–200 km^2) Subwatersheds (40-120 km^2) Counties	- habitat - population - threats	Relative to dataset (25/75th percentiles)	Patch size, interior habitat area	3 different prioritization objectives
Land conservation suitability (Humphries et al. 2008)	Interior Columbia River Basin (56,000 km^2)	17,227 biophysical units (3-6,943 km^2)	35 naturally vegetated cover types	Absolute cover type targets	Polygon size, neighborhood suitability	15 land cost scenarios
Sierra Nevada Checkerboard (White et al. 2005)	Central Sierra Nevada mixed ownership landscape (6,192 km^2)	Public land survey sections (2.59 km^2)	$\sim$50 pertaining to biodiversity, mature forest connectivity, recreation, and threats	Relative to dataset (Min/Max)	5 km buffer for neighborhood characteristics	None
Western OR High Conservation Value Forest (Staus et al. 2010)	BLM Western OR Plan Revision area (77,000 km^2)	Public land survey sections (average size of 2.13 km^2)	17 indicators pertaining to aquatic and terrestrial values, including spp. presence and vegetation characteristics	Absolute where available, relative otherwise ($\pm$1 SD)	Distance to nearest old growth patch Index of fragmentation	None

evaluation unit and its immediate vicinity, using a 5 km buffer. These themes also included inputs explicitly chosen to reflect their potential values in the future (e.g., current low development density indicating greater future conservation potential). Since the modeling objective was land prioritization for conservation, the majority of the evaluation criteria were set using the minima and maxima of each data set.

Staus et al. (2010) similarly looked at both aquatic and terrestrial habitats in a study to identify forest areas of high conservation value in western Oregon, which were not specifically managed for conservation. They used a number of fine-filter biologically-based indicators, such as presence of NSOs, murrelets and other rare species, as well as coarse-filter metrics of percent old-growth forest and spatial metrics of its distribution (fragmentation).

The Integrated Restoration and Protection Strategy of US Forest Service Region 1 (Jensen et al. 2009, Chap. 5) also included significant wildlife components. Similar to the Sierra Nevada study, they included topics for both aquatic and terrestrial biodiversity, each of which incorporated coarse- and fine-scale metrics of current condition and potential threats. Their analysis units varied by resource and data resolution, and vegetation and aquatic resource conditions were evaluated at the ecological subsection level (100–200 km^2, Bailey 1996) and subwatershed level (4,000–12,000 ha, Seaber et al. 1987, http://nhd.usgs.gov/), respectively. The assessment only needed a relative rating of areas, so it used the 25th and 75th percentiles of the data to set the lower and upper bounds on evaluation criteria.

EMDS has also been used to evaluate the condition of forest resources at the national level in the US. Evaluation included a number of indicators related to biodiversity. The 1992 United Nations Conference on Environment and Development (UNCED) produced a Statement of Forest Principles, which in turn has led to regional efforts to promote and harmonize reporting on national forest conditions. The US participates in the Montreal Process, Criteria and Indicators group, and produced its first major national report in 2004 (Guldin and Kaiser 2004). Nine indicators were grouped under a criterion for biodiversity. The US report generally attempted to quantify these indicators, but did not attempt to evaluate acceptable levels of each indicator. Reynolds et al. (2003) argued that such evaluation was an essential step, and demonstrated how this could be done using EMDS (see also Reynolds et al. 2008). The units of analysis varied with the indicators, which ranged from coarse-filter changes in forest types to fine-filter metrics of endangered species listings and population counts. Few reference conditions were available for setting evaluation criteria. In some cases, relatively recent historical data were used (1960s, 1970s) and the general strategy for the effort was to focus on trends starting from the initial report.

Most similar in scope to the single-species application detailed in this chapter, Heaton et al. (2008) used EMDS to identify the best translocation sites for desert tortoises (*Gopherus agassizii*) given a planned expansion of military training grounds in southern California, USA. In contrast to other wildlife habitat assessments, more is known about threats to the tortoise than about its biophysical habitat preferences, so many of the indicators used in the model represented threats (e.g., roads, urbanization) and opportunities (i.e., land ownership). Evaluation criteria were assigned by an expert panel and weighted by grouping them into two tiers; a top tier combined criteria with an AND operator (limiting factor approach) and a lower tier combined with a UNION (averaging) operator. Many of the indicators were spatial in nature, involving proximity to features (roads, urban areas). The analysis did not involve any measure of landscape configuration (i.e., landscape metrics), but the team made an especially effective use of alternative scenarios to incorporate differing scientific opinions. They constructed seven scenarios, with each altering the use of one or more indicators under contention. The results of all scenarios were overlaid and areas with high suitability values in every scenario were identified as the preferred translocation sites.

From the examples above, we clearly show that EMDS has been used successfully for a number of wildlife-related studies, usually dealing with large landscapes (>5000 km^2). Similarly, analysis units used in evaluations have been large (>2 km^2) and often based on ownership/management divisions than on ecological considerations. Indicators used have ranged from exclusively coarse-filter cover types to fine-filter metrics related to individual species, and often both have been combined in a single model. While EMDS is not designed to have any population modeling capabilities, population numbers and species presence/absence have been incorporated into a number of the assessments. Nor does EMDS have built-in spatial analyses, yet a number of studies have incorporated spatial proximity and fragmentation metrics through preprocessing, and assigned these values to individual analysis units.

Finding published, objective criteria for use in data evaluation appears to be a challenging modeling aspect. Many studies simply used reference statistics from the datasets themselves (e.g., minimum/maximum, 25th/75th percentiles), which provide a relative comparison but say little about habitat values in any absolute sense. Some studies applied more externally-relevant criteria, generated from analyses of historical conditions or drawing on established habitat values. By and large, uncertainty/sensitivity was not addressed in these studies. Jensen et al. (2009), however, demonstrated how model structures could be reconfigured to address different management objectives (management uncertainty), and Humphries et al. (2008) combined their outputs with 15 land-cost scenarios to evaluate sensitivity to these costs. Heaton et al. (2008) went the farthest in addressing uncertainty in their outputs by running five alternative model formulations and then prioritizing areas scoring highly on all five.

3 Motivation for Using EMDS in This Context

The DNR's principal need for the South Puget NSO analysis was to rate dispersal habitat conditions across a set of management alternatives. This focus meant that a habitat rather than a population model was the most appropriate approach, and, since little data were available on the needs of dispersing owls, the model would need to be largely informed by expert opinion. The DNR informally reviewed a number of approaches and chose EMDS for a number of reasons. EMDS had been used in a number of habitat assessments already (as described above), and its basis in fuzzy logic provided a mechanism well-suited to the integration of expert knowledge. The fuzzy logic approach also provided a simple solution to problems identified with the strict threshold approach to habitat definition; EMDS provides a method to integrate a variety of indicators and can produce combined evaluations on a continuous scale. Finally, the polygon-based framework used by EMDS fit well with the DNR's stand-based forest inventory data, and with the division of these stands into finer-grained analysis polygons.

4 What Worked Well

4.1 Data Structure

The first step our project team took was to develop a prototype model based on DNR's current inventory data. These data were organized by stand polygons, so it was quite easy to link the data into an EMDS model. There were also a considerable number of stands lacking certain indicators or inventory data altogether. EMDS accommodates such data gaps by assigning them an evaluated score of zero (no evidence for or against the proposition). This feature allowed the team to produce a prototype model quickly, which in turn helped team members and owl experts to understand the approach and see its advantages. The fact that EMDS is scale independent was also a significant advantage, since, for the forest modeling alternatives, the original stand data were further subdivided by intersecting a number of features relevant to the planning process, such as stream and road buffers.

4.2 Expert Engagement

Wildlife models designed to answer management questions must often rely on wildlife experts to synthesize available information to provide the model structure and key parameters. The processes used to engage experts can involve significant time and effort, but they positively influence model quality (Andelman et al. 2001; Gordon and Gallo 2011). Here, we describe our process and experiences based on detailed notes from a dedicated note-taker and a written post-workshop review compiled by the project team.

The project benefitted from a number of previously conducted reviews of NSO habitat characteristics (Courtney et al. 2004; Hanson et al. 1993; Thomas et al. 1990); however, as mentioned above, there was little empirical data on the habitat needs of dispersing owls. DNR managers recognized that there was a significant effort needed to synthesize the available data in a manner that took advantage of

EMDS modeling capabilities. To do this synthesis, the project team identified a group of NSO experts: two biologists from the DNR (in addition to the project lead), and one each from the Washington Department of Fish and Wildlife and Oregon State University. To quickly produce a model, which would go out for broader review as part of the draft forest plan, a small group of experts most immediately involved in the state forest planning process was chosen.

We scheduled a two-day workshop to begin the expert engagement process. Given the limited amount of time available, the team prepared for the workshop carefully, including the development of a written plan comprising the following elements: equipment, materials (presentations, handouts, computer reference files), roles (manager, facilitator, data provider, and note-taker), and a detailed agenda of modeling phases. 2 weeks prior to the workshop, we provided a packet of background materials on the EMDS modeling process, the available data, and a set of indicator values extracted from the literature. The availability of a science synthesis report (Hanson et al. 1993), prepared for the 1997 HCP, provided a strong base for beginning discussions of the habitat model.

We began the workshop with three presentations on the background and objectives, EMDS modeling process, and available data. We then moved into model specification, using the following order to proceed from overall model structure to the individual indicators of owl dispersal habitat (steps are described below):

1. Sketch out model structures
2. Choose the specific indicators
3. Choose evaluation (scaling) criteria for each indicator
4. Document follow-up tasks and concerns.

The experts appeared to have little difficulty understanding the EMDS approach to modeling, and the choice of indicators proceeded rapidly. Instead of building one habitat model, they choose to build three, in order to reflect different life history requirements of dispersing owls: foraging, roosting, and movement. This division was somewhat unusual for a habitat suitability model, and it was important in recognizing that not all portions of the landscape must meet all functions in order to support dispersal. Conceptual diagrams of the models are presented in Fig. 2a–c. Model structures were helpful in deconstructing individual life history requirements into indicators, such as thermoregulation. At the finest level of disaggregation, indicators had to be directly or indirectly measurable in the field. For example, while thermoregulation itself is difficult to measure, indirect indicators that provide thermoregulatory habitat (e.g., canopy cover) can be quantified by taking measurements in forested stands. The process slowed when we reached the third stage: setting the evaluation criteria values for each indicator (which we discuss further in the "Limitations" section). This slowing was understandable, since it represented the operational decisions regarding numerical threshold values of some significance, and the most appropriate datasets to use. Not all aspects of our model were specified in a single workshop—questions

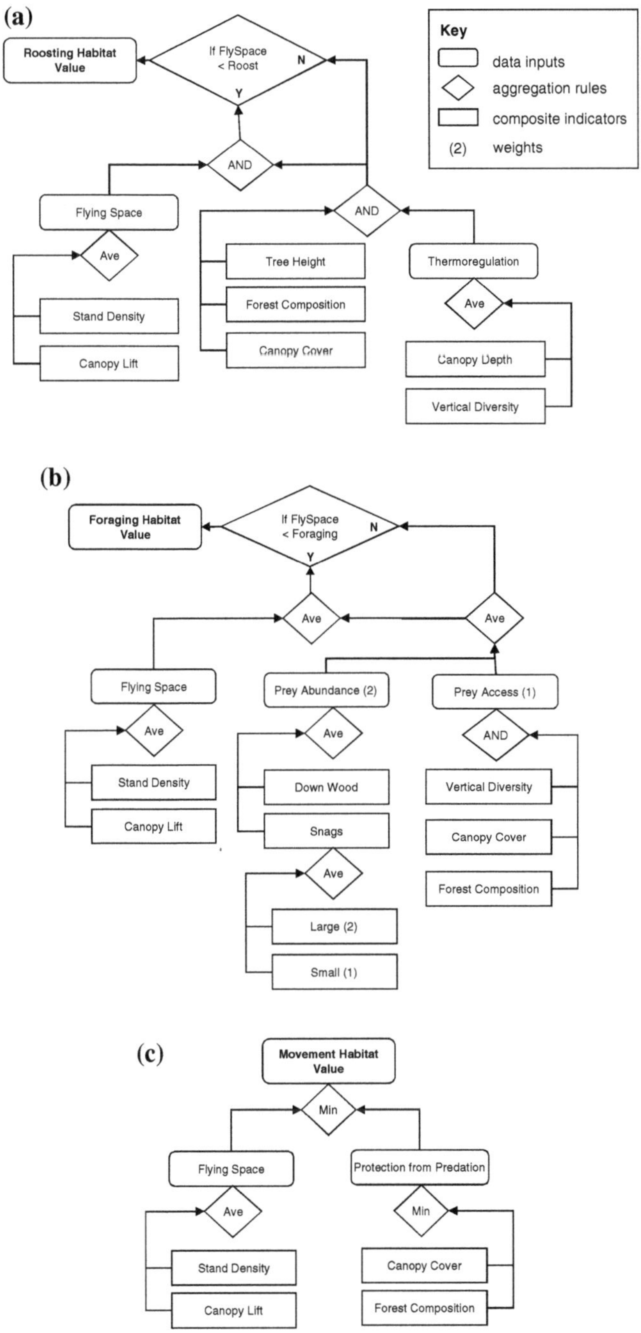

Fig. 2 EMDS model structure diagrams. **a** roosting habitat model; **b** foraging habitat model; **c** movement habitat model

requiring further research often arose. These were recorded in a special section of the model—Notes for Follow-up.

4.3 Data Evaluation and Integration

One of DNR's major objectives was to move away from a simple binary approach to habitat management, where each indicator either passes or fails. It was frustrating to DNR managers that this pass/fail system would not distinguish a recent clearcut from a stand approaching (but not quite meeting) threshold conditions for dispersal habitat. The fuzzy-logic in EMDS provided a solution to this problem. Once evaluation criteria were chosen, each indicator was scored on a standardized, continuous scale; stands closer to optimum conditions received higher scores.

Developing evaluation criteria is often the most challenging aspect of EMDS model development. For example, in the case of the NSO, a key question centers on the tree diameter threshold that differentiates "no support" from "support" for good habitat. A number of EMDS applications discussed above were able to solve this problem by basing the criteria on the distribution of the data being analyzed (e.g., all stands >80th percentile are scored +1). However, a relativizing approach assumes that either desirable conditions can be captured by a known percentile of the current conditions, or that only a relative comparison is needed. Since the DNR wished to reassess the very definition of dispersal habitat, it was unknown what percentile of current conditions was acceptable. Although the ultimate use of the spotted owl evaluation was in fact a relative comparison between management options, relative criteria (i.e., chosen simply based on statistics, such as maximum and minimum) were insufficient for a few reasons. More absolute criteria (i.e., based on external considerations, such as actual habitat needs) had already been established by the HCP and mapped, and any new habitat assessment method would be viewed with a similar expectation. The HCP required certain amounts of habitat to be produced, rather than just choosing from among the best options.

Since single indicator evaluation thresholds had already been established by the HCP, the project team started with these indicators in the development of their prototype. The binary threshold of each indicator was spread into an upper and lower bound by applying a percentage multiplier, e.g., the 11-inch tree diameter was multiplied by $\pm 20\ \%$ resulting in lower and upper bounds of this criterion of $8.8''$ (-1, evaluates to no support) and $13.2''$ (+1, evaluates to full support). This approach enabled the quick generation of a prototype, but the DNR saw the need to have these thresholds reviewed and updated by an expert group. The experts appeared to readily understand and appreciate the fuzzy logic approach to indicator evaluation.

Another shortcoming of HCP binary habitat definitions was that any stand could fail to meet habitat conditions by narrowly missing a threshold for a single indicator. The EMDS approach of producing a standardized score for each indicator, and aggregating scores within a hierarchical (or network) structure, provided a solution to this problem. In some cases, the owl experts maintained a limiting factor approach similar to the HCP (a MINIMUM aggregation function), but in others they allowed partial compensation between indicators (AVERAGE operator). Indicators were also weighted in a few cases. Interestingly, one of the first steps taken by the experts was to divide the habitat assessment into distinct models for three different life history needs of owls: roosting, foraging, and movement. It was felt that previous thresholds had been facilitated safe movement of owls through the forest, without considering concurrent roosting and foraging needs. The models generally contained one intermediate level of aggregation based on more detailed aspects of life history needs ("Prey Abundance" in Fig. 2). As mentioned above, the experts did not appear to have much difficulty in devising the hierarchies for the three models.

Initial results from the models produced a pattern counterintuitive to the project lead's expectations: the movement model, which was expected to be less demanding, rated more stands poorly than either the roosting or foraging models. By comparing the models, we determined that the lower scores were caused chiefly by a Flying Space node (a combination of stand density and canopy lift), which was in the movement model, but not in the others. The expert group felt that flying space was an important consideration for foraging and roosting also, and the modularity of the EMDS model enabled us to quickly copy this structure into the other models.

4.4 Model Validation

Model validation as usually understood in the natural sciences means testing to see if a model produces empirically accurate results with respect to independent, real world observations (Oreskes et al. 1994). However, knowledge-based systems are often built for situations in which such empirical tests are neither possible nor affordable. For this reason, validation in the expert systems sense is often done by comparing model processes and results back to the judgments of experts (Turban and Aronson 2001). Sufficiently detailed data on habitat use during dispersal were not available, so our validation strategy was limited to consideration and confirmation of model results by the expert panel.

Our first step was to send out the model results to the expert group via email. We received some feedback via email, but it was piecemeal and driven by a few initial questions from one of the participants. To get a more comprehensive assessment, we scheduled a follow-up model review workshop. Because of a difficulty experts had expressed in visualizing the stand-level inventory data during the first workshop, we decided to spend the first day of the review workshop in the field, examining a few sample stands and their indicator scores, and comparing them with associated data values.

We tested a number of handouts designed to enhance the visit. The first was simply an aerial photo of the stand, which provided the big-picture view, since we would only be visiting one point in the larger stand. Second, we provided a stand statistics page with the indicator values from the stand inventory. We also included a graphic picture of the stand produced with the Stand Visualization System application (McGaughey 1997), to test how well such visualizations might compare with actual conditions. Third, we provided a "visual stand scoring sheet," on which we asked each expert to independently score stand indicators based on their visual assessment after observing each stand. Experts were then asked to share their scores, and discussions ensued about the numeric evaluation points for each indicator. Finally, we prepared a set of questions to ask at each stand concerning the most problematic indicators. The experts found the handouts quite useful. The visual scoring sheets and prepared questions helped keep the conversations on topic, but we were not ultimately able to collect enough scores to compare to the EMDS result scores.

A few months after the second workshop, we sent the revised stand model results out by email for a third round of review, along with a first draft of results from the landscape model. The expert group had a few remaining questions on the stand model but seemed largely satisfied with the results.

5 What Didn't Work Well

5.1 Data Structure

Our EMDS model adapted well to the DNR data structure; however, managing the fourth dimension (time) was more challenging. The environmental impact statement (EIS) process involved estimating the impacts of three alternative management options over the next 100 years. A 10-year time-step was chosen by the EIS team, which meant we would process 30 (3 alternatives × 10 periods) data tables as input to the owl model. Given that multiple runs were needed to test and refine the operation of the model, this represented a considerable analytical burden. Initially, we reduced the load by simply limiting our analysis to three of the 10 periods (beginning, middle, and end). Ultimately, however, we wished to see the full trend over time.

ArcMap and EMDS did not have a facility for handling a time series of data by assigning more than one value to a single polygon, so our initial approach was non-spatial. We developed Python® (http://www.python.org/)code to stack the 30 input tables into one master table, with new fields to track which "alternative" and "time period" each record was associated with. Then we ran this combined table through the EMDS-NetWeaver module directly, which does not require spatial objects to be associated with the data.

This approach worked reasonably well for the stand-level model done in the EMDS framework. However, as our landscape model evolved (described further below in Data Evaluation), we began to see the potential advantages of integrating the two scales. Without such integration, a model run involved multiple human interventions: an initial preparation of the data for NetWeaver, the NetWeaver run, and finally the initiation of the landscape model run. During the draft EIS review process, after comments were tendered requesting additional model runs to test the sensitivity of various parameters, we decided to fully integrate the stand and landscape models. We replicated the functionality of the basic NetWeaver fuzzy logic operators in a Python script, so that the complete model and all iterations could be performed as a single program run. The basic functionality was not particularly difficult to mimic: nevertheless, the integration and testing required considerable time of an expert programmer over several weeks. Another advantage of this approach was that it could be more closely integrated with the DNR data structures, drawing on and writing to them directly without manual intervention. Disadvantages were the loss of other EMDS functionality (e.g., quick creation of output maps and scenarios) and the programming expertise necessary to maintain the model.

5.2 Expert Engagement

As described above, the owl experts expressed difficulty in visualizing stand conditions from the statistics provided (e.g., trees per acre). They referred to stand photos shown in one of the introductory presentations, which suggested the need for additional photos representative of stand conditions and classification of stands into a few basic types. They also requested seeing frequency distributions of the indicators to aid them in setting thresholds. We were able to provide some displays "on the fly," but preparation of a handout would be more helpful in the future. In our post-workshop review, we also agreed that beginning the assessment of each indicator by explicitly reviewing our literature summary (from the background materials) would have been helpful.

In general the experts appeared to quickly understand the EMDS modeling framework. Where we encountered the most difficulty was in sharing the modeling results with the group as part of the validation process. We could not expect the participants to run GIS or other parts of the modeling software, and thus needed to encapsulate the results in a more accessible format. Simply exporting the EMDS result tables to a spreadsheet produces a mass of variable names and numbers that were not ordered according to the model hierarchy. We worked around this by using database software (Microsoft Access) to reformat and export the results to a spreadsheet format (Microsoft Excel). Additional manual formatting and writing of explanatory text was also done. We manually exported EMDS map results and imported these into PowerPoint. Through these steps we were able to make the results accessible; however, the EMDS modeling process would have greatly expanded utility if these needs could be better addressed by the software.

Our second major problem with expert engagement was not due to, but rather highlights, the utility of EMDS. The second day of the initial expert workshop focused on modeling landscape connectivity. We presented only a few general ideas on the various scales that might be relevant to this task (e.g., neighborhood, patch, landscape), but the value of a chosen analytical framework, such as EMDS, became quickly evident, because with no specific roadmap to follow, the resulting discussion was unfocused and alternated between broad biological principles and narrow computational challenges. A number of these broad principles were eventually useful to frame the EMDS model (e.g., risk is determined by the size of roosting/foraging patches and the connectivity between them, owl dispersal needs vary between transience and colonization stages), but little overall progress was made in assembling a functional model specification.

5.3 Data Evaluation and Integration

EMDS worked well for the basic stand-level habitat assessment. The experts appeared to readily understand and appreciate the fuzzy logic approach to indicator evaluation; however, based on progress made during the initial workshop, choosing the evaluation criteria values seemed considerably more difficult than choosing the indicators and structuring them into a model. This is not a limitation of EMDS, but a limitation of expert knowledge, which EMDS exposes.

The fact that we were looking outside the EMDS framework for a solution to the landscape connectivity question does indicate one of its limitations. A number of other EMDS models have integrated spatially-aware attributes: Bourgeron et al. (2000) used polygon size and the suitability of neighboring units; White et al. (2005) also used neighborhood suitability scores; and Heaton et al. (2008) used proximity to other spatial attributes, such as roads. All of these metrics were based on the proximity of attributes to each polygon. Although EMDS does not perform this “windowing” itself, it is easy enough since it sits inside a GIS application and the results are directly assignable to the individual members of the existing polygon set. In contrast, our overarching question was the connectivity of the landscape as a whole, which required a more holistic assessment.

We ultimately settled on a graph theoretic approach (Urban and Keitt 2001), based on identifying habitat patches and the distances between them, which we then summarized using landscape metrics (Saura and Pascual-Hortal 2007). Constructing the patches required combining the EMDS model scores from the roosting and foraging models (both had to meet a minimum threshold) and then aggregating these polygons from the EMDS stand-level assessment. In using a patch-based approach we lost some of the finer gradation of scores produced by the fuzzy logic, because an in/out patch threshold had to be chosen, but we were able to preserve some of this information by assigning a patch suitability score based on size and quality.

Whereas previous owl modeling efforts generally have assumed that habitat between patches is uniform, our focus on dispersal led us to give this aspect more thought. Cost distance is a well-established technique in GIS, in which the distance from one point to another is calculated, based not only on distance, but also on the difficulty of traversing the intermediate terrain. A number of researchers had applied it to animal habitats (e.g., see Bunn et al. 2000; Singleton et al. 2002). Problems with applying cost distance approaches to habitat assessment include broadening its application to patches instead of points, and working around its GIS implementation, which focuses on finding the least-cost path, an unrealistic assumption for species without access to GIS. We based our model on recent work by Theobald et al. (2006), which addresses both of these issues. Cost distance is generally implemented using a raster rather than vector data format, which required us to convert the data to raster as it was moved from the stand to landscape sections of the model. Adding the capability of handling raster data to EMDS would increase its applicability to a wider set of analyses.

5.4 Model Validation

A few issues related to model validation have already been discussed above: providing the model output in an accessible format and keeping the expert team engaged enough to understand the technical aspects of the model. With the publication of the draft EIS, the validation emphasis shifted focus from the expert group to the broader public. In terms of the validity of the model to assess owl habitat, the only major comment received actually came from a member of our expert panel, but this was more a reflection of his role as the representative of the state wildlife regulatory agency. The comment centered on the need to acknowledge and incorporate uncertainties into the analysis.

Discussions of uncertainty in the fuzzy logic literature tend to focus on linguistic uncertainty—how it addresses language concepts such as high, medium and low (Adriaenssens et al. 2004; Regan et al. 2002). We found little explanation of how such systems address uncertainties in model inputs and relationships, which was the focus of the comment. Such uncertainties seem more easily explained in the use of Bayesian belief networks, which are often used for similar applications, because they explicitly combine probability distributions. However, these distributions tend to be greatly simplified, often into bins such as high, medium and low. Uncertainty in the partition points for these bins resembles that of inflection points on fuzzy logic functions, but the latter appear harder to explain simply using the language of certainty because they are tied to fuzzy set theory instead of probability theory. However, it could be argued that fuzzy logic naturally incorporates a degree of uncertainty in inputs, which is reflected in the shape of the fuzzy evaluation functions.

Because we did not find sufficient discussion about input uncertainties in the literature and the concept is not intuitively familiar, we employed a standard sensitivity analysis approach. We discussed the perceived degree of uncertainty around each of the model parameter inputs with the regulatory agency representative and chose alternate plausible values for four aspects perceived to be the most uncertain. Three of these aspects were related to parameters in the stand model (down wood, snags, and their combination) and one was in the landscape model (the costs of moving through unfavorable cover types). We re-ran the model for each of these four different assumptions, which resulted in a range of final landscape scores that varied from our baseline assumptions by −26 to +41 % in the initial period to −26 to +15 % in the final planning period (WADNR 2010a).

6 Conclusions

EMDS has been shown to be an effective tool in the wildlife modeling domain. Its primary uses have been in habitat suitability assessment. Its basis in fuzzy logic is compatible with techniques developed for habitat suitability indexing when using ramp functions to evaluate specific habitat attributes and when combining these attribute scores using common mathematical operators to obtain overall suitability scores. EMDS is not a simulation system, so it is not suitable for population simulations; however, some applications have successfully integrated population estimates (from other sources) into EMDS assessments and prioritization models. In the same vein, EMDS does not include spatial neighborhood functions, but a number of cases have included such metrics in assessments by pre-calculating them in a GIS. In sum, the EMDS framework is most appropriate for species dependent on habitat variables that can be mapped to specific spatial units; it is less suitable for analyzing influences based on population size or spatial behaviors, such as herding or migration. However, the simplicity and flexibility of the EMDS framework have led to its use in large landscape assessments that attempt to integrate diverse factors (e.g., habitat, population, threats).

Our analysis of NSO dispersal habitat benefitted greatly from the analytical framework provided by EMDS. The software enabled the rapid development of a prototype model, and it facilitated the use of expert knowledge to define habitat for dispersing NSOs. We did encounter limitations though, especially related to the spatial modeling of owl dispersal and the automation of multiple model runs.

Many analyses of wildlife habitat in forested regions begin with polygon-based maps from forest stand inventories. The EMDS polygon-based data structure fits this type of data well, and can easily manage polygon subdivision produced by combining data layers. However, many wildlife analyses also use raster-based functions, which are generally superior for analyzing proximity and connectivity issues. Some EMDS applications have worked around this by using regular grids of polygons (Heaton et al. 2008; White et al. 2005), but this approach does not provide access to many raster-based analysis functions in GIS software. Dispersal modeling places a special emphasis on habitat connectivity, and we found that this aspect of the analysis was better done in a raster environment. The ability to work with raster data would be a particularly useful addition for wildlife modeling with EMDS. Integrating EMDS with subsequent raster analysis was also a challenge, which may be solved by exposing some of the basic EMDS analytical functions as tools within the ArcGIS environment. Exposing such functions could further help with the challenge of automating model runs for multiple time periods or sensitivity analyses.

The fuzzy logic engine in EMDS was effective at moving habitat scoring from a strict threshold-based system to a finer, multi-valent gradation of habitat values. This change in itself was quite significant in the eyes of many DNR staff, who had long been frustrated with the insensitive, black and white definition of habitats specified in the HCP. Fuzzy logic also provided a well-accepted method for using expert knowledge to set habitat evaluation criteria. This aspect was crucial given the lack of direct empirical data on the habitat needs of dispersing owls—a situation which occurs all too frequently in wildlife modeling. The new polygon-based habitat maps were a useful product themselves. Such maps were a familiar format for DNR staff and the EMDS methodology was relatively easy to explain. Although the EMDS framework appears easy to grasp, we believe the software could benefit from including functions to help package and share the results with experts, managers, and other stakeholders who cannot be expected to run the program themselves.

Finally, quantifying model uncertainties appears to be increasingly important to model users. In addition to our work, at least two other EMDS wildlife analyses have addressed uncertainty through the use of scenarios (Bourgeron et al. 2000; Heaton et al. 2008). Such analyses have simply involved altering input parameters or model structure and re-running the model. The EMDS scenario capability provides good support for this type of sensitivity analysis, but support could be improved by providing more automation tools to run a range of values in batch mode. More fundamental discussion and analysis in the literature of how fuzzy logic manages uncertainty would also be useful for analysts and stakeholders.

References

Adriaenssens V, Baets BD, Goethals PLM, Pauw ND (2004) Fuzzy rule-based models for decision support in ecosystem management. Sci Total Environ 319(1–3):1–12

Akçakaya HR, Raphael MG (1998) Assessing human impact despite uncertainty: viability of the northern spotted owl metapopulation in the northwestern USA. Biodivers Conserv 7:875–894

Andelman SJ, Beissinger S, Cochrane JF, Gerber L, Gomez-Priego P, Groves C, Haufler J, Holthausen R, Lee D, Maguire L, Noon B, Ralls K, Regan H (2001) Scientific standards for conducting viability assessments under the National Forest Management Act: report and recoomendations of the NCRAS working group. National Center for Ecological Analysis and Synthesis, Santa Barbara

Applied Biomathematics (2011) RAMAS analytical and educational software Applied Biomathematics, Inc. http://www.ramas.com/. Accessed 20 April 2011

Bailey RG (1996) Ecosystem geography. Springer, New York

Bourgeron PS, Humphries HC, Reynolds KM (2000) Conducting large-scale conservation evaluation and conservation area selection using a knowledge-based system. Paper presented at the proceedings of the 4th international conference on integrating geographic information systems and environmental modeling: problems, prospectus, and needs for research, banff, (AB) Canada

Buchanan JB (2004) Managing habitat for dispersing northern spotted owls–are the current management strategies adequate? Wildl Soc Bull 32(4):1333–1345

Bunn AG, Urban DL, Keitt TH (2000) Landscape connectivity: a conservation application of graph theory. J Environ Manage 59(4):265–278

Caughley G, Sinclair ARE (1994) Wildlife ecology and management. Blackwell Scientific Publications, Boston

Courtney SP, Blakesley JA, Bigley RE, Cody ML, Dumbacher JP, Fleisher RC, Franklin AB, Gutierez RJ, Marzluff JM, Sztukowski L (2004) Scientific evaluation of the status of the northern spotted owl. Sustainable Ecosystems Institute, Portland

Cushman SA, McKelvey KS, Noon BR, McGarigal K (2010) Use of abundance of one species as a surrogate for abundance of others. Conserv Biol 24(3):830–840. doi:10.1111/j.1523-1739.2009.01396.x

Dai JJ, Lorenzato S, Rocke DM (2004) A knowledge-based model of watershed assessment for sediment. Environ Model Software 19(4):423–433

Dawson WR, Ligon JD, Murphy JR, Myers JP, Simberloff D, Verner J (1987) Report of the scientific advisory panel on the spotted owl. The Condor 89(1):205–229

Doak D (1989) Spotted owls and old growth logging in the Pacific Northwest. Conserv Biol 3(4):389–396

ESA (1973) Endangered Species Act of 1973

Fall A, Fall J (2001) A domain-specific language for models of landscape dynamics. Ecol Model 141(1–3):1–18

Gallo K, Lanigan SH, Eldred P, Gordon SN, Moyer C (2005) Northwest Forest Plan—the first 10 years (1994–2003): preliminary assessment of the condition of watersheds. USDA Forest Service Pacific Northwest Research Station, Portland

Gordon SN, Gallo K (2011) Structuring expert input for a knowledge-based approach to watershed condition assessment for the Northwest Forest Plan, USA. Environ Monit Assess 172(1):643–661

Guldin RW, Kaiser HF (2004) National report on sustainable forests–2003. USDA Forest Service, Washington

Hanson E, Hays D, Hicks L, Young L, Buchanan JB (1993) Spotted owl habitat in Washington: a report to the Washington Forest Practices Board. Washington Forest Practices Board, Spotted Owl Advisory Group, Olympia

Harrison S, Stahl A, Doak D (1993) Spatial models and spotted owls: exploring some biological issues behind recent events. Conserv Biol 7(4):950–953

Heaton J, Nussear K, Esque T, Inman R, Davenport F, Leuteritz T, Medica P, Strout N, Burgess P, Benvenuti L (2008) Spatially explicit decision support for selecting translocation areas for Mojave desert tortoises. Biodivers Conserv 17(3):575–590

Hirzel AH, Hausser J, Perrin N (2004) Biomapper 3.0. Laboratory for Conservation Biology, University of Lausanne, Division of Conservation Biology, University of Bern

Hulse D, Gregory S, Baker J (eds) (2002) Willamette River basin planning atlas: trajectories of environmental and ecological change. Oregon State University Press, Corvallis

Humphries H, Bourgeron P, Reynolds K (2008) Suitability for conservation as a criterion in regional conservation network selection. Biodivers Conserv 17(3):467–492

Jensen M, Reynolds K, Langner U, Hart M (2009) Application of logic and decision models in sustainable ecosystem management. Paper presented at the 42nd Hawaii international conference on system sciences, Hawaii, HI, 5–8 Jan 2009

Johnson DH, O'Neil TA (eds) (2001) Wildlife-habitat relationships in Oregon and Washington. Oregon State University Press, Corvallis

Lacy RC (2000) Structure of the VORTEX simulation model for population viability analysis. Ecological Bulletin 48:191–203

Lamberson RH, Noon BR, Voss C, McKelvey R (1994) Reserve design for territorial species: the effects of patch size and spacing on the viability of the Northern Spotted Owl. Conserv Biol 8(1):185–195

Lande R (1988) Demographic models of the northern spotted owl (Strix occidentalis caurina). Oecologia 75(4):601–607
Lang S (2004) Subdivision analysis with subdiv.avx extension for ArcView 3.x. Spatial Indicators for European Nature Conservation Project, Salzburg, Austria
Lee PL (1985) History and current status of spotted owl (Strix occidentalis) habitat management in the Pacific Northwest Region, US Department of Agriculture Forest Service. In: Gutierrez RJ, Carey AB (eds) Ecology management of the Spotted Owl in the Pacific Northwest, vol GTR-PNW-185 USDA Forest Service. PNW Research Station, Portland, pp 5–9
Leslie PH (1945) The use of matrices in certain population mathematics. Biometrika 33(3):183–212
Lindenmayer DB, Manning AD, Smith PL, Possingham HP, Fischer J, Oliver I, McCarthy MA (2002) The focal-species approach and landscape restoration: a critique. Conserv Biol 16(2):338–345. doi:10.1046/j.1523-1739.2002.00450.x
Lint J (2005) Northwest Forest Plan—the first 10 years (1994–2003): status and trends of northern spotted owl populations and habitat. USDA Forest Service, Pacific Northwest Research Station, Portland
MacNally R, Fleishman E (2004) A successful predictive model of species richness based on indicator species. Conserv Biol 18(3):646–654. doi:10.1111/j.1523-1739.2004.00328_18_3.x
McComb WC, McGrath MT, Spies TA, Vesely D (2002) Models for mapping potential habitat at landscape scales: an example using northern spotted owls. For Sci 48(2):203–216
McGarigal K, Marks B (1995) FRAGSTATS: spatial pattern analysis program for quantifying landscape structure. USDA Forest Service, Pacific Northwest Research Station, Portland
McGaughey RJ (1997) Visualizing forest stand dynamics using the stand visualization system. In: Proceedings of ACSM/ASPRS, vol 4, no 2, pp 48–57
Morrison ML, Marcot B, Mannan RW (2006) Modeling wildlife-habitat relationships. In: Morrison ML, Marcot B, Mannan RW (eds) Wildlife-habitat relationships: concepts and applications. Island Press, Washington, pp 320–376
Noss RF (1987) From plant communities to landscapes in conservation inventories: a look at the Nature Conservancy (USA). Biol Conserv 41:11–37
Oreskes N, Shrader-Frechette K, Belitz K (1994) Verification, validation, and the confirmation of numerical models in the earth sciences. Science 263(5147):641–646
Possingham HP, Davies I (1995) ALEX: a population viability analysis model for spatially structured populations. Biol Conserv 73:143–150
Reeves GH, Hohler DB, Larsen DP, Busch DE, Kratz K, Reynolds K, Stein KF, Atzet T, Hays P, Tehan M (2004) Aquatic and riparian effectiveness monitoring plan for the Northwest Forest Plan. Gen. Technicl Report U.S. Department of Agriculture, Forest Service, Pacific Northwest Research Station, Portland, OR
Reynolds K, Peets S (2001) Integrated assessment and priorities for protection and restoration of watersheds. Paper presented at the Proceedings of the IUFRO 4.11 conference on forest biometry, modeling and information science, University of Greenwich, UK, 26–29 June
Regan HM, Colyvan M, Burgman MA (2002) A taxonomy and treatment of uncertainty for ecology and conservation biology. Ecol Appl 12(2):618–628. doi:10.1890%2F1051%2D0761%282002%29012%5B0618%3AATATOU%5D2.0.CO%3B2
Rempel RS, Carr AP (2003) Patch analyst extension for ArcView: version 3
Reynolds K, Gordon SN, Johnson KN (2008) Using logic to evaluate forest ecosystem sustainability. In: Paper presented at the forest criteria and indicators analytical framework and report workshop, Joensuu, Finland, 18–21 May 2008
Reynolds KM, Jensen ME, Andreasen J, Goodman I (2000) Knowledge-based assessment of watershed condition. Comput Electron Agric 27:315–333
Reynolds K, Johnson KN, Gordon SN (2003) The science/policy interface in logic-based evaluation of forest ecosystem sustainability. Forest Policy Econ 5:433–446
Reynolds KM, Rodriguez S, Bevans K (2002) Ecosystem Management Decision Support 3.0 user guide. USDA Forest Service Pacific Northwest Research Station, Corvallis

ROD (1994) Record of decision for amendments to Forest Service and Bureau of Land Management Planning Documents within the range of the Northern Spotted Owl; GPO 1994–589–111/0001, U.S. Government Printing Office, Region No. 10: Washington, DC, USA
Roloff GJ, Wilhere GF, Quinn T, Kohlmann S (2001) An overview of models and their role in wildlife management. In: Johnson DH, O'Neil TA (eds) Wildlife-Habitat relationships in Oregon and Washington. Oregon State University Press, Corvallis, pp 512–536
Saura S, Pascual-Hortal L (2007) A new habitat availability index to integrate connectivity in landscape conservation planning: comparison with existing indices and application to a case study. Landscape Urban Plann 83(2–3):91–103
Schumaker NH (1996) Using landscape indices to predict habitat connectivity. Ecology 77(4):1210–1225
Schumaker NH, Ernst T, White D, Baker J, Haggerty P (2004) Projecting wildlife responses to alternative future landscapes in Oregon's Willamette basin. Ecol Appl 14(2):381–401
Seaber PR, Kapinos PF, Knapp GL (1987) Hydrologic unit maps. Water-supply paper 2294. U.S. Department of Interior, U.S. Geological Survey, Washington
Simberloff D (1998) Flagships, umbrellas, and keystones: Is single-species management passé in the landscape era? Biol Conserv 83(3):247–257
Singleton PH, Gaines WL, Lehmkuhl JF (2002) Landscape permeability for large carnivores in Washington: a geographic information system weighted-distance and least-cost corridor assessment USDA Forest Service. Pacific Northwest Research Station, Portland
Staus NL, Strittholt JR, Dellasala DA (2010) Evaluating areas of high conservation value in western Oregon with a decision-support model. Conserv Biol 24(3):711–720
Sutherland GD, O'Brien DT, Fall SA, Waterhouse FL, Harestad AS, Buchanan JB (2007) A framework to support landscape analyses of habitat supply and effects on populations of forest-dwelling species: a case study based on the northern spotted owl. British Columbia Ministry of Forests and Range, Victoria
Theobald DM, Norman JB, Sherburne MR (2006) FunConn v1 user's manual: ArcGIS tools for functional connectivity modeling. Natural Resource Ecology Lab, Colorado State University, Fort Collins CO
Thomas JW, Forsman ED, Lint J, Meslow EC, Noon BR, Verner J (1990) A conservation strategy for the northern spotted owl. Report of the interagency scientific committee to address the conservation of the northern spotted owl. USDA Forest Service, Portland
Turban E, Aronson JE (2001) Decision support systems and intelligent systems. Prentice Hall, Upper Saddle River
U.S. Department of Agriculture, Forest Service [USDA FS] (1984) Final regional guide and final environmental impact statement for the Pacific Northwest region. Portland
U.S. Fish and Wildlife Service [USFWS] (1981) Ecological services manual 103. Standards for the development of habitat suitability index models. Division of Ecological Services, Washington
Urban D, Keitt T (2001) Landscape connectivity: a graph-theoretic perspective. Ecology 82(5):1205–1218. doi:10.1890%2F0012%2D9658%282001%29082%5B1205:LCAGTP%5D 2.0.CO%3B2
Walker R, Keithley C, Henly R, Downie S, Cannata S (2007) Ecosystem Management Decision Support (EMDS) applied to watershed assessment on California's North Coast. In: Standiford RB, Giusti GA, Valachovic Y, Zielinski WJ, Furniss MJ (eds) Proceedings of the redwood region forest science symposium: what does the future hold? Gen. Technical Report GTR-PSW-194. Pacific Southwest Research Station, Forest Service, U.S. Department of Agriculture, Albany California
Washington Department of Natural Resources [WADNR] (2004) Final environmental impact statement on alternatives for sustainable forest management of state trust lands in western Washington and for determining the sustainable harvest level. Olympia, WA

Washington Department of Natural Resources [WADNR] (2010a) Forest land plan for the South Puget Habitat Conservation Plan (HCP) planning unit: final environmental impact statement. Olympia, WA

Washington Department of Natural Resources [WADNR] (2010b) Science behind the choices: biodiversity, DNR and what does an indicator species have to do with it? http://washingtondnr.wordpress.com/2010/10/22/science-behind-the-choices-biodiversity-dnr-and-what-does-an-indicator-species-have-to-do-with-it/. Accessed 22 Oct 2011

White MD, Heilman GE Jr, Stallcup JA (2005) Science assessment for the Sierra Checkerboard Initiative. Conservation Biology Institute, Corvallis

Wiens JA, Hayward GD, Holthausen RS, Wisdom MJ (2008) Using surrogate species and groups for conservation planning and management. Bioscience 58(3):241–252. doi:10.1641/b580310

Zabel CJ, Dunk JR, Stauffer HB, Roberts LM, Mulder BS, Wright A (2003) Northern spotted owl habitat models for research and management application in California (USA). Ecol Appl 13(4):1027–1040

Planning for Urban Growth and Sustainable Industral Development

Mª Carmen Ruiz Puente

Abstract This chapter describes a planning application designed in EMDS to evaluate land areas for their suitability to industrial park siting. The search for sustainable development approaches to park siting and for improving future viability of existing parks necessitated a structured decision-making process, which took into account all the phases of development, and anticipated uses of the neighboring land and resources. This chapter analyzes key factors in siting, and proposes a multi-criteria decision tool that is useful to considering the viability of present and future industrial areas, and their integration with their surroundings. The inherently spatial characteristics of our planning problem necessitated the use of an application development tool like EMDS to assist us in planning. EMDS was an appropriate platform for developing this application and for assessing the relative suitability of different land areas to siting. Our application was developed for a 646.2 km^2 district in the Cantabria region of northern Spain. We specified and integrated the sustainability criteria and the resulting maps clearly identified suitable zones for industrial park siting. The tool we developed is also extensible to evaluating existing industrial parks and opportunities for retrofitting.

Keywords Urban planning · Sustainable development · Spatial decision support · EMDS · Industrial park

1 Historical Background

Territorial and urban planning greatly influences the social, economic and environmental conditions of many regions of Spain. The siting (location) of industrial parks is a critical part of this process. Industrial parks are defined as "land which is

Mª. C. Ruiz Puente (✉)
Transport and Projects Technology and Processes Department, School of Industrial and Telecommunication Engineers, University of Cantabria, Santander, Spain
e-mail: ruizpm@unican.es

K. M. Reynolds et al. (eds.), *Making Transparent Environmental Management Decisions*, Environmental Science and Engineering,
DOI: 10.1007/978-3-642-32000-2_11, © Springer-Verlag Berlin Heidelberg 2014

divided and developed for the simultaneous use of various economic activities with shared infrastructures and proximity between the enterprises" (Trinder et al. 1993). Industrial park planning and promotion have been a fundamental part of urban land use plans. Furthermore, industrial park development is an important component of economic development strategies employed worldwide since 1970, particularly among developed nations (Smith 1971). Nevertheless, industrial parks present a significant environmental risk because they spatially concentrate the environmental problems of each of the component enterprises, and the park's infrastructure and services greatly modify native ecosystem structure and processes. Moreover, the lack of a comprehensive environmental management approach often leads to impacts resulting from waste generation, air and water pollution, and unsafe working or environmental conditions (UNEP 1997), which can interfere with and create conflicts with adjacent or concurrent urban, tourism, and recreation zones.

In urban planning, industrial areas are usually located far from residential areas to minimize environmental impact and land use conflicts. However, with time, urban growth can bring residential and industrial areas into greater proximity, making it increasingly difficult to separate land uses. The integration of industry with the environment rather than the transformation of environments by industry is a necessary strategy for minimizing conflicts associated with disparate land uses in close proximity, especially as available land area for such uses declines.

According to the guidelines of European Directive 2001/42/EC (2001), integration of environmental concerns during early industrial planning and project design phases greatly reduces impacts and the overall footprint of industrial parks on the natural environment. Likewise, pollution may be reduced where industrial operations are designed to mimic ecosystem structure and process, a principle of industrial ecology.

Industrial ecology proposes the search for common ground between industrial operations and urban or exurban surroundings, wherein industrial production and ecological processes are considered interdependent elements. The main objective is to promote symbiosis and synergy between various human and ecological activities in a given area through the carefully planned use or exchange of materials and energy, knowledge use, and development of shared facilities or initiatives (Graedel and Allenby 2003). Urban planning, production system design, and environmental management have always been considered separately in the planning and design of industrial areas. Industrial ecology seeks integrated solutions using an approach combining economic and environmental advantages.

The first publication to explicitly address industrial ecology was by Frosh and Gallopoulos (1989), although scientific publications on the basic ideas of industrial ecology date back to the mid-1950s (Erkman 1997). There are many examples of the use of industrial ecological principles in production systems at the corporate level through material, energy and water exchange, and recycling cycles (Lowe and Evans 1995; Frosh 1995; UNEP 1995). These principles also emerge at the international level for the planning and design of all the lifecycle phases of industrial parks (Kirschner 1995; Rosenfeld 1997; UNEP 1997; Roberts 2004; Tudor et al. 2007).

The conditions and evolution of these industrial ecosystems have been presented in various scientific works published in the last decade (Côté and Cohen-Rosenthal 1998; Lambert 2002; Gibbs 2005; Korhonena and Snäkin 2005; Gibbs 2007). Initiatives are found mainly in Western Europe (United Kingdom, Netherlands, and France), North America, and East Asia (China, Japan, Philippines), in developed countries and those with current high economic growth. Nevertheless, the true level of implementation is limited when compared with the number of planned projects. Furthermore, a great number of the studied cases involve transforming already functioning industrial areas into eco-industrial parks.[1]

The analysis of industrial siting factors in developed nations shows that the proximity to transportation infrastructures, workforce availability, and market or raw material proximity are still the principal driving variables (Leitham et al. 2000; Figueiredo et al. 2002; Somlev and Hoshino 2005). In contrast to the goal of reducing establishment and working costs found in classic location (siting) theories (Hoover 1948; Isard 1954; Lösch 1954), the current challenge is to find new competition factors. Environmental responsibility and proximity of businesses to infrastructure and R&D centers are recent examples.

According to Aalborg (1994), agreements on balanced development, the relationship between an urban environment with optimal social services, and the search for a quality physical environment to be conserved, have taken on a key role in the combining of industrial activity and the environment. Thus, it is essential to carry out integrated planning that includes the different policies of various sectors (e.g., transportation, energy, land use) when combining socio-economic development and environmental protection. Under the opinion of the author, a lack of foresight in the planning of spaces for urban development and economic activity leads to frequent changes in spatial organization and frequently re-building alignment and typology. Rational planning should consider and forecast, to a practical extent, industrial siting factors that make socio-economic and environmental interests compatible.

It is therefore crucial to consider new criteria in classical industrial location theory that help achieve integrated planning of the use of resources and minimize negative impacts to the environment created by establishment of new land uses. With the current state of development, there is no doubt that industrial parks should be sited to simultaneously maximize production capacity and minimize adverse effects on the environment. Reaching this level of integration is complex and requires the development of models and tools that allow for interrelating and ranking the numerous variables in a coherent and structured manner, to aid in the

[1] An eco-industrial park (EIP) is an industrial park in which businesses cooperate with each other and with the local community in an attempt to reduce waste and pollution, efficiently share resources (such as information, materials, water, energy, infrastructure, and natural resources), and help achieve sustainable development, with the intention of increasing economic gains and improving environmental quality. An EIP may also be planned, designed, and built in such a way that it makes it easier for businesses to co-operate, and that results in a more financially sound, environmentally friendly project for the developer (Lowe 2001).

decision making process for the agents who are directly involved in sustainable regional and urban design.

The combination of spatial analysis with multi-criteria decision analysis techniques (MCDA) has allowed for the creation of Multi-Criteria Spatial Decision Support Systems (MC-SDSS) with the objective to formulate and resolve spatial decision problems. MC-SDSS differ from conventional multi-criteria analysis techniques by including a spatially explicit geographic component (Malczewski 1999). In contrast with MCDA, spatial MCDA requires information on the values of the evaluation criteria, the geographic locations of the alternatives, and the preferences of the decision-makers regarding the values of the criteria and their locations.

The geographical information system (GIS) component of MC-SDSS may include statistical and mathematical modeling to explore existing data sources and spatially extrapolate the results. These tools can be also be used to explore alternative decisions and evaluate uncertainties associated with the available data, and the sensitivity of decisions to errors associated with the data or their extrapolation.

A well-designed user interface component can aid decision makers by helping them to visualize the key components of decisions and the supporting data, the chief differences among alternatives, and how these are displayed geographically and by criteria.

MC-SDSS applications have been developed in a wide variety of field applications. For example, in environmental planning and management, there are applications aimed at waste management (Maniezzo et al. 1998; Chang et al. 2008); natural resource management (Kallali et al. 2007), and territorial and urban planning (Dai et al. 2001; Herbst and Herbst 2006; Hernández et al. 2004). Depending on the specifications of applications and their system requirements, each project presents its own unique integration of a GIS and an MCDA model.

One application in particular, designed for land suitability evaluation, employs an expert system (ES) (Kalogirou 2002; Tabeada et al. 2006). ESs are rule, fact, and/or procedural knowledge-bases that have been employed to solve specialized problems, where significant expert-based information may exist. They provide several advantages: (1) they can easily handle qualitative and quantitative knowledge in the same utility; (2) can readily organize the implicit knowledge of experts; (3) are structurally flexible, and (4) are adaptable for different applications. As such, ESs may be highly useful to adaptive management, especially in capturing a snapshot of the relevant knowledge that is applied to a management concern, and the specific ways in which it was used to formulate alternative management approaches (e.g., see Humphries et al. 2008).

Here, we present a new application of an MC-DSS; an industrial area siting (location) model that incorporates the core principles of industrial ecology and sustainable development (Fernández and Ruiz 2009). Our model includes location and environmental sustainability factors, and evaluates prospective land areas at two levels: the first identifies the best regional, district, or district group siting areas for industrial development; the second evaluates specific aspects of localized land areas identified in the first level.

Due to the variety of criteria applied in evaluation, spatial analysis and presentation of the variables is necessary to aid in decision making by the local or regional agents involved. To accomplish this, a fuzzy rule-based expert system was constructed with NetWeaver software using the Ecosystem Management Decision Support (EMDS) system (Reynolds 2002, 2005). The result was a spatial urban planning MC-SDSS that employs MCDA. In our application, we created digital summary maps that show the suitability of different zones for locating potential industrial areas. Additionally, the tool is extensible to evaluating existing industrial areas.

2 Methods

2.1 Study Area

The study area is located in the Cantabria region of northern Spain, in one of the region's principal river corridors. The Cantabria region covers 18 municipalities with a total land surface area of 646.2 km^2, and a population of 121,629 inhabitants (Fig. 1).

2.2 Structure of the Model

Our model applies theory and methods appropriate to industrial area location developed by Fernández and Ruiz (2009). The model includes traditional as well as new value-added factors representing contributions to the global economy. Given the wide variety in potentially available productive spaces, this model focuses on mixed industrial parks, which are typified by small and mid-sized enterprises from secondary and tertiary economic sectors of local and regional importance. In these types of parks, there is little interdependence between enterprises, which are highly varied; there is often much turnover in tenancy, leading to a dynamic rather than fixed cross-section of business enterprises. The conceptual framework of the model is shown in Fig. 2, and described as follows:

- *Structuring the logic model.* This first stage establishes the geographic application scale of the siting model. Each of the levels in the model represents a sorting and ranking of the objectives by the application scale. Once established, the variables and their respective evaluation criteria are defined, categorized, and subcategorized within levels. The place they occupy within the model structure represents the order in which variables are analyzed.

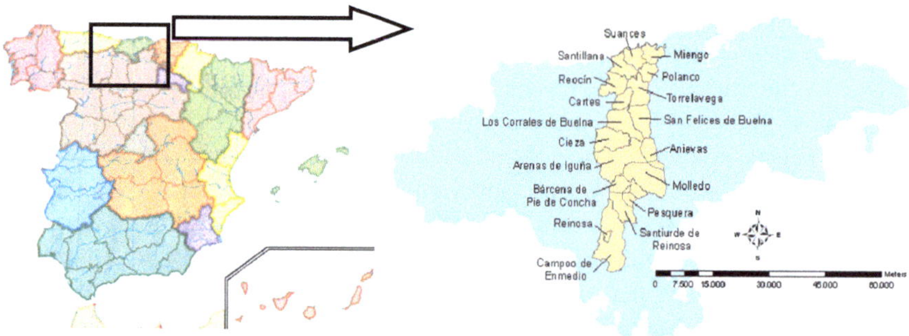

Fig. 1 Location of the study area in the region of Cantabria, Spain

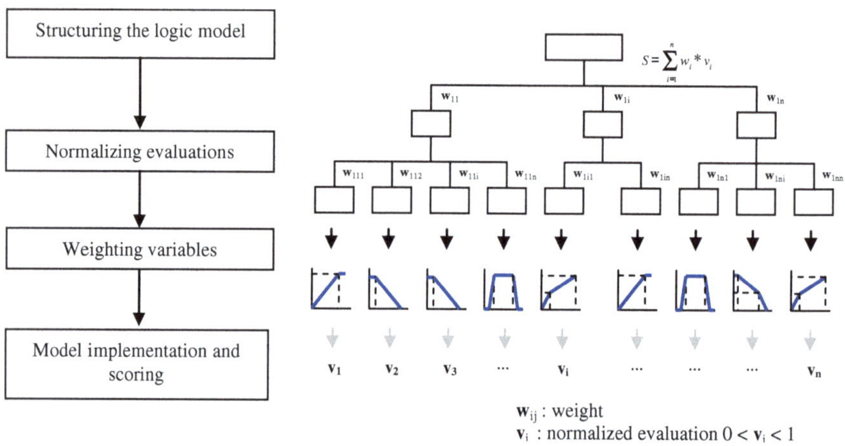

Fig. 2 Overall flow of the logic modeling effort

- *Normalizing evaluations.* Each of the defined variables in the structure is evaluated according to criteria and subcriteria, and reference ranges of values for each criterion and sub-criterion via functions that normalize the results to values that fall between 0 and 1. The evaluation functions are represented using fuzzy logic (Mukaidono 2004) and are implemented in the NetWeaver component of EMDS.
- *Weighting variables.* This is the third stage of the evaluation method and consists of assigning the degree of importance using weights for each level, category, subcategory and indicator. Weights are derived using the Analytical Hierarchy Process (Saaty 1980, 2005). Each of the parties involved in the project acts as decision maker and individually obtains weighting factors. Each decision maker builds comparison matrices using pairs of variables that are present in each node of the hierarchy in descending order. From the comparison matrices, the weight vector is obtained by means of the geometric sum method

and it is checked that the consistency ratio for each trial is less than 10 % so that therefore the obtained weights are consistent. The final weights are calculated using the geometric average of the weighting factors obtained individually and through the consensus of the group. In order to carry out this weighting process the software Expert Choice® was used.

- *Model implementation and scoring*. The scoring of sites is accomplished using a simple weighted scoring algorithm that considers each level, criterion, sub-criterion, and variable, and its weighted contribution, in ascending order in the data structure. The final score represents the estimated degree of compliance of any given site with the established criteria, considering their relative influence. Partial results are obtained for all criteria and sub-criteria, and the final map result is the evaluated area divided into zones and colorized according to the final score, which ranges between 0 and 1.

Figure 3 shows the structure and flow of the siting model and the evaluation sequence. Model structure is presented in two phases: Phase 1) Evaluation and selection of appropriate regional and district zones, and Phase 2) Evaluation of specific lots within regions or districts. The first phase applies to regions or districts covering one or more municipalities, and considers socio-economic, physical-environmental, and infrastructure and urban development criteria. Legal and technical restrictions (sub-criteria) are considered later for areas selected in Phase 1.

Legal restrictions protect zones with outstanding natural beauty, environmental sensitivity, or cultural interest. Technical restrictions are applied to avoid construction or development in zones of high quality soils for agricultural uses, or to avoid physical or substrate hindrances to construction, such as zones with a steep grade or occurring over solid bedrock.

After legal and technical restrictions are considered, potential and probable risks are evaluated, including risks of flooding, landslides, subsidence, and the like. Final lot selection is done in Phase 2 at a local level, using the detailed analysis resulting from Phase 1. In Phase 2, we evaluate available resources and infrastructure linked to water, energy, waste management facilities and ICTs and the specific costs associated with acquiring each lot. The economic feasibility of developing the land is determined as well, along with cost estimates of construction and development of each new industrial park. These factors vary considerably and employ detailed technical reports that are unique to each lot.

2.3 Design of the Spatial Decision Support System

Five methodological steps were followed in developing the logic model:

- *Obtaining the information*. Collect necessary data to parameterize and validate the model.
- *Data entry*. ArcGIS was chosen as the GIS platform (ArcGIS 1997), and all map layers were projected in vector format. In this step, all map layers were set to a

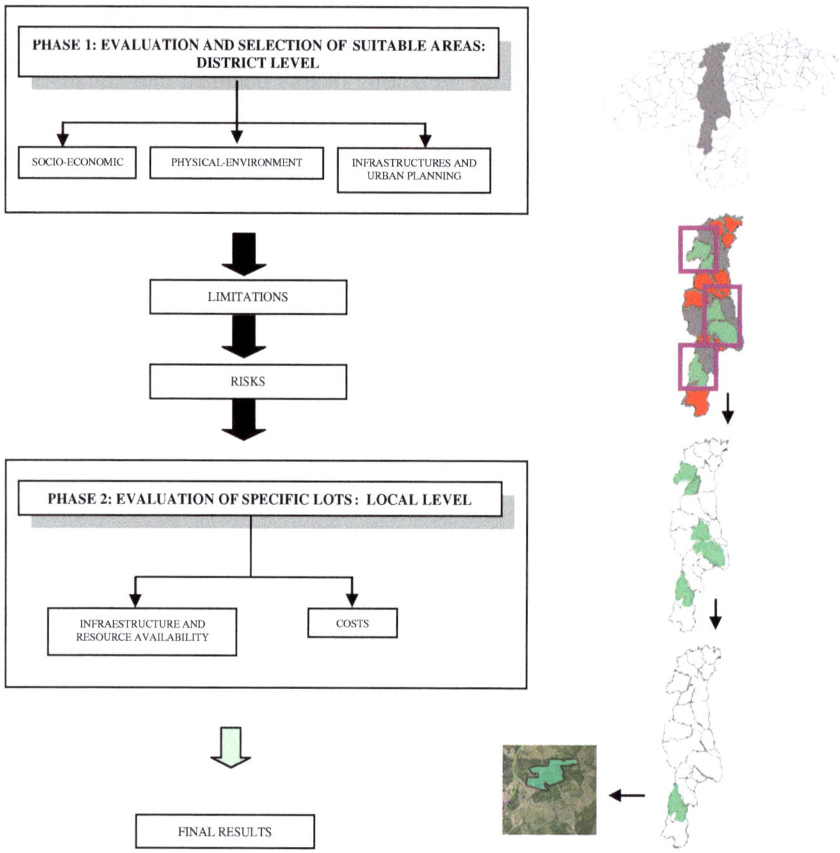

Fig. 3 Structure and flow of the logic model

common map projection (Universal Transverse Mercator System), edited, and edge matched.

- *Data preparation for NetWeaver modeling.* All map layers were grouped into a common file geodatabase and readied for access as a structured flat file that was readable by the NetWeaver model.
- *Multi-criteria evaluation.* Fuzzy evaluation functions and weights were applied to evaluation variables to calculate partial scores.
- *Output.* Results were graphically presented in maps and a dynamically linked data table, which allowed decision makers and users to visualize contributions of variables, subcategories, and categories to the overarching goal.

Linear and trapezoidal fuzzy functions were developed for each variable included in the MC-SDSS (Figs. 3 and 4). The function type (see examples in Fig. 3), parameters of each variable and weighting factor for the socio-economics, physical-environmental, infrastructure, and urban development primary criteria and sub-criteria are shown in Tables 1, 2 and 3 respectively.

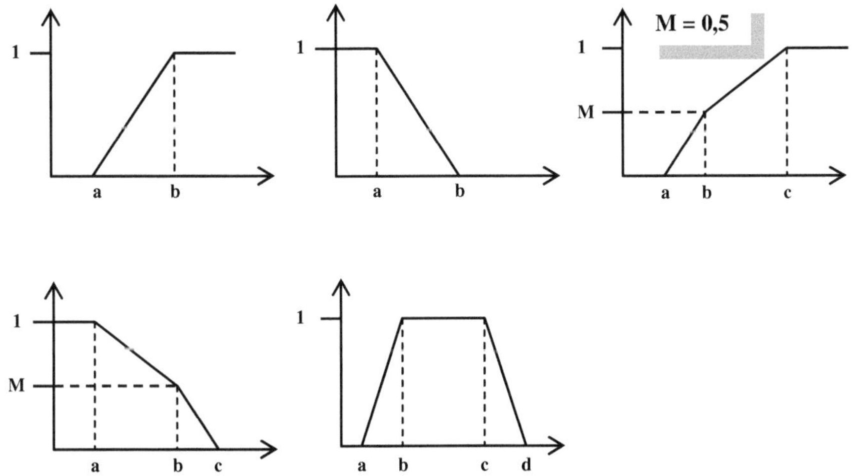

Fig. 4 Evaluation functions used were: *L1* positive linear function; *L2* negative linear function; *L3* positive linear function with changes in slope; *L4* negative linear function with changes in slope; *T1* trapeziodal function

The socio-economic criterion evaluated the potential viability of developing new industrial areas in the context of other economic sector alternatives (Table 1). For example, analysis was completed to determine the contributions of local economies, populations (e.g., demographics, occupations), existing workforces (e.g., age, education level) and locally implemented economic strategies to new industrial activities (Amiti and Pissarides 2005; Kilkenny and Thies 1999).

The physical-environmental primary criterion evaluated the capacity of the natural environment to absorb the impacts resulting from new industrial activities (Table 2). To this end, issues such as water and land availability and quality were analyzed as natural resources. A natural environment can be severely damaged by population density or industrial activity, or by poor management (Ferrarini et al. 2001). Avoiding damage requires knowing the impact absorption capacity and carrying out a thorough planning of the use of resources (Fabbri 1998).

The infrastructure and urban development criterion evaluated the infrastructures and facilities needed for viable operation of a new industrial area using analyses of existing transportation infrastructure, water and electrical supply, among other factors (Table 3). Presence of adequate infrastructure and services of emergency or information and communication is essential for the development of new production activities, and increases the likelihood that a good standard of living is associated with project implementation (Martin 1999; Ferrarini et al. 2001).

Model Builder, a standard utility in ArcGIS, was used to manipulate data and compute values for criteria, sub-criteria, and variables (Fig. 5). NetWeaver software was used to reflect the logic of the ES in a logic dependency network (NetWeaver Developer 1987; EMDS 2004).

Table 1 Evaluation functions and weights of the socio-economic factors

Indicator	Normalized evaluation		Weights
Socio-economical	Type	Parameters	0.143
Social			0.269
Population change (%)	L1	**a** = 0; **b** = 30	0.134
Unemployment rate (%)	L1	**a** = 5; **b** = 10	0.433
Rate of aging (%)	L2	**a** = 1; **b** = 1.25	0.117
Population age structure			0.171
Ages between 16–44 (%)	L1	**a** = 25; **b** = 50	0.667
Ages between 45–49 (%)	L1	**a** = 10; **b** = 20	0.333
Educational level			0.145
Population with no education (%)	L2	**a** = 0; **b** = 11	0.088
Population with primary education (%)	L2	**a** = 25; **b** = 100	0.199
Population with secondary education (%)	T1	**a** = 0; **b** = 23; **c** = 60; **d** = 100	0.478
Population with university education (%)	T1	**a** = 0; **b** = 21; **c** = 30; **d** = 100	0.235
Economic			0.731
Economic sector efficiency			0.120
Efficiency of primary sector (*% population/ % GVA*)	L1	**a** = 0.7; **b** = 1.3	0.157
Efficiency of secondary sector (*% population/ % GVA*)	L2	**a** = 0.7; **b** = 1.3	0.572
Efficiency of tertiary sector (*% population/ % GVA*)	L2	**a** = 0.7; **b** = 1.3	0.271
Occupancy levels in industrial areas (%)	L1	**a** = 0; **b** = 80	0.354
Costs			0.526
Land prices (*normalized cost estimate*)	L1	**a** = 1; **b** = 0	0.651
Workforce costs (*€/worker/month*)	L2	**a** = 1297; **b** = 1297.1	0.198
Housing costs (%)	L2	**a** = 0; **b** = 15.75	0.151

3 Results and Discussion

This section presents and discusses the results obtained from applying the MC-SDSS to the Cantabria region. Model results are presented by the primary criteria–socio-economic factors, physical-environmental factors, and infrastructures and urban development factors. Combining partial scores leads to an overall result that provides diagnostically important information about the most appropriate zones for the development of a new industrial area using sustainability criteria. Legal and technical restrictions that could limit the resulting available land areas are applied to the final results.

Table 2 Evaluation functions and weights of the physical-environmental factors

Indicator	Normalized evaluation		Weights
Physical-environmental	Type	Parameters	0.327
Atmosphere			0.193
Air			0.500
Zones of atmospheric safety (m)	L1	**a** = 500; **b** = 2000	0.276
Environmental pollutant concentration (n° times)	L4	**a** = 30; **b** = 60; **c** = 100	0.473
Stationary pollution sources (m)	L1	**a** = 1000; **b** = 2000	0.252
Noise (dBA)	L4	**a** = 35; **b** = 50; **c** = 65	0.250
Environmental electromagnetic pollution ($\mu W/cm^2$)	L2	**a** = 0; **b** = 450	0.125
Light pollution (*type of luminosity in the sky*)	L1	**a** = 0; **b** = 1	0.125
Land			0.623
Construction specifications			0.443
Land use capacity (*soil type*)	L1	**a** = 0; **b** = 1	0.360
Grade (*grade*)	L2	**a** = 5; **b** = 10	0.640
Soil quality			0.557
Soil agricultural capacity (*scale used in Cantabria*)	L1	**a** = 0; **b** = 1	0.290
Prior industrial use (*yes/no*)	L1	**a** = 0; **b** = 1	0.710
Water			0.184
Presence of surface water			0.255
Presence of rivers (m)	T1	**a** = 250; **b** = 500; **c** = 1000; **d** = 2000	0.682
Presence of lakes, reservoirs (m)	T1	**a** = 250; **b** = 500; **c** = 1000; **d** = 2000	0.318
Surface water quantity			0.454
River quantity (m^2)	L1	**a** = 25000; **b** = 2576100	0.619
Lake, reservoir quantity (m^2)	L1	**a** = 25000; **b** = 2576100	0.381
Surface water quality			0.290
River quality (*water type*)	L1	**a** = 0; **b** = 1	0.619
Lake, reservoir quality (*water type*)	L1	**a** = 0; **b** = 1	0.381

3.1 Socio-Economic Factors

Results from evaluating sub-criteria under the socio-economic factors primary criterion (Fig. 6c) were derived from a weighted scoring of social (26.9 %, Fig. 6a) and economic sub-criteria (73.1 %, Fig. 6b). The presence of zones with moderate suitability was due to lower than average land prices, distance from protected spaces and from important economic activity centers, the presence of more efficient economic sectors, and a higher unemployment rate. Inversely, zones with lowest suitability were characterized by higher than average land prices, poor efficiency of economic sectors, and a low unemployment rate.

Table 3 Evaluation functions and weights of the infrastructures and urban development factors

Indicator	Normalized evaluation		Weights
Infrastructures and urban development	Type	Parameters	0.53
Infrastructures and facilities			0.550
Transportation			0.400
Land transport			0.422
Presence of land transport infrastructures			0.518
Turnpikes and motorways (m)	L4	**a** = 2000; **b** = 5000; **c** = 10000	0.685
Main roads (m)	L4	**a** = 500; **b** = 1500; **c** = 4000	0.224
Secondary roads (m)	L4	**a** = 100; **b** = 1000; **c** = 2000	0.091
Fluidity of land transport infrastructures			0.482
Turnpikes and motorways (*intensity/capacity*)	L2	**a** = 0.5; **b** = 0.9	0.685
Main roads (*intensity/capacity*)	L2	**a** = 0.5; **b** = 0.8	0.231
Secondary roads (*intensity/capacity*)	L2	**a** = 0.5; **b** = 0.7	0.084
Railway transport			0.156
Presence of railway transport infrastructures			0.518
Presence of transfer stations (m)	L4	**a** = 5000; **b** = 10000; **c** = 20000	0.543
Presence of railway (m)	L4	**a** = 3000; **b** = 10000; **c** = 15000	0.457
Railway transport infrastructures quality			0.482
Transfer station category (*station category*)	L1	**a** = 0; **b** = 1	0.568
Railway category (*railway type*)	L1	**a** = 0; **b** = 1	0.432
Sea transport			0.074
Presence of a port (m)	L2	**a** = 20000; **b** = 50000	0.623
Port category (*port category*)	L1	**a** = 0; **b** = 1	0.377
Air transport			0.052
Presence of an airport (m)	L4	**a** = 10000; **b** = 18000; **c** = 30000	0.623
Airport category (*airport category*)	L1	**a** = 0; **b** = 1	0.377
Access to personal transport			0.112
Public transport			0.819
Road transport (m)	L4	**a** = 400; **b** = 1000; **c** = 1200	0.500
Railway transport (m)	L4	**a** = 400; **b** = 1000; **c** = 1200	0.500
Possibility of travelling on foot or by bicycle (m)	L2	**a** = 500; **b** = 2000	0.181
Junctions (m)	L2	**a** = 10000; **b** = 25000	0.184
Energy			0.123
Electricity			0.688
High tension lines (m)	L4	**a** = 100; **b** = 500; **c** = 1000	0.435
Low tension lines (m)	L4	**a** = 50; **b** = 200; **c** = 300	0.122
Power stations and substations (m)	L2	**a** = 500; **b** = 1500	0.444
Natural gas (m)	L2	**a** = 1000; **b** = 5000	0.202
Combustible liquids (m)	L2	**a** = 5000; **b** = 30000	0.110

(continued)

Table 3 (continued)

Indicator	Normalized evaluation		Weights
Water			0.216
Water supply			0.755
Water supply network (m)	L4	**a** = 500; **b** = 1500; **c** = 2000	0.400
Other large infrastructures (water highway) (m)	L2	**a** = 1000; **b** = 10000	0.600
Wastewater treatment			0.245
Wastewater network (m)	L4	**a** = 500; **b** = 1500; **c** = 2000	0.800
Wastewater treatment stations (m)	L4	**a** = 1000; **b** = 7000; **c** = 10000	0.200
Dumping sites (m)	L4	**a** = 10000; **b** = 20000; **c** = 50000	0.050
Waste treatment management and facilities			0.050
Urban solid waste management (m)	L4	**a** = 10000; **b** = 20000; **c** = 50000	0.233
Non-hazardous waste management (m)	L4	**a** = 10000; **b** = 20000; **c** = 50000	0.403
Hazardous waste management (m)	L4	**a** = 10000; **b** = 20000; **c** = 50000	0.364
Landfills			0.060
Presence of landfills			0.400
Urban waste landfills (m)	L4	**a** = 10000; **b** = 20000; **c** = 50000	0.179
Inert waste landfills (m)	L4	**a** = 10000; **b** = 20000; **c** = 50000	0.165
Non-hazardous waste landfills (m)	L4	**a** = 10000; **b** = 20000; **c** = 50000	0.377
Hazardous waste landfills (m)	L4	**a** = 10000; **b** = 20000; **c** = 50000	0.279
Infrastructure capacity			0.600
Urban waste landfills (*years*)	L3	**a** = 10; **b** = 25; **c** = 50	0.254
Inert waste landfills (*years*)	L3	**a** = 10; **b** = 25; **c** = 50	0.152
Non-hazardous waste landfills (*years*)	L3	**a** = 10; **b** = 25; **c** = 50	0.312
Hazardous waste landfills (*years*)	L3	**a** = 10; **b** = 25; **c** = 50	0.282
Information and communication technologies (m)	L4	**a** = 500; **b** = 750; **c** = 2000	0.030
Emergency services (min)	L2	**a** = 10; **b** = 60	0.030
Urban development			0.450
Land use classification and zoning (*zoning*)	L1	**a** = 0; **b** = 1	0.500
Industrial land use (%)	L3	**a** = 65; **b** = 75; **c** = 100	0.100
Territorial land policy			0.400
Coastal law (POL) (*land use policies*)	L1	**a** = 0; **b** = 1	0.568
Special plans (*yes*)	L1	**a** = 0.5; **b** = 1	0.432

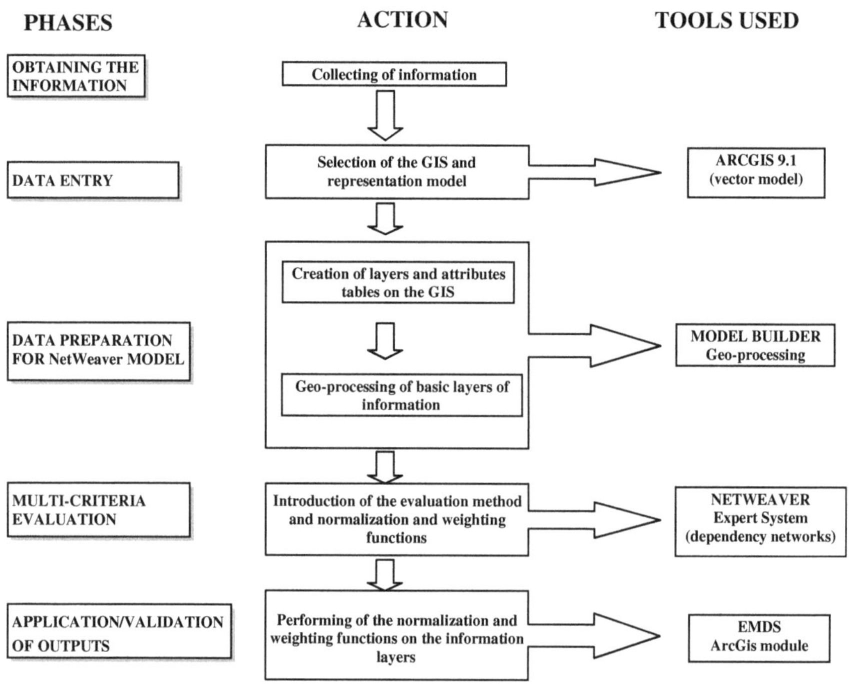

Fig. 5 Implementation methodology for the assessment model

3.2 Physical-Environmental Factors

Results from evaluating subcriteria under the physical-environmental factors primary criterion (Fig. 7d) were derived from a weighted scoring of atmosphere (19.3 %, Fig. 7a), land (62.3 %, Fig. 7b), and water (18.4 %, Fig. 7c) sub-criteria. Model results in Fig. 7d show that none of the eighteen municipalities had homogeneous suitability; rather that suitability diminished as one moved down the Besaya River corridor from south to north, varying from a moderately positive to a very low suitability result.

Regions with low overall physical and environmental suitability displayed low suitability scores in the land and water sub-criteria. For the majority of the municipalities studied, low land suitability was driven by inherently steep grades and a lack of land area already degraded by prior industrial use. Even though the *quantity* of available water (calculated as 50 % of the difference between the annual average volume and the ecological volume) indicated overall suitability throughout the Besaya River basin, the poor *quality* of the water in the tributaries of the lower and central basin led to a negative score for the water sub-criterion.

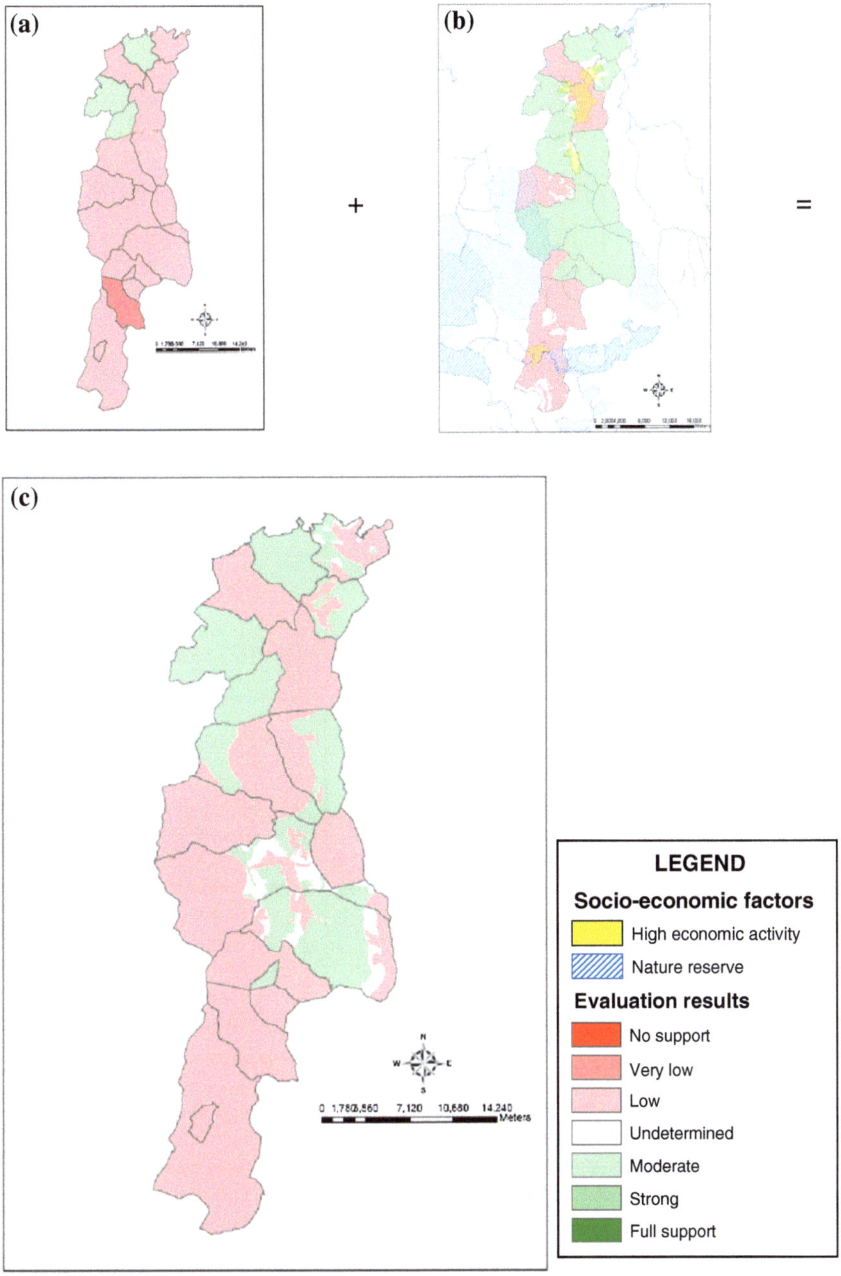

Fig. 6 Results from the evaluation of socio-economic factors: **a** social factors; **b** economic factors; **c** overall result of socio-economic factors

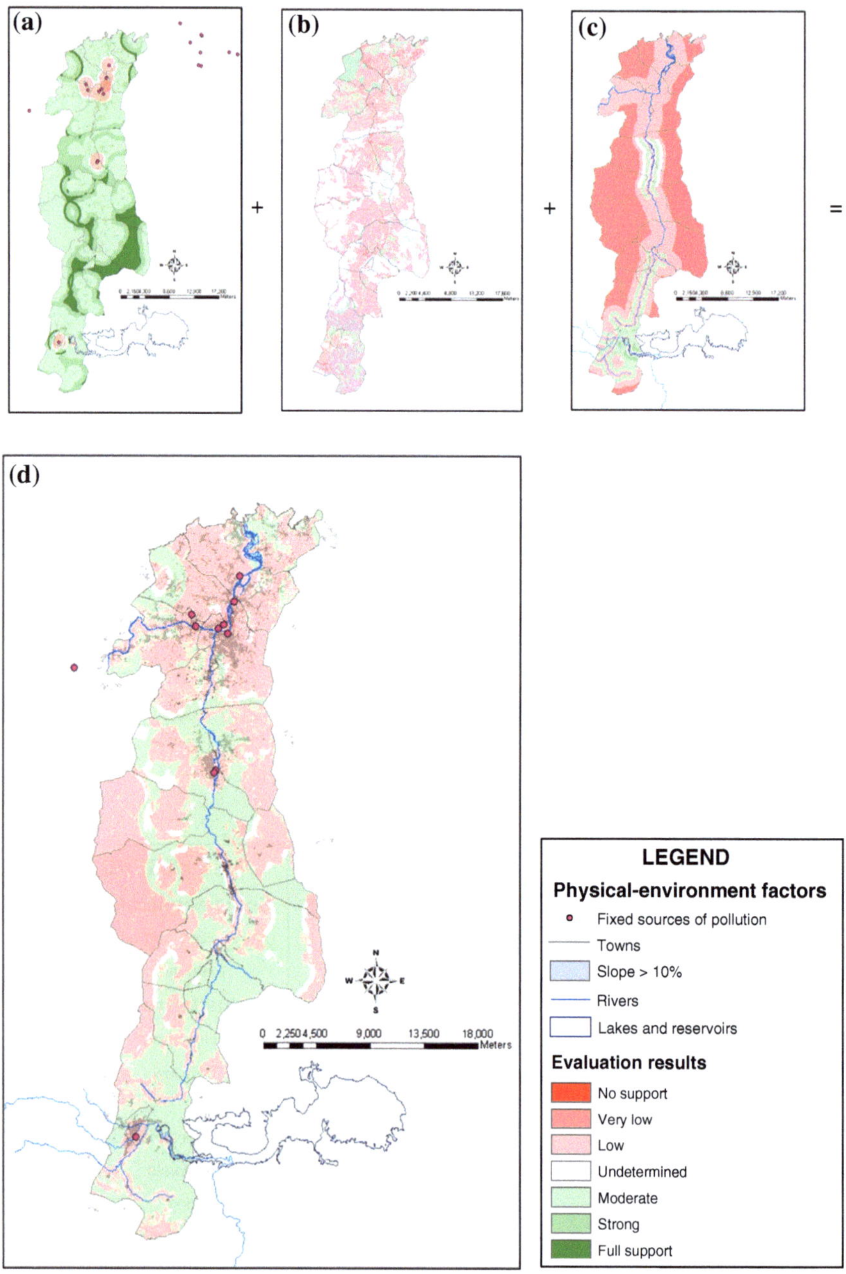

Fig. 7 Results from the evaluation of physical-environmental factors: **a** atmosphere; **b** land; **c** water; **d** overall results of the physical-environmental factors

In contrast, regions that were scored as moderate suitability displayed relatively high air quality, due to distance from point pollution sources, low concentration of pollutants in the environment, good land characteristics for development due to low grades and high carrying capacity, and readily available water resources from nearby rivers and lakes. The moderate suitability displayed in municipalities of zone S-E were exceptional; difficult terrain for development of this zone was compensated for by overall suitability offered by adequate air quality.

3.3 Infrastructures and Urban Development Factors

The map resulting from the evaluation of the infrastructure and urban development criterion is shown in Fig. 8c. Results were derived from a weighted scoring of the infrastructures and facilities (45.7 %, Fig. 8a) and urban development (54.3 %, Fig. 8b) sub-criteria. The central and upper portions of the Besaya River basin exhibited low or very low suitability for development of industrial areas. The zones displaying positive suitability were located in the central and lower basin near road, railway, water and energy infrastructures. Low and very low suitability were primarily driven by low partial scores among urban development factors (45 % of the overall weight), which hinders stimulus and development of new areas of economic activity.

3.4 Full Model and Limitations

The weighted sum of the socio-economic (weight of 14.3 %), physical-environmental (weight of 32.7 %) and infrastructures and urban development (weight of 53 %) criteria shown above provided the total evaluation and final results of the model (Fig. 9).

The resulting land surface area displaying moderate suitability (not considering legal and technical restrictions) was 62,824,157 m^2, 12.77 % of the regional area analyzed. Of the 18 municipalities evaluated, only one exhibited suitability in each of the three primary criteria, five showed low suitability for establishment of an industrial area, and the remaining municipalities produced a mixed distribution of zones with moderate and poor suitability.

The legal and technical restrictions applied to the model in Phase 2 are reflected in the overall map shown in Fig. 10. The majority of the zones were affected by restrictions from Coastal laws, urban zoning, protected spaces and steep grades. The land surface area suitable for industrial development after legal and technical

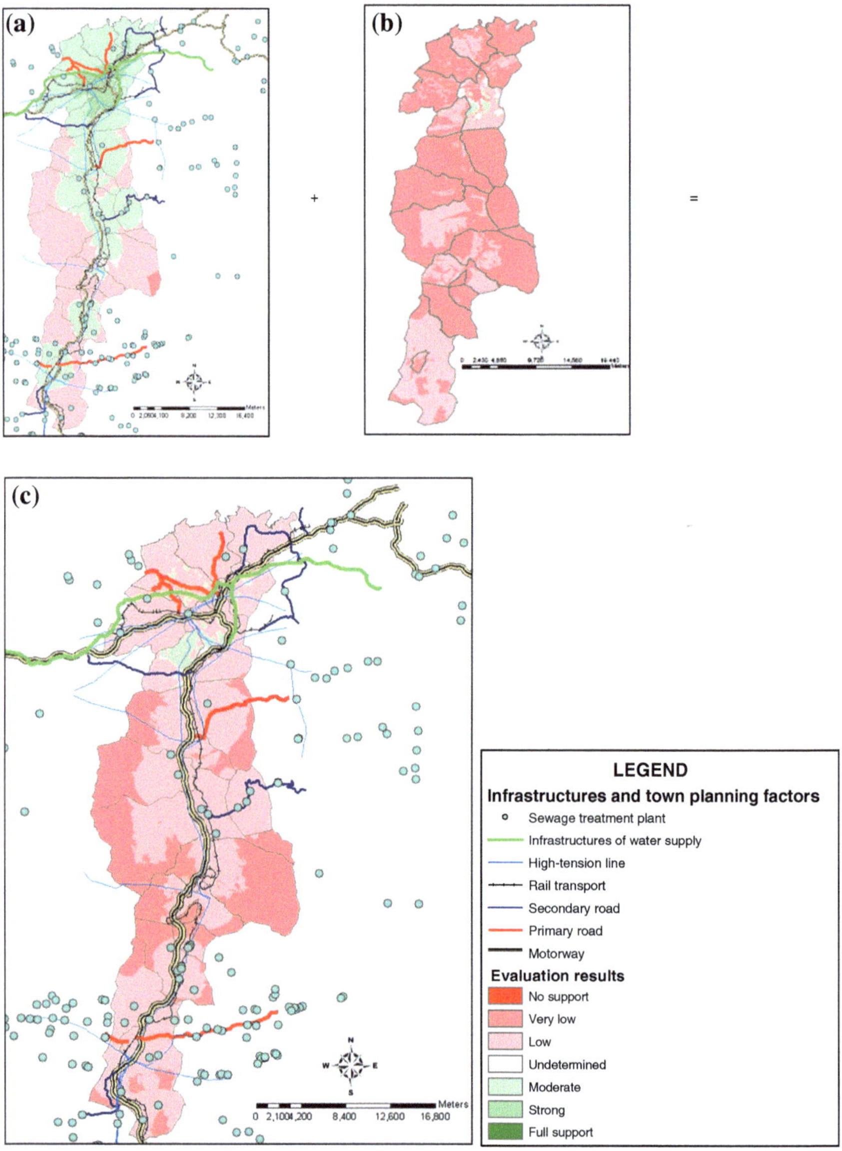

Fig. 8 Results from the evaluation of the infrastructures and urban development factors: **a** infrastructures and facilities; **b** urban development; **c** overall results for infrastructures and urban development factors

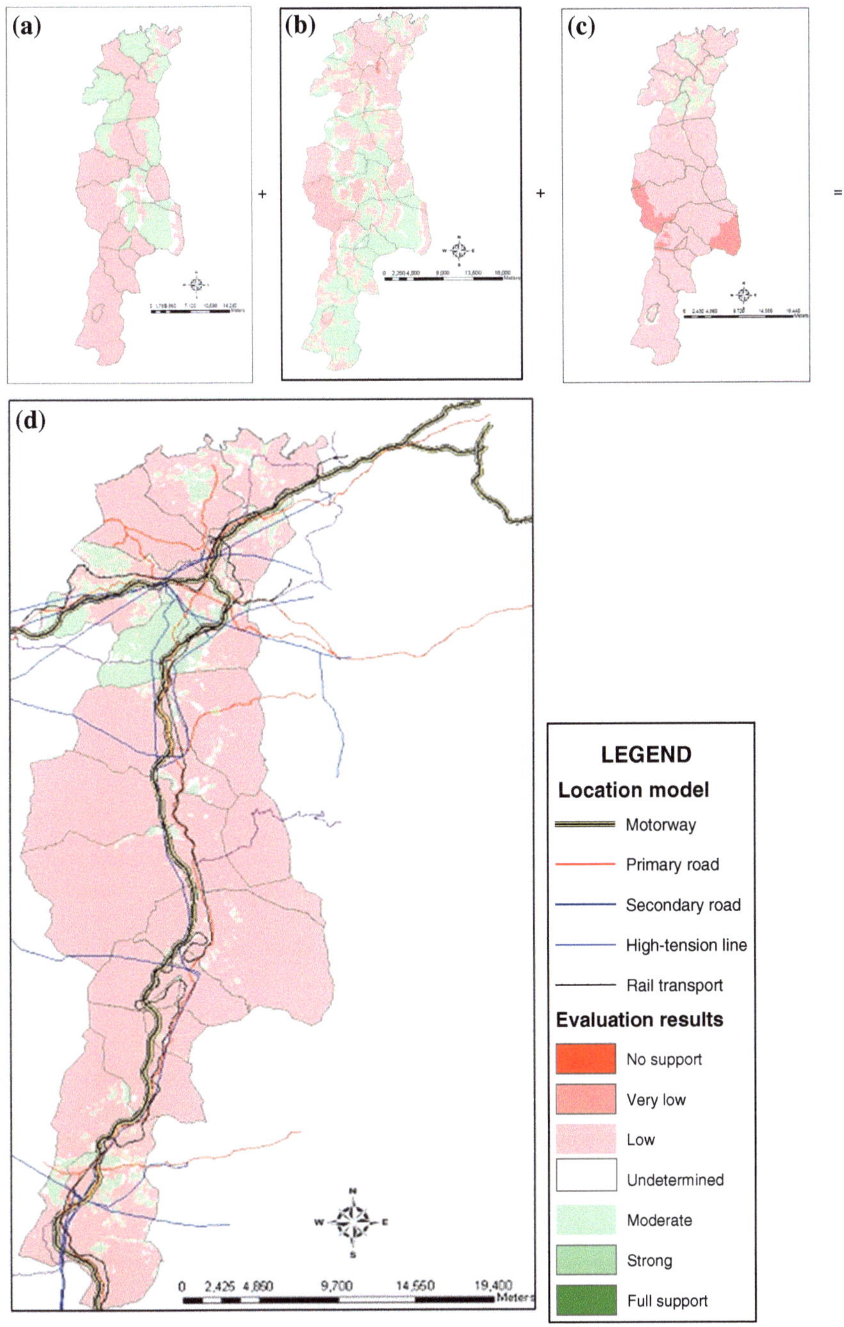

Fig. 9 Complete model: **a** socio-economic factors; **b** physical-environmental factors; **c** infrastructures and urban development factors; **d** overall evaluation

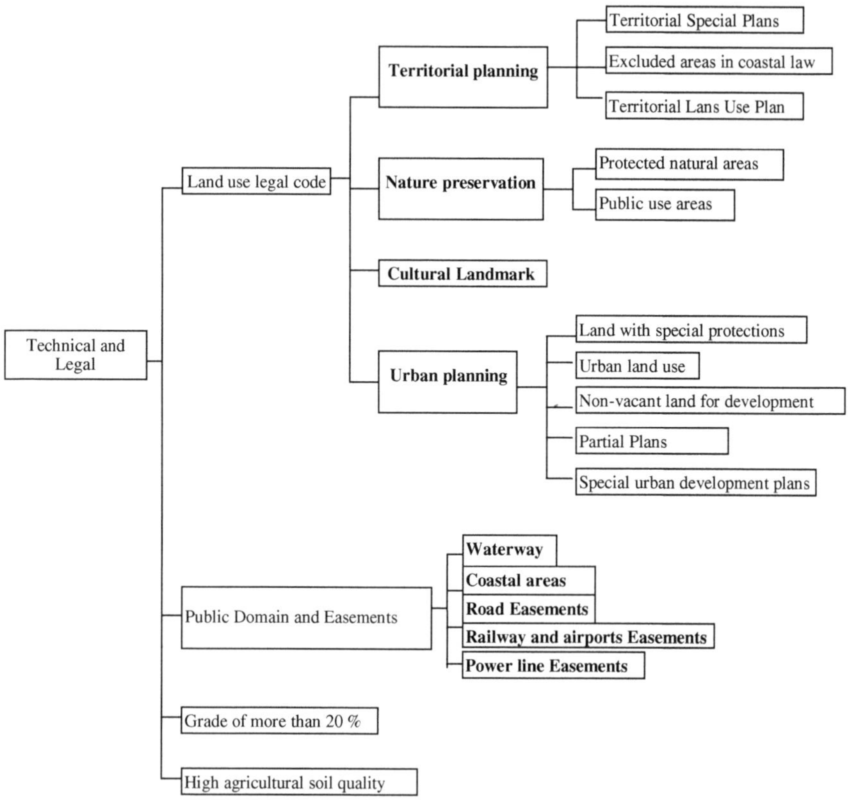

Fig. 10 Overview of land-use restrictions implemented in the model

restrictions were applied was reduced to 33,877,132 m^2, or 5.24 % of the area analyzed. The results for municipalities with moderate suitability before and after applying the restrictions are shown in Table 4. Using these results, one can establish a preference for municipalities where land development for industrial production is socially and environmentally sound, technically feasible, and legally responsible. With few exceptions, we were able to complete evaluations of all municipalities using the same model evaluation sequence before and after applying the legal and technical restrictions.

Table 4 Data on the surface areas of municipalities with moderate suitability

Municipality	Municipal surface area (m^2)	Moderate suitability surface area (m^2)		
		Total	Restricted	Free of restrictions
Cartes	19,000,000	18,049,397.79	10,782,012.10	7,267,385.73
Reocín	32,100,000	13,611,702.49	3,805,280.80	9,806,421.69
Campoo de Enmedio	91,100,000	8,425,884.87	4,683,060.48	3,742,824.39
Torrelavega	35,500,000	8,018,268.10	3,535,653.02	4,482,615.08
Suances	24,600,000	3,588,650.88	665,911.45	2,922,739.43
Polanco	12,700,000	2,888,330.47	502,827.43	2,385,503.04
Miengo	21,500,000	2,159,872.72	1,212,092.60	947,780.12
Los Corrales de Buelna	45,400,000	1,930,014.93	1,788,373.91	141,641.02
Molledo	71,100,000	1,263,473.09	496,964.85	766,508.24
Arenas de Iguña	86,800,000	1,095,616.39	965,045.72	130,570.67
Reinosa	4,100,000	961,770.86	61,731.82	900,039.04
San Felices de Buelna	36,200,000	717,451.11	416,990.55	300,460.56
Pesquera	8,900,000	113,723.34	31,080.45	82,642.89

4 Conclusions

Planning and location of industrial areas is a strategic decision with powerful implications for existing and future activities and human communities within a given area. Typically, industrial location or siting models have only considered the presence of basic infrastructures and proximity to raw materials or the market place. However, past failures to design the built environment to be in greater harmony with the natural environmental have generated a new awareness of the need to create a more sustainable industrial development model. Such a model requires attention to a wider range of environmental criteria, legal and technical restrictions, which can add complexity to any decision making. The MC-SDSS set forth in this chapter demonstrates a clear application of this new awareness.

Integration of socio-economic and environmental information, as in the present model, is a valuable and practical approach to decision making and the elaboration of strategic policies. The present model shows how one can integrate the principles of industrial ecology with an operational siting of industrial areas.

Of the available options for the design and construction of an MC-SDSS, the combined NetWeaver and EMDS-ArcGIS platform met the needed requirements. The NetWeaver tool allowed us to propose in the design stages, a variety of different criteria and structures for grouping variables and evaluating data, which facilitated the planning process for decision makers and provided considerable flexibility in model design and implementation. Furthermore, the use of an expert system approach implemented with fuzzy logic allowed for the relatively simple application of a theoretic model with 75 variables.

The MC-SDSS is an open and changeable system. The logic structure, data considered, and evaluation criteria may all be readily adapted to unique territorial needs and the scientific and technical advances at the time. But to be clear, a new user of the MC-SDSS system would need to spend several days with the model and data to become adequately familiar with its structure and data requirements, such that they could adapt it to their own uses. However, this is generally true when a user is gaining familiarity with any new modeling system—there is start-up time.

One final aspect when considering transfer and application of an MC-SDSS, such as this one, to another environment is the critical nature of the data collection stage. In general, at least 75–90 % of the overall effort is collecting and organizing the needed data structure for access by the MC-SDSS. In our application, the SDSS was applied to a region of Spain located in the corridor of one of its principal rivers. The surface area analyzed was 646.2 km^2, or roughly 12.38 % of the entire region. To facilitate a rapid data capture and development process, we created an integrated work environment in ArcGIS, which reduced the time and effort needed to generate useful information. We also spent considerable time evaluating available data sources for the highest quality data because the reliability and quality of the information used has a critical influence on the accuracy of the results. Low quality data lead to low quality modeling and industrial area planning results.

In conclusion, we constructed a decision-support system for regional and urban planning using the functionality of EMDS. This application of EMDS led to the creation of digital maps that were highly useful to distinguishing different zones and their appropriateness for sustainable industrial area siting.

References

Aalborg Chart (1994) First conference or sustainable cities and Towns, Aalborg

Amiti M, Pissarides CA (2005) Trade and industrial location with heterogeneous labour. J Int Econ 67:392–412

ArcGis v9.1 (1997) Environmental Systems Research Institute ESRI, Inc., Redlands. http://www.esri-es.com

Chang N, Parvathinathan G, Breeden JB (2008) Combining GIS with fuzzy multicriteria decision-making for landfill sitting in a fast-growing urban region. J Environ Manage 87:139–153

Côté RP, Cohen-Rosenthal E (1998) Designing eco-industrial parks: a synthesis of some experiences. J Clean Prod 6:181–188

Dai FC, Lee CF, Zhang XH (2001) GIS-based geo-environmental evaluation for urban land-use planning: a case study. Eng Geol 61:257–271

Directive 2001/42/EC (2001) Directive of the European Parliament and of the Council on the assessment of the effects of certain plans and programmes on the environment, Luxembourg

EMDS v3.1 (2004) Ecosystem Management Decision Support. Redlands Institute, Redlands. http://www.institute.redlands.edu/emds/index.htm

Erkman S (1997) Industrial ecology: an historical view. J Clean Prod 5(1–2):1–10
Fabbri KP (1998) A methodology for supporting decision making in integrated coastal zone management. Ocean Coast Manage 39:51–62
Fernández I, Ruiz MC (2009) Descriptive model and evaluation system to locate sustainable industrial areas. J Clean Prod 17:87–100
Ferrarini A, Bodini A, Becchi M (2001) Environmental quality and sustainability in the province of Reggio Emilia (Italy): using multi-criteria analysis to assess and compare municipal performance. J Environ Manage 63:117–131
Figueiredo O, Guimaraes P, Woodward D (2002) Home-field advantage: location decisions of Portuguese entrepreneurs. J Urban Econ 52:341–361
Frosh R (1995) The industrial ecology of the 21st century. Sci Am 273(8):178–181
Frosch RA, Gallopoulos NE (1989) Strategies for manufacturing in managing planet earth. Sci Am 261(3):144–153
Gibbs D, Deutz P (2005) Implementing industrial ecology? Planning for eco-industrial parks in the USA. Geoforum 36:452–464
Gibbs D, Deutz P (2007) Reflections on implementing industrial ecology through eco-industrial park development. J Clean Prod 15:1683–1695
Graedel TE, Allenby BR (2003) Industrial ecology. Prentice Hall, New Jersey
Herbst H, Herbst V (2006) The development of an evaluation method using a geographic information system to determine the importance of wasteland sites as urban wildlife areas. Landsc Urb Plan 77:178–195
Hernández J, García L, Ayuga F (2004) Assessment of the visual impact made on the landscape by new buildings: a methodology for site selection. Landsc Urb Plan 68:15–28
Hoover M (1948) The location of economic activity. McGraw-Hill, New York. http://www.archive.org/details/locationofeconom 02987 4mbp. Accessed 20 Oct 2008
Humphries HC, Bourgeron PS, Reynolds KM (2008) Suitability for conservation as a criterion in regional conservation network selection. Biodivers Conserv 17:467–492
Isard W (1954) Location and space-economy. The M.I.T Press, Cambridge
Kallali H, Anane M, Jellali S, Tarhouni J (2007) GIS-based multi-criteria analysis for potential wastewater aquifer recharge sites. Desalination 215:111–119
Kalogirou S (2002) Expert systems and GIS: an application of land suitability evaluation. Comput Environ Urban Syst 26:89–112
Kilkenny M, Thies JF (1999) Economics of location: a selective survey. Comput Oper Res 26:1369–1394
Kirschner E (1995) Eco-industrial parks find growing acceptance. Chem Eng News 73:15
Korhonena J, Snäkin J (2005) Analysing the evolution of industrial ecosystems: concepts and application. Ecol Econ 52:169–186
Lambert AJD, Boons FA (2002) Eco-industrial parks: stimulating sustainable development in mixed industrial parks. Technovation 22:471–484
Leitham S, McQuaid RW, Nelson JD (2000) The influence of transport on industrial location choice: a stated preference experiment. Transp Res 34:515–535
Lösch A (1954) The economics of location. Yale University Press, New Haven
Lowe EA, Evans LK (1995) Industrial ecology and industrial ecosystems. J Clean Prod 3(1–2):47–53
Lowe EA (2001) Eco-industrial park handbook for Asian developing countries. A Report to Asian Development Bank, Environment Department, Indigo Development, Oakland
Malczewski J (1999) GIS and multicriteria decision analysis. Wiley, New York
Maniezzo V, Mendes I, Paruccini M (1998) Decision support for sitting problems. Decis Support Syst 23:273–284
Martin P (1999) Public policies, regional inequalities and growth. J Public Econ 73:85–105
Mukaidono M (2004) Fuzzy logic for beginners. World Scientific, New Jersey
NetWeaver Developer (1987) Rules of Thumb, Inc., Pennsylvania. http://rules-of-thumb.com

Reynolds KM (2002) Landscape evaluation and planning with EMDS 3.0. 2002 ESRI user conference, San Diego, CA, 9–12 July 2002. Environmental Systems Research Institute, Redlands

Reynolds KM (2005) Integrated decision support for sustainable forest management in the United States: fact or fiction? Comput Electron Agric 49:6–23

Roberts BH (2004) The application of industrial ecology principles and planning guidelines for the development of eco-industrial parks: an Australian case study. J Clean Prod 12:997–1010

Rosenfeld SA (1997) Bringing business clusters into the mainstream of economic development. Eur Plan Stud 5:3–23

Saaty TL (1980) The analytic hierarchy process: planning, priority setting, resource allocation. McGraw-Hill, New York

Saaty TL (2005) Theory and applications of the analytic network process: decision making with benefits, opportunities, costs and risks. RWS Publications, Pittsburgh

Smith D (1971) Industrial location: an economic geographical analysis. Wiley, New York

Somlev IP, Hoshino Y (2005) Influence of location factors on establishment and ownership of foreign investments of foreign investments: the case of the Japanese manufacturing firms in Europe. Int Bus Rev 14:577–598

Tabeada J, Matías JM, Araújo M, Ordóñez C (2006) Assessing the viability of underground slate mining by combining an expert system with a GIS. Eng Geol 87:75–84

Trinder B, Fohl A, Shayt DH et al (1993) The blackwell encyclopedia of industrial archaeology. Blackwell Publishers, Oxford

Tudor T, Adam E, Bates M (2007) Drivers and limitations for the successful development and functioning of EIPs (Eco-Industrial Parks): a literature review. Ecol Econ 61:199–207

UNEP, United Nations Environment Programme (1995) Cleaner production: a guide to sources of information. UNEP, Paris

UNEP, United Nations Environment Programme (1997) The environmental management of industrial estates. UNEP, Paris

Measuring Biological Sustainability via a Decision Support System: Experiences with Oregon Coast Coho Salmon

Thomas C. Wainwright, Peter W. Lawson, Gordon H. Reeves, Laurie A. Weitkamp, Heather A. Stout and Justin S. Mills

Abstract Conservation of Pacific salmon (*Oncorhynchus* spp.) has become increasingly important as major populations have declined in abundance to the point of being listed under the U.S. Endangered Species Act. The complex life-history of Pacific salmon species and the diversity of habitats they occupy require multifaceted recovery efforts, and the metrics needed to evaluate species status and progress toward recovery are necessarily complex. Formal decision support systems (DSS) are designed to assist decision-makers in integrating and evaluating many factors. We describe a knowledge-based DSS for evaluating the biological status of Oregon coast coho salmon (O. *kisutch*). We then compare our DSS to similar tools and consider its advantages and disadvantages. We show how the DSS can provide a transparent and logical framework linking multiple criteria across geographic scales for a unified assessment. Once constructed, the DSS can serve as an institutional knowledge base, codifying the pathways from data to criteria evaluation and supporting consistent future status evaluations with a path to incorporating new knowledge over time. The DSS was not trivial to implement, nor is it easy to explain to resource managers, and we offer suggestions to address these problems. The DSS was particularly helpful in providing a logical and reproducible way to quantify multiple risks and assess progress toward recovery across multiple spatial and temporal scales. Development of this DSS is an important step in the evolution of assessment tools for salmon conservation.

T. C. Wainwright (✉) · P. W. Lawson · L. A. Weitkamp · H. A. Stout
Northwest Fisheries Science Center, National Oceanic and Atmospheric Administration, 2032 Southeast OSU Drive, Newport, OR 97365, USA
e-mail: thomas.wainwright@noaa.gov

G. H. Reeves
U.S. Department of Agriculture, Forest Service, Pacific Northwest Research Station, Forestry Sciences Laboratory, 3200 SW Jefferson Way, Corvallis, OR 97331, USA
e-mail: greeves@fs.fed.us

J. S. Mills
Frank Orth and Associates, 4040 Lake Washington Boulevard NE, Suite 208, Kirkland, WA 98033, USA

K. M. Reynolds et al. (eds.), *Making Transparent Environmental Management Decisions*, Environmental Science and Engineering,
DOI: 10.1007/978-3-642-32000-2_12, © Springer-Verlag Berlin Heidelberg 2014

Keywords Coho salmon · Biological conservation · Decision support · Sustainability · Recovery planning · Extinction risk

1 Introduction

Decision-support systems (DSS) are widely advocated as a means of structuring diverse information to support natural resource decisions. In conservation, they are used for spatial planning and prioritization of conservation actions (e.g., Steel et al. 2008; Wilson et al. 2011; Beechie et al. 2012, other chapters, this volume) and for tracking species at risk of extinction (Wong et al. 2007). Here, we consider their use in assessing biological sustainability of Pacific salmon (*Oncorhynchus* spp.) listed under the U.S. Endangered Species Act (ESA).

Our work focuses on the Oregon Coast Coho Salmon (OCCS) Evolutionarily Significant Unit (ESU). This ESU is composed of a distinct group of coho salmon (*O. kisutch*) populations that spawn in rivers and streams along the Oregon Coast (Weitkamp et al. 1995). Abundance of OCCS formerly exceeded one million adults (Lichatowich 1989), but declined in abundance by about 90 % during the 20th century, reaching historic low abundances in the 1990s (Weitkamp et al. 1995, 2000). Concern over these declines led the National Marine Fisheries Service (NMFS) to list OCCS as a threatened species under the ESA (NMFS 1998, 2008).

Coho salmon is an anadromous species that ranges along the west coast of North America from central California to Point Hope, Alaska and along the coast of Asia from Kamchatka to the Sea of Japan/East Sea (Sandercock 1991). There is much latitudinal variation in life-history patterns, but along the Oregon Coast, coho salmon typically spawn in freshwater during winter. Offspring rear for approximately a year and a half in freshwater before entering the ocean, where they rear for another year and a half before returning to freshwater to spawn (Sandercock 1991; Weitkamp et al. 1995).

In 2002, NMFS convened an OCCS Work Group to describe the biological conditions required for recovery of the ESU. OCCS have a complex metapopulation structure, with populations spawning and rearing in a range of coastal freshwater systems including ephemeral streams, large rivers, lakes, and tidal lowlands. In describing the population structure of this ESU, the work group identified five "biogeographic strata" representing major genetic and ecological diversity within OCCS, and identified 21 major independent populations within these strata (Lawson et al. 2007; Fig. 1).

In addition to identifying population structure, the work group developed a suite of biological sustainability criteria as a formal DSS (Wainwright et al. 2008). The term "biological sustainability" implies that a population is able to survive prolonged periods of adverse environmental conditions, while maintaining its genetic legacy and long-term adaptive potential. Sustainability also implies that habitat

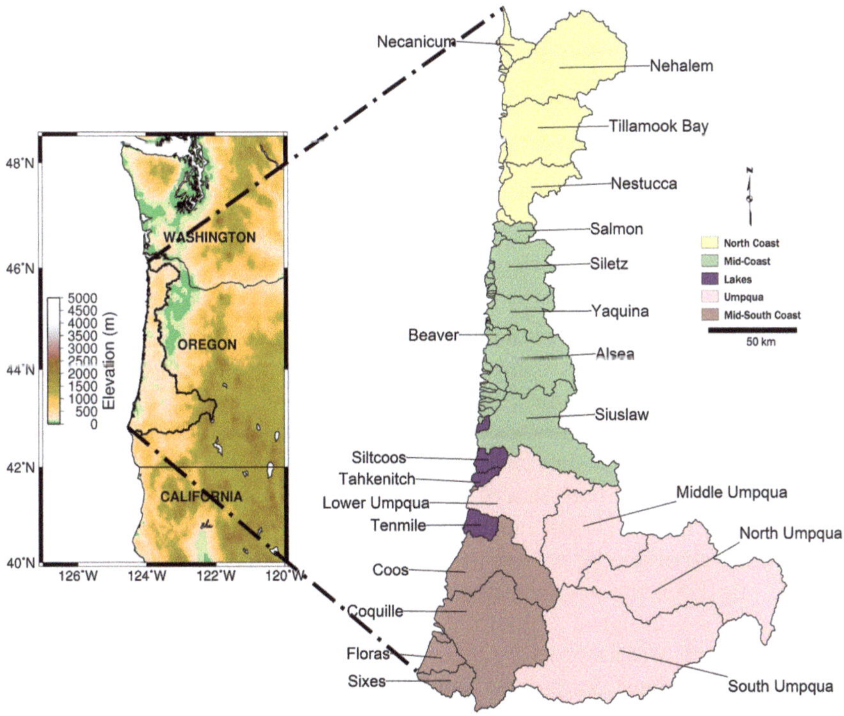

Fig. 1 Map of the range of the Oregon Coast Coho Salmon ESU, showing the locations of 21 major populations (named river basins) and numerous smaller populations (unnamed basins). The individual populations are aggregated into 5 biogeographic strata, indicated by different colors. Redrawn from Lawson et al. (2007)

conditions needed to support the full expression of the populations' life history diversity will be available into the foreseeable future (Wainwright et al. 2008).

A DSS is a computer-based tool that can analyze and compare numerous pieces of data, producing results that assist managers in making decisions (Turban and Aronson 2001). These systems allow decision makers to perform complex evaluations quickly, present a consistent assessment that draws from a variety of data sources, and accurately track large sets of information, thus improving decisions by supplementing and supporting human judgment (Rauscher 1999). In addition, a DSS can incorporate substantial uncertainty about the precise conditions that are optimal for the target organisms. Uncertainties may result from gaps in information and the lack of perfect knowledge about the interrelationships among relevant factors (Reynolds and Hessburg 2005). The OCCS DSS has been used as part of recovery planning by the state of Oregon (Chilcote et al. 2005) and by NMFS as part of an updated ESA status review (Stout et al. 2012) and ongoing recovery planning activities (NMFS 2013).

To our knowledge, such systems have not previously been applied directly to the evaluation of status for species at risk. Here, we summarize the OCCS DSS structure, illustrate its application, and discuss its utility for conservation management. In doing so, we consider whether such a DSS can improve practical decisions regarding the status of species at risk.

2 Methods

A complete assessment of the biological condition of any species is necessarily multifaceted, including a variety of interrelated criteria, with varying data quantity and quality. For OCCS, these criteria relate to biological processes at a variety of temporal and spatial scales. Time scales vary from a single salmon generation (ca. 3 yr) to evolutionary time (100–1000s of years), and spatial processes vary from individual stream reaches (ca. 1 km) to the domain of the ESU (ca. 400 km). To track this large suite of data and criteria in a transparent and logically consistent framework, the work group decided that some form of DSS was essential.

The OCCS DSS was designed to provide an integrated suite of biological sustainability criteria to be used as part of ESA recovery planning. As a "support system," the DSS was not intended to "make" a decision on ESA listing or de-listing—it was designed to inform a decision-maker's judgment, not to replace it. Our approach was inspired by that used by the U.S. Forest Service for the Aquatic and Riparian Effectiveness Monitoring Plan component of the Northwest Forest Plan (Reeves et al. 2004). In defining the component criteria, the work group sought to meet the following characteristics of good recovery criteria:

- Measurable—Criteria are quantifiable based on obtainable data
- Comprehensive—Criteria include all important aspects of sustainability
- Sound—Criteria are based on best available science
- Transparent—Criteria are clearly defined and consistent
- Reproducible—Criteria can be consistently re-applied
- Useful—Criteria provide practical advice in a form useful to decision makers

2.1 DSS Structure

Here, we present an abridged summary of the DSS's more complex structure; a full description of which can be found in the work group's report (Wainwright et al. 2008).

The OCCS biological recovery criteria encompass a variety of biological requirements that contribute to ESU sustainability. To incorporate these into a DSS, the various criteria were expressed as a network of clearly-defined logical propositions whose truth could be evaluated from data. The DSS thus consisted of a hierarchical set of individual propositions describing various aspects of sustainability (Table 1). At the lowest level, propositions were evaluated based on

Table 1 Definitions of sustainability propositions evaluated in the decision support system (DSS)

Title	Definition	Code
ESU-Scale propositions		
ESU sustainability	The ESU is self-sustaining into the foreseeable future	ES
All strata sustainable	All biogeographic strata are sustainable (see Stratum Sustainability)	ES-1
ESU-scale diversity	The ESU has sufficient broad-scale diversity to maintain its ecological and evolutionary functions into the foreseeable future	ES-2
Genetic diversity	ESU-scale genetic diversity is sufficient for long-term sustainability of the ESU	ED-1
Genetic structure	Genetic diversity within the ESU is comparable to healthy coho salmon ESUs and forms the basis for life-history diversity	ED-1a
Effects of selection	Human-driven selection is not sufficient to decrease genetic diversity	ED-1b
Effects of migration	Genetic diversity is not compromised by changes in the movements of fish	ED-1c
Phenotypic and habitat diversity	ESU-scale phenotypic and habitat diversity are sufficient for long-term sustainability of the ESU	ED-2
Phenotypic diversity	Phenotypic diversity is present within the ESU at levels comparable to healthy ESUs or the historical template.	ED-2a
Habitat diversity	Habitats are sufficiently productive, diverse, and accessible to promote phenotypic plasticity	ED-2b
Small populations	Dependent populations within the ESU are not permanently lost	ED-3
Biogeographic Stratum-Scale propositions		
Stratum sustainability	The stratum is self-sustaining (in terms of both diversity and functionality) into the foreseeable future	SS
Stratum diversity	Most of the historically independent populations in the stratum are at present sustainable (see Population Sustainability)	SD
Stratum functionality	All of the historically independent populations in the stratum are functional (see Population Functionality)	SF
Population-Scale propositions		
Population sustainability	The population is able to sustain itself into the future. Requires both Population Persistence and Population Diversity	PS
Population persistence	The population will persist for the next 100 years	PP
Population productivity	Productivity at low abundance is sufficient to sustain the population through an extended period of adverse environmental conditions	PP-1

(continued)

Table 1 (continued)

Title	Definition	Code
Probability of persistence	The population has a high likelihood of persisting over the next 100 years, as estimated from PVA models	PP-2
Critical abundance	Population abundance is maintained above levels where small-population demographic risks are likely to become significant	PP-3
Population diversity	The population has sufficient diversity and distribution to ensure continued fitness in the face of environmental change	PD
Spawner abundance	The population has sufficient naturally produced spawners to prevent loss of genetic variation due to random processes over a 100-year time frame	PD-1
Artificial influence	The abundance of naturally spawning hatchery fish will not be so high as to be expected to have adverse effects on natural populations	PD-2
Spawner distribution	On average, the historically occupied watersheds in the population's range have spawners occupying the available spawning habitat (see Watershed Spawner Occupancy)	PD-3
Juvenile distribution	On average, the historically occupied watersheds in the population's range have juveniles occupying the available juvenile habitat (see Watershed Juvenile Occupancy)	PD-4
Population functionality	Habitat quality and quantity are adequate to support sufficient abundance to maintain long-term genetic integrity of the population	PF
Watershed-Scale propositions		
Watershed spawner occupancy	Spawners occupy a high proportion of the available spawning habitat within the watershed	W-Sp
Watershed juvenile occupancy	Juveniles occupy a high proportion of the available juvenile habitat within the watershed	W-Ju

Propositions are testable assertions; they are organized by metapopulation scale. "Code" is the abbreviation used to identify each proposition as it is evaluated in the DSS, shown graphically in Fig. 2

data collected at the *population* or *watershed* scale (Fig. 2). Various low-level propositions were combined at the *population* scale, then aggregated upward to the *stratum* scale and finally to the entire *ESU* (flow chart in Fig. 2). This approach views OCCS as a metapopulation, with substructure at a continuum of scales, from individual spawning grounds, up to the entire OCCS range. Along this continuum, the work group identified four discrete scales as important for defining attributes related to ESU status: (1) watersheds (defined as fifth-field hydrologic units, aka 5th HUCs, Seaber et al. 1987. Regional Ecosystem Office 2002), (2) populations

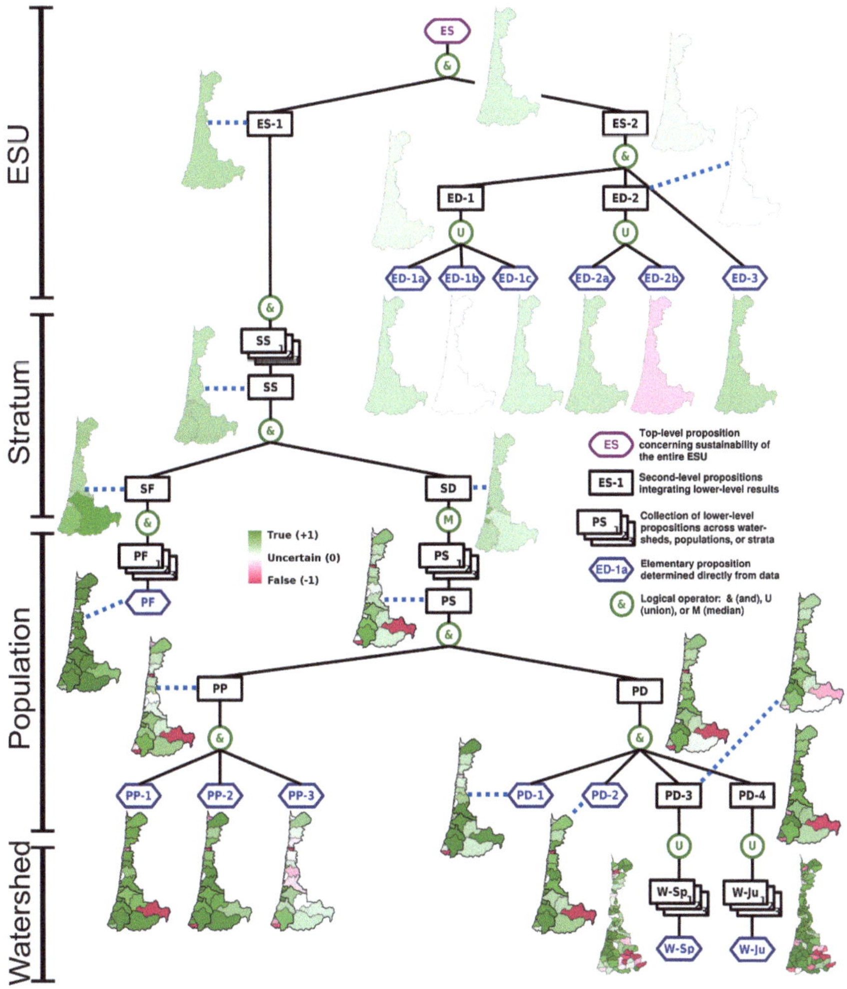

Fig. 2 The logic network for the decision support system, showing the information flow from lower-level propositions to the top-level ESU Sustainability ("ES") proposition, along with maps of resulting strength of evidence in support of each proposition. Proposition abbreviations are defined in Table 1

(fish residing in major river basins or subbasins), (3) biogeographic strata (regional population aggregates), and (4) the entire ESU (Lawson et al. 2007). Data used to assess OCCS status included juvenile salmon densities, adult spawner (a spawning fish) abundance and distribution, the proportion of spawners that were of hatchery origin, and professional judgment regarding ESU-wide genetic, phenotypic, and habitat diversity (Table 1). In our judgment, a multi-scale, multi-parameter approach would best support maintaining natural process variability, an important component of salmon restoration planning (Bisson et al. 2009).

Watershed-scale Attributes. Many ecological processes occur at small geographic scales. For the most part, these considerations have been subsumed into population-scale attributes, but two watershed-scale propositions (Juvenile Occupancy and Spawner Occupancy; Table 1) were used to examine within-population diversity and distribution.

Population-scale Attributes. There are a variety of population-scale attributes that contribute to population health. For NMFS recovery plans, these attributes are typically categorized into four parameters: population size, population growth rate, spatial structure, and diversity (McElhany et al. 2000). These same four parameters appear in the Oregon Native Fish Conservation Policy (Oregon Department of Fish and Wildlife 2003). The four are not independent of one another and their relationship to sustainability depends on a variety of interdependent ecological processes. Hence, the work group included 11 parameters in our analysis at this scale (Table 1).

Biogeographic Stratum-scale Attributes. In evaluating OCCS status, biogeographic strata play two important roles. First, they represent the largest pieces of habitat, genetic, and life history diversity, thus ensuring that they preserve much of the among-population diversity. Second, by ensuring that all the strata are preserved, a hedge is provided against loss of the whole ESU from large-scale catastrophic disturbance(s). The important attributes of strata are primarily the number and status of their constituent populations and associated habitats. Thus, in this analysis, the four stratum-scale propositions were simply combinations of the propositions for their component populations.

ESU-scale Attributes. On the ESU scale, concern focuses on catastrophes (infrequent large scale disturbances), long-term demographic processes, and long-term evolutionary potential (McElhany et al. 2000). At this scale, the propositions focus on ensuring that (1) all strata are independently sustainable ("All Strata Sustainable," Table 1), and (2) the ESU as a whole maintains sufficient genetic, phenotypic, and habitat diversity ("ESU-Scale Diversity").

Network Structure. The various propositions at these four scales form the nodes in a logical dependency network (Fig. 2). The links in this network take the form of logical operators that define relations among the input values. In traditional Boolean logic (which evaluates propositions as either absolutely true or false), propositions are "knife-edged," meaning the result of an operation has only one of two values—true or false—and, when the input values are near the edge, the result is very sensitive to small changes in the input. In the knowledge-based system used here, a type of approximate logic (referred to as "fuzzy logic"; Zadeh 1965, Adriaenssens et al. 2004) allows incorporating and assessing the imprecise knowledge available for the system. The advantage of using fuzzy logic is that it allows the degree of uncertainty to be evaluated and expressed in an outcome, ranging from certainly false, to uncertain, to certainly true. Being able to work with levels of uncertainty enables decision makers to evaluate the degree of risk inherent in decision-making.

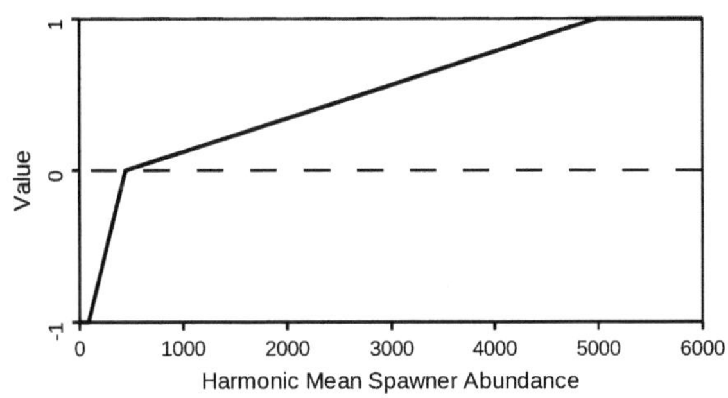

Fig. 3 Example of a fuzzy membership function for the Spawner Abundance (PD-1) proposition, showing how observations of harmonic mean spawner abundance (*horizontal axis*) correspond to the strength of evidence (*vertical axis*) in support of the proposition. Strength of evidence values range from –1 (no support) to +1 (full support) with 0 indicating unknown support. Redrawn from Wainwright et al. (2008)

The low-level metrics are evaluated on a scale from +1.0 (completely true) to –1.0 (completely false) through the application of "fuzzy membership functions" (see Chap. 2). These functions are used to evaluate the degree of evidence in support of specific propositions about current conditions. They were developed through analysis and application of the work group's best professional judgment, informed by relevant scientific literature. These functions allow intermediate values between true and false to occur when the strength of evidence supporting a proposition is intermediate, representing uncertainty in knowledge. For example, the "Spawner Abundance" proposition (Fig. 3) evaluates the degree of support for the proposition that "a population has sufficient numbers of naturally produced spawners to prevent the loss of genetic variation due to random processes over a 100-year time frame" (Wainwright et al. 2008, p. 44). Based on genetic theory (e. g., Waples 2006) and coho salmon life-history characteristics, the work group derived a range of long-term harmonic mean[1] abundance values that define the membership function for the proposition. Values ranged from 85 spawners (assigned a value of –1, = no support for the proposition) to 5000 spawners (assigned a value of +1, = full support for the proposition), with unknown support (assigned a value of 0) when spawner abundance is equal to 450.

Proposition combinations (within scales) and aggregations (across scales) are performed using three logical operators:

- "AND" (abbreviated "&" in Fig. 2)—evaluates to true only if evaluations of all of antecedent propositions are true (+1), and to false if any of the antecedents are false (–1). Between +1 and –1, the operator returns a weighted average of the antecedent values, with the weighting favoring the minimum value.

[1] Use of a harmonic mean to characterize spawner abundance reflects the nature of genetic variation in populations with fluctuating abundance (Waples 2006).

For example, in order for "Population Persistence" ("PP" in Fig. 2) to be true, all subpropositions ("PP-1", "PP-2", "PP-3") must be true.
- "UNION" ("U")—treats its antecedents as contributing compensating evidence in support of a proposition, thereby averaging the strength of evidence values of its antecedents. This operator is used where trade-offs among subpropositions fall within an acceptable range with respect to the fuzzy membership function (i.e., values falling between no support and full support, –1 and +1). For example, for "Spawner Distribution" ("PD-3" in Fig. 2) a high value from one watershed could compensate for a low value from another.
- "MEDIAN" ("M")—returns the median level of support for the proposition derived from evaluating its antecedents. Use of this operator implies that it is desirable that a majority of the subpropositions are true. For example, "Stratum Diversity" ("SD") requires that a majority of the populations within a stratum are sustainable ("Population Sustainability," "PS").

Our DSS was initially developed using the Ecosystem Management Decision Support (EMDS) system (Reynolds et al. 2002, Chap. 1). However, we were unable to fully implement our propositions within the EMDS framework, owing to unique features of our application. We developed alternative implementations of the DSS logic in both R (R Development Core Team 2013) and Microsoft Excel[2] that emulated the logic model. Further details on methods and data sets (including computer code implementing the DSS) can be found in Wainwright et al. (2008).

2.2 Application

To date, three versions of the DSS have been used by agencies in decision-making relevant to managing OCCS. The work group provided an initial example application of the DSS in its report (Wainwright et al. 2008), and that analysis was included by NMFS as part of the information considered in on-going listing decisions and recovery planning (NMFS 2008). A second version of the DSS was used by the State of Oregon in developing its Oregon Plan for Salmon and Watersheds (Chilcote et al. 2005). Most recently, NMFS and Oregon Department of Fish and Wildlife staff updated the data sets and applied the DSS to inform an updated status review for OCCS (Stout et al. 2012). Below, we briefly summarize that most recent application as a concrete example of the approach.

[2] Reference throughout this document to trade names does not imply endorsement by the National Marine Fisheries Service, NOAA.

3 Results

Running the DSS produces a systematic evaluation of the degree of support associated with each proposition, for all the population units for the geographic scale at which the proposition applies. This results in a large number of individual proposition values, which are summarized in the color-coded maps in Fig. 2. After combining propositions within scales and aggregating values across scales, the ultimate result for "ESU Sustainability" ("ES," which evaluates the proposition that "the ESU is self-sustaining into the foreseeable future"), was 0.24 (on a scale of –1–+1), corresponding to the light-green map at the top of Fig. 2.

This overall ESU value summarizes a great deal of variability in population and stratum-scale information, as can be seen in the maps for subpropositions in Fig. 2. For example, the "Population Persistence" values for individual populations ranged from –1 (Salmon River, Sixes River) to +0.98 (Tenmile Lakes), and approximately two-thirds of the populations had persistence values greater than 0.25. "Population Sustainability" values ranged from –1.0 in two populations to a high of 0.85 in the Coquille River. The values for "Stratum Sustainability" were less variable, in the narrow range of 0.39–0.48.

3.1 Communication and Interpretation

The DSS provides quantitative measures, i.e., strength of evidence in support of propositions, concerning the biological condition of OCCS at population, biogeographic-stratum, and whole-ESU scales. These values can inform decisions about ESA listing and recovery, as well as more general conservation planning; but tables of strength of evidence values are not very "user friendly," especially for decision makers unfamiliar with the technique. To ease the communication and interpretation of results, the work group took two approaches: verbal interpretations of numerical values, and visual display of results.

First, the work group addressed the verbal description of results, particularly with regard to the ESU sustainability propositions. For the extreme values, a value of –1.0 indicates a particular proposition is not met; meaning both the risk of inaction and the expected restoration cost may be high. Similarly, a value of +1.0 means that the proposition is fully met, and both risks and costs approach zero. However, between these extremes, the values do not translate directly to yes or no conclusions, and interpretation of intermediate values may prove difficult for decision makers. To aid in interpretation, the work group developed a set of verbal descriptions of the degree of certainty associated with a given strength of evidence value (Fig. 4). Values near zero are considered "uncertain," those near ± 0.1 are considered to indicate "low" certainty, those near ± 0.3 "moderate" certainty, those near ± 0.6 "high" certainty, and values near ± 1.0 are considered fully certain. Combining the sign (which indicates truth or falsity) with the magnitude of the value allows statements to be made about the propositions. Applying this

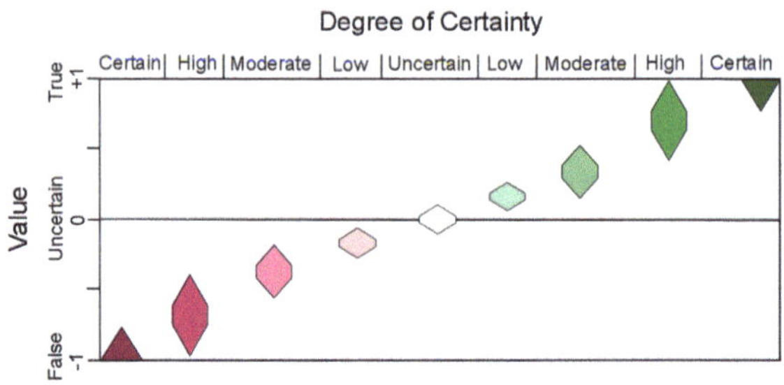

Fig. 4 Verbal interpretation of strength of evidence values produced by the decision support system. Adjectives describing degree of certainty are related to imprecise degree of support values, with color-coded levels corresponding to those used on the maps in Fig. 2. Redrawn from Wainwright et al. (2008)

semantic tool, the ESU sustainability value of 0.24 could be expressed as "low-to-moderate certainty" that the ESU is sustainable for the foreseeable future.

The second way the work group communicated results was visually, as a set of maps of the values for each proposition within the DSS network (Fig. 2). The maps provided an easy visual guide to populations and habitat areas that likely limit the sustainability of the ESU as a whole. The network of maps provided a tool for understanding how observed conditions at population and watershed scales contributed to overall population sustainability, and how populations contributed to the condition of the biogeographic strata and ESU. In addition, the maps provided a guide to areas where recovery efforts might be prioritized.

4 Discussion

The OCCS DSS represents an important step in the evolution of approaches to Pacific salmon status evaluations. Beginning around 1990, early status reviews (e.g., Waples et al. 1991) relied on biological information evaluated by a Biological Review Team, with a decision reached using professional best judgment without any defined set of criteria. In the mid-1990s, NMFS initiated a series of coordinated coast-wide reviews of Pacific salmon from California, Oregon, Washington, and Idaho; this led to a standardized set of metrics for population abundance, trends, and hatchery influence (Weitkamp et al. 1995), while the decision still relied on group professional judgment. This system was replaced by a more formal risk-matrix approach (Wainwright and Kope 1999). The specific risks were categorized as abundance, trends/productivity, genetic integrity, and other risks, and review team members voted on five levels (ranging from very low to high) within each risk category. While decisions still relied on team member professional judgment, this new approach clarified which risk categories were of greatest concern, and documented the degree of uncertainty and disagreement within review teams.

Following the development of NMFS' "Viable Salmonid Populations" (VSP) approach (McElhany et al. 2000), the risk matrix was further refined to provide consistent evaluations of the four VSP risk categories (e.g., Good et al. 2005). However, even with this added structure, status assessments strictly relied on professional judgment, and were neither transparent nor easily repeatable. The approach summarized here provides a suite of criteria that are based on well-defined data, have explicit statements of their scientific basis, and are connected via logical operations that provide a repeatable and explicit methodology to evaluate overall ESU status. The DSS still involved professional judgment in defining propositions, designing the logic network, and in drawing management conclusions from the results. However, that judgment was constrained and organized into documented steps such that persons outside the review process could follow (and potentially repeat) the logic of any management decisions based on the DSS.

The manner in which the work group implemented the DSS involved a number of subjective choices that could influence decision outcomes; several of these merit further discussion (immediately below).

4.1 Alternative Approaches

The work group considered a number of alternatives before choosing the DSS approach. First, we considered an approach with no formal DSS, where each criterion would be evaluated separately, with ESU status decided by subjective integration of the various criteria. This approach was rejected for two primary reasons: (1) by leaving the integration of criteria and the aggregation from populations to ESU open to subjective interpretation, it would not provide an objective means to evaluate ESU status, and (2) it lacks an explicit description of the logical connections among the criteria, thus reducing the reproducibility and transparency of analysis.

Second, we considered a logic framework with strict pass/fail propositions. This approach would ignore uncertainties in the science and available data, and would require a priori determination of acceptable risk prior to assessment. No such determination was available, and the work group believed the choice of acceptable risk should be a sociopolitical decision, and thus outside the scope of work. Another problem with pass/fail propositions would arise where results approach threshold values, in which case decisions could flip-flop with minor changes in data, potentially reducing the credibility of the process. Based on these considerations, the work group sought a method that expresses the degree of biological risk facing an ESU, rather than a pass/fail result.

Finally, other approaches to evaluating risk within the DSS framework were considered, notably Bayesian probabilistic risk assessment (Marcot et al. 2006). A Bayesian approach could provide an integrated measure of risk, resulting in a final status evaluation that may be easier to interpret. However, it would have required a number of complex assumptions regarding conditional probabilities, which can hide important details of analysis. The work group concluded that a logic-based DSS provided greater transparency than probabilistic approaches, while retaining a full description of the logical connections leading from low-level data to the integrated ESU evaluation.

4.2 Advantages and Disadvantages

As part of the initial DSS design, the work group focused on a knowledge-based system that codified the best available scientific knowledge. Our approach to assessing biological sustainability has several advantages over other approaches:

1) The method organizes quantitative and qualitative knowledge about numerous and diverse factors into a single, coherent logical framework that reflects the current state of knowledge about processes affecting biological sustainability, how they are logically connected, and how they operate at different spatial scales.
2) The DSS allows us to explicitly disclose how the aggregate of various indicators was evaluated; not only the grouping of indicators, but the nature of the interdependence among them as they represent biological sustainability.
3) The DSS clearly represents the data used, assumed relations among the data, information used as the basis for assumed relations, and how conclusions were drawn.

The DSS design incorporates a hierarchy of geographic population structure representing different scales of genetic and life-history diversity. Our ability to display this geographic structure as maps of values for the various propositions (Fig. 2) provides two distinct advantages. First, it provides a quick basin-by-basin overview of areas with good and poor conditions, which provide a starting point for prioritizing recovery actions. Second, when the proposition maps are viewed alongside the logic network, they provide a means of directly visualizing how the parts contribute to the whole.

Knowledgebases provide consistent data interpretation in a specified framework that is clear and unambiguous. Many current sustainability assessment efforts do not have an accompanying logic framework providing consistent, integrated interpretation of data. This lack may lead to differences of opinion about what the data mean and inconsistent conclusions over time. Knowledgebases can clearly articulate relations between indicators and biological condition as well as relations among indicators. As a result, calculations of aggregate indices are consistent and less likely influenced by successive assessments of individual evaluators.

Another advantage of the knowledge-based framework is that its transparency and logical rigor can provide a clear road map for revision and adaptation. The DSS can serve as a "snapshot" of institutional knowledge, codifying the scientific knowledge and data at the time of its creation. This means there is no need to re-create this process at each subsequent decision point, so updates are easier.

However, the DSS also provides a structure for incorporating new information; as new knowledge becomes available, changes in relationships can be redefined without losing prior knowledge. Similarly, new propositions can be added if new data types become available.

While a DSS such as ours works well in data-rich environments, the approach may also be robust to missing data. Although full results require that consistent data series be available for all primary propositions, where data are lacking, the DSS defaults to a result of zero, indicating complete uncertainty regarding that proposition. This provides an indication of the consequences of missing data to decisions.

The approach is of course not without disadvantages. One disadvantage is the time required to design and implement a DSS, which involves extensive literature review and discussions to elucidate interacting ecological processes that define the logic network. Once the basic propositions have been linked in the network, the next task is developing specific metrics (statistical summaries of data) and fuzzy membership functions for them. In our application, this process took over a year of subcommittee work. A second disadvantage is the difficulty of communicating DSS results, and how biological conditions affect results, to policy-makers and interested public, particularly explaining the meaning of numerical results. While the DSS provides transparency in the sense that all parts of the process are documented and the software to run the analysis is available, the knowledge network is not simple to explain. As an analogy, describing the DSS is like describing a clock with a transparent case: all the parts can be seen, but understanding how it works requires a bit more effort, and perhaps a copy of *The Way Things Work* (MacCaulay 1988). This difficulty is to some degree inherent in any decision model—some effort and expertise is required to understand the workings of a complex system.

4.3 Lessons Learned

A number of issues were important during development of the DSS, and led to modifications in either the overall approach, or how results were communicated. These included (1) consistency with other Pacific salmon recovery plans, (2) the role played by professional judgment, (3) the use of fuzzy logic, and (4) the level of resources needed to implement the approach.

Consistency. To ensure that the biological recovery criteria were comprehensive and consistent with approaches used by other Pacific salmon recovery plans, the work group incorporated the aforementioned VSP approach (McElhany et al. 2000). VSP identifies four key population parameters that influence the sustainability of populations.

- Population size—the abundance of all life stages of the species
- Population growth rate (productivity)—production over the entire life cycle

- Spatial structure—fish distribution among habitats and habitat connectivity
- Diversity—variation in phenotypic and genetic traits among individuals in a population and among OCCS populations.

While recognizing the importance of VSP parameters to population sustainability, problems were identified while using them as organizing principles for a consistent set of recovery criteria. Two primary concerns were (1) the lack of direct connection with habitat characteristics, and (2) strong interdependency among the four parameters. The goal of recovery and restoration is not merely to meet a set of artificial criteria, but to restore or repair ecological processes that contribute to long-term sustainability of the resource. As the propositions used in these analyses were developed, the work group focused on connections between the criteria and fundamental processes, specifically those that contribute to sustainability at various scales.

Professional Judgment. All ESA evaluations rely to some degree on professional judgment, and this judgment is often implicit and intractable. In a DSS approach however, judgments can be well-defined and constrained by a logical structure, and the pathways by which judgment influences evaluations are traceable. In addition, because the judgment is contained within the logic structure of the DSS, it is codified in the model and cannot be influenced by analytical results.

In our DSS, judgment entered three ways, when the work group defined the structure of the logic network, established fuzzy membership functions, and developed metrics to evaluate ESU-scale diversity propositions. The first two of these are common to (and unavoidable in) all decision-support applications; it is how scientists apply expert knowledge and judgment formed through experience to the art of decision making. The third case—basing some propositions on professional judgment rather than objective data—was a design choice made reluctantly by the work group. Evaluating diversity at the ESU scale proved to be somewhat intractable. While there was some theory as to the processes important to maintaining broad-scale diversity, there were no quantitative models of those processes, no clear relationships between diversity measures and sustainability, and little available data. Because the use of professional judgment introduced great uncertainty in these propositions, their values were given a reduced weight relative to the population and stratum-scale propositions (Wainwright et al. 2008).

Fuzzy Logic. Of all the parts of the DSS, the concept of fuzzy logic was probably the most difficult to communicate to non-technical audiences. While fuzzy logic is widely recognized (though perhaps not well understood) in the scientific community (Adriaenssens et al. 2004), with policy-maker and public audiences we found the term to be an unfortunate name for a useful approach. In fact, during a stakeholder review meeting, one of the review panel members quipped: "Fuzzy logic? That sounds like the federal government!" Because of widespread misunderstanding of the term, for public venues the work group adopted the informal term "approximate logic" to reduce initial negative reactions.

Beyond the problem of terminology, another problem with using fuzzy logic is that it results in numerical strength of evidence values that are difficult to directly interpret. The structure of fuzzy membership functions with numeric values between "false" and "true" is unfamiliar and uncommonly used. Interested parties want to ask "is the ESU sustainable?" and get a simple yes or no answer. Instead, we evaluate the statement "the ESU is sustainable" and get a numeric value between true (+1) and false (–1). A value of "0" (complete uncertainty) can result from two different cases: either the input value is 0 on the fuzzy membership function (e.g., 450 spawners in Fig. 3), or there is a complete lack of information. This frequently requires further explanation to avoid confusion.

Similarly, it is not always clear how to interpret numerical DSS results for an ESA listing or de-listing (yes/no) decision. Verbal interpretation of the final value 0.24 for the ES proposition ("The ESU is self-sustaining into the foreseeable future") suggests a qualified answer: "Yes, but with a low to moderate degree of certainty" (Fig. 4). Whether that value leads to de-listing would of course depend on other factors, including consideration of the desired level of precaution in the decision, legal interpretations of the phrase "likely to become endangered" in the ESA, and evaluations of other ESA listing factors such as habitat conditions, exploitation, or the adequacy of regulatory mechanisms.

Resources Needed. For most members of the work group, this was their first exposure to constructing a knowledge-based DSS, and this DSS was the most complicated that any of us had designed. The experience provided a few lessons that should be useful to other teams that try this approach. Development of both the recovery criteria (propositions to be evaluated) and the fuzzy membership functions took far longer than anyone expected. Because propositions and membership functions form the foundation of analysis, it is essential that all are carefully worded and structured to faithfully capture the attributes of interest, and ensure that propositions, functions, and metrics are based on the best available science. Thus, although the process was time consuming, it was critical for success.

The process was also iterative. Once the DSS was complete, the work group revisited several propositions and membership functions that were not representing the attribute under evaluation as intended. In addition to the appropriate scientific expertise, successful implementation of the DSS required support staff with expertise in scientist herding, database management, GIS, and programming.

Even though the OCCS is one of the most data-rich Pacific salmon ESUs, there were still major data gaps. For example, the work group used professional judgment to assess genetic and phenotypic diversity, hardly satisfying when compared to extremely good data sets on juvenile or spawner abundance. Application of this method to other ESUs that are less data-rich would be more difficult, and would require more professional judgment. Clearly, the utility of developing a formal DSS system should be established early, especially where data are limited.

To reduce the possibility of implementation errors, the work group relied on two independent software implementations of the DSS: one spreadsheet-based (in MS Excel), and the other programmed in the R language (R Development Core Team 2009). This extra effort turned out to be essential to a reliable analysis. During initial development, test runs of the two implementations identified several ambiguities in verbal definitions of propositions and metrics. Even during the analysis phase, a few errors were caught in data alignments that resulted from the complexity of the spatially structured time series data sets. Had there been only one version of the DSS and one person implementing it, reliability of the analysis would have been sacrificed and, worse, we probably wouldn't have known there were problems.

4.4 Future Directions

The DSS fell somewhat short of initial expectations. We suggest a number of improvements that could be made within the existing framework, then a few new directions.

Improvements. The work group was unsatisfied with several of the propositions. Most problematic was the set of ESU-scale diversity propositions evaluated solely based upon expert opinion. We would like to revisit these in the future and suggest specific monitoring that could directly inform them, or, failing that, define a rigorous future method of eliciting expert opinion (e.g., Aspinall 2010). Other propositions that should be revisited include "Population Functionality" and "Stratum Functionality", which are based on habitat models rather than direct measures of available habitat, and "Population Persistence", which is based on a suite of four models that are time-consuming to update and run.

Another area where there is room for improvement is in software implementation. As noted above, we started DSS development in the EMDS system, but found that our needs were better met by implementing the DSS logic in both R and MS Excel. This decision was made because of software limitations and portability issues. First, we could not implement our full logic network (in particular, the median operator used in the Stratum Sustainability proposition) in EMDS, nor could we easily conduct desired sensitivity analyses (Appendix A in Wainwright et al. 2008). Second, EMDS is not portable, in the sense that it is designed for a single operating system and requires additional proprietary commercial software. The work group felt that it was important that our decision system be accessible to government decision-makers and stakeholders so that they could recreate and potentially modify our results without specialized commercial software.

One effect of this decision was that we no longer had a fully integrated system that could be used from network design through presentation of results. Although the work group used two versions of the model, neither could be considered user-friendly, particularly in the level of effort needed to update and cross-check the many data series that drive the analysis. This could be improved by developing direct database linkages from agency-maintained data sets to the DSS software. Additional improvements could be made in the user interface, which currently requires script programming or intimate familiarity with multi-page spreadsheet management, as well as GIS expertise to translate the results into maps. A more integrated system would better support frequent and rapid data updates integrated with direct GIS results display. Such improvements would make it much more likely that the DSS would play an active role in ESA recovery scenario planning.

New Directions. The work group was unable to address two important issues in the current version of the DSS: freshwater habitat issues, and climate variation and trends. Habitat availability and quality are important statutory considerations in ESA recovery planning, but this DSS does not directly capture habitat quality in any meaningful way. The Oregon Department of Fish and Wildlife conducts extensive annual salmon habitat surveys, so it should be possible to incorporate habitat data into the population diversity criteria, or possibly to form a separate branch of the DSS to evaluate habitat and other statutory listing factors (ESA Sect. 4a). Recently developed habitat trend models (Ward et al. 2012) suggest a possible approach.

Much of the concern that led to the ESA listing of OCCS arose in the 1990s during a prolonged period of marine conditions that led to poor juvenile survival for northwest coastal coho and other salmon stocks (Stout et al. 2012). One concern of the work group was to define criteria that would ensure that OCCS would survive another period of similar conditions without substantial loss of diversity. Presently, the DSS considers only the most recent 12 years of data (except for long-term means in the "Spawner Abundance" proposition). This is problematic when one considers that climatic fluctuations often persist longer than 12 years, and it also does not account for long-term trends in climate conditions. Climate change is an emerging issue for conservation planning for this and other ESUs (Stout et al. 2012; Wainwright and Weitkamp 2013).

5 Conclusions

The Oregon Coast coho salmon DSS is a unique approach to informing ESA listing and recovery decisions by contributing an integrated and comprehensive analysis of important biological sources of risk. This DSS meets goals for measurable, transparent, and reproducible criteria. Although initially time-consuming to build, the DSS has served a variety of useful functions. The formal process of developing the DSS network provided a basis for finding common ground and consensus building among scientists with diverse backgrounds and priorities. It summarized a variety of objective, measurable information, creating statistics that were comparable among multiple populations. Based on these statistics, reported in table and map form, patterns in the status of a listed species and the underlying causes of these patterns became evident. Through periodic updates the DSS has quantified incremental changes in biological status. Likewise, the DSS has provided a unifying platform for discussion of OCCS status and has supported the decision-making process.

Acknowledgments In addition to the authors, the Oregon Coast Coho Salmon Work Group included Mark Chilcote, Tom Nickelson, Chuck Huntington, and Kelly Moore, all of whom contributed valuable ideas and discussion. Kelly Christensen provided additional GIS expertise.

This work was supported by the National Oceanic and Atmospheric Administration and the Oregon Department of Fish and Wildlife. We also thank George Pess, Chris Jordan, and Joanne Butzerin for invaluable comments on earlier drafts of this manuscript.

References

Adriaenssens V, De Baets B, Goethals PLM, De Pauw N (2004) Fuzzy rule-based models for decision support in ecosystem management. Sci Total Environ 319:1–12

Aspinall W (2010) A route to more tractable expert advice. Nature 463:294–295

Beechie T, Imaki H, Greene J, Wade A, Wu H, Pess G, Roni P, Kimball J, Stanford J, Kiffney P, Mantua N (2012) Restoring salmon habitat for a changing climate. River Res Appl. doi:10.1002/rra.2590

Bisson PA, Dunham JB, Reeves GH (2009) Freshwater ecosystems and resilience of Pacific salmon: habitat management based on natural variability. Ecol Soc 14(1):45. http://www.ecologyandsociety.org/vol14/iss1/art45/

Chilcote M, Nickelson T, Moore K (2005) Oregon coastal coho assessment. Part 2: viability criteria and status assessment of Oregon coastal coho. Oregon Plan for Salmon and Watersheds. State of Oregon, Salem

Good TP, Waples RS, Adams P (eds) (2005) Updated status of federally listed ESUs of West Coast salmon and steelhead. NOAA Technical Memorandum NMFS-NWFSC-66. U. S. Department of Commerce, Northwest Fisheries Science Center, Seattle

Lawson PW, Bjorkstedt EP, Chilcote MW, Huntington CW, Mills JS, Moore KMS, Nickelson TE, Reeves GH, Stout HA, Wainwright TC, Weitkamp LA (2007) Identification of historical populations of coho salmon (*Oncorhynchus kisutch*) in the Oregon Coast Evolutionarily Significant Unit. NOAA Technical Memorandum NMFS-NWFSC-79. U. S. Department of Commerce, Seattle

Lichatowich JA (1989) Habitat alteration and changes in abundance of coho (*Oncorhynchus kisutch*) and Chinook salmon (*O. tshawytscha*) in Oregon's coastal streams. In: Levings CD, Holtby LB, Henderson MA (eds) Proceedings of the national workshop on effects of habitat alteration on salmonid stocks. Canadian special publications in fisheries and aquatic sciences 105. Department of Fisheries and Oceans, Ottawa, pp 92–99

MacCaulay D (1988) The way things work. Houghton Mifflin, Boston

Marcot BG, Hohenlohe PA, Morey S, Holmes R, Molina R, Turley MC, Huff MH, Laurence JA (2006) Characterizing species at risk II: using Bayesian belief networks as decision support tools to determine species conservation categories under the Northwest Forest Plan. Ecol Soc 11:12. http://www.ecologyandsociety.org/vol11/iss2/art12/

McElhany P, Ruckelshaus MH, Ford MJ, Wainwright TC, and Bjorkstedt EP (2000) Viable salmonid populations and the recovery of Evolutionarily Significant Units. NOAA Technical Memorandum NMFS-NWFSC-42. U.S. Department of Commerce, Seattle

National Marine Fisheries Service (1998) Endangered and threatened species: threatened status for the Oregon Coast Evolutionarily Significant Unit of Coho Salmon. Fed Reg 63:42587–42591

National Marine Fisheries Service (2008) Endangered and threatened species: Final threatened listing determination, final protective regulations, and final designation of critical habitat for the Oregon Coast Evolutionarily Significant Unit of Coho Salmon. Fed Reg 73:7816–7873

National Marine Fisheries Service (2013) Endangered and threatened species; notice of intent to prepare a recovery plan for Oregon Coast Coho Salmon Evolutionarily Significant Unit. Fed Reg 78:38011–38012

Oregon Department of Fish and Wildlife (2003) Native fish conservation policy (revised). Oregon Department of Fish and Wildlife, Salem. http://dfw.state.or.us/fish/CRP/nfcp.asp. Accessed 19 Mar 2013

R Development Core Team (2013) R: a Language and environment for statistical computing. R Foundation for Statistical Computing, Vienna. http://www.R-project.org. Accessed 20 july 2013

Rauscher HM (1999) Ecosystem management decision support for federal forests in the United States: a review. Forest Ecol Manag 114:173–197

Reeves GH, Hohler DB, Larsen DP, Busch DE, Kratz K, Reynolds K, Stein KF, Atzet T, Hays P, Tehan M (2004) Effectiveness monitoring for the aquatic and riparian component of the Northwest Forest Plan: conceptual framework and options. General Technical Report PNW-GTR-577. U.S. Forest Service, Corvallis

Regional Ecosystem Office (2002) Hydrologic unit boundaries for Oregon, Washington, and California. http://www.reo.gov/gis/projects/watersheds/reohucv13.zip. Accessed 28 Feb 2008

Reynolds KM, Hessburg PF (2005) Decision support for integrated landscape evaluation and restoration planning. Forest Ecol Manag 207:263–278

Reynolds KM, Rodriguez S, Bevans K (2002) Building EMDS models and applications: the ecosystem management decision support system version 3.0. U.S. Forest Service. http://www.spatial.redlands.edu/emds/manuscripts/docs/EMDS%20Training%20Manual.pdf. Accessed 29 Jul 2013

Sandercock FK (1991) Life history of coho salmon (*Oncorhynchus kisutch*). In: Groot C, Margolis L (eds) Pacific Salmon life histories. UBC Press, Vancouver, pp 395–445

Seaber PR, Kapinos PF, Knapp GL (1987) Hydrologic unit maps. Water-Supply Paper 2294. U.S. Department of Interior, U.S. Geological Survey, Washington, DC

Steel EA, Fullerton A, Caras Y, Sheer MB, Olson P, Jensen D, Burke J, Maher M, McElhany P (2008) A spatially explicit decision support system for watershed-scale management of salmon. Ecol Soc 13:50. http://www.ecologyandsociety.org/vol13/iss2/art50

Stout HA, Lawson PW, Bottom DL, Cooney TD, Ford MJ, Jordan CE, Kope RG, Kruzic LM, Pess GR, Reeves GH, Scheuerell MD, Wainwright TC, Waples RS, Weitkamp LA, Williams JG, Williams TH (2012) Scientific conclusions of the status review for Oregon Coast coho salmon (*Oncorhynchus kisutch*). NOAA Technical Memorandum NMFS-NWFSC-118. U. S. Department of Commerce, Northwest Fisheries Science Center, Seattle

Turban E, Aronson JE (2001) Decision support systems and intelligent systems, 6th edn. Prentice Hall, Upper Saddle River

Wainwright TC, Kope RG (1999) Methods of extinction risk assessment developed for US West Coast salmon. ICES J Mar Sci 56:444–448

Wainwright TC, Weitkamp LA (2013) Effects of climate change on Oregon Coast coho salmon: habitat and life-cycle interactions. Northwest Sci 87(3):219–242

Wainwright TC, Chilcote MW, Lawson PW, Nickelson TE, Huntington CW, Mills JS, Moore KMS, Reeves GH, Stout HA, Weitkamp LA (2008) Biological recovery criteria for the Oregon Coast Coho Salmon Evolutionarily Significant Unit. NOAA Technical Memorandum NMFS-NWFSC-91. U. S. Department of Commerce, Northwest Fisheries Science Center, Seattle

Waples RS (2006) Seed banks, salmon, and sleeping genes: effective population size in semelparous, age-structured species with fluctuating abundance. Am Nat 167:118–135

Waples RS, Johnson OW, Jones RP Jr (1991) Status review for Snake River sockeye salmon. NOAA Technical Memorandum NMFS F/NWC-195. U. S. Department of Commerce, Northwest Fisheries Science Center, Seattle

Ward EJ, Pess GR, Anlauf-Dunn K, Jordan CE (2012) Applying time series models with spatial correlation to identify the scale of variation in habitat metrics related to threatened coho salmon (*Oncorhynchus kisutch*) in the Pacific Northwest. Can J Fish Aquat Sci 69:1773–1782

Weitkamp LA, Wainwright TC, Bryant GJ, Teel D, Kope R (2000) Review of the status of coho salmon from Washington, Oregon, and California. In: Knudsen EE, Steward CR, MacDonald DD, Williams JE, Reiser DW (eds) Sustainable fisheries management: Pacific salmon. Lewis Publishers, New York, pp 111–118

Weitkamp LA, Wainwright TC, Bryant GJ, Milner GB, Teel DJ, Kope RG, Waples RS (1995) Status review of coho salmon from Washington, Oregon, and California. NOAA Technical Memorandum NMFS-NWFSC-24. U. S. Department of Commerce, Northwest Fisheries Science Center, Seattle

Wilson KA, Lulow M, Burger J, Fang Y, Andersen C, Olson D, O'Connell M, McBride MF (2011) Optimal restoration: accounting for space, time and uncertainty. J Appl Ecol 48:715–725

Wong IW, Bloom R, McNicol DK, Fong P, Russell R, Chen X (2007) Species at risk: data and knowledge management within the WILDSPACE decision support system. Envir Model Softw 22:423–430

Zadeh LA (1965) Fuzzy sets. Inform Control 8:338–353

Part III
The Next Version and Final Thoughts

EMDS 5.0 and Beyond

Steve Paplanus, Bruce Miller, Philip J. Murphy, Keith Reynolds and Michael Saunders

Abstract The EMDS Consortium plans, designs, and oversees software development and integration for the EMDS suite of tools. This chapter presents an overview of the Consortium's plans for the next version of EMDS. Our objective is to release an EMDS 5.0 that will support web-services and Microsoft's workflow foundation. The 5.0 release will include major updates to the NetWeaver® engine, the Priority Analyst® engine, and the EMDS core.

Keywords Architecture · Workflows · Web services · Engine services · Data services · Business logic · NetWeaver · Criterium decisionplus · VisiRule

---

S. Paplanus (✉)
University of Redlands, 1200 East Colton Avenue, Redlands, CA 92373, USA
e-mail: Steve_Paplanus@spatial.redlands.edu

B. Miller · M. Saunders
Rules of Thumb, Inc., 11817 Cedar Mill Road, North East, PA 16428, USA
e-mail: bjmiller@gmail.com

M. Saunders
e-mail: mcs5@psu.edu

P. J. Murphy
InfoHarvest, Inc., PO Box 25155Seattle, WA 98165-2055, USA
e-mail: philip_murphy@infoharvest.com

K. Reynolds
U.S. Department of Agriculture, Forest Service, Pacific Northwest Research Station, 3200 SW Jefferson Way, Corvallis, OR 97331, USA
e-mail: kreynolds@fs.fed.us

M. Saunders
Department of Entomology, Penn State University, 501 Agricultural Science and Industries Building, University Park, PA 16802, USA

K. M. Reynolds et al. (eds.), *Making Transparent Environmental Management Decisions*, Environmental Science and Engineering,
DOI: 10.1007/978-3-642-32000-2_13,

1 Introduction

Development work on the Ecosystem Management Decision Support (EMDS) system was begun by the Pacific Northwest Research Station (U.S. Department of Agriculture, Forest Service) in 1995. For the initial versions of EMDS through version 3, system implementation was done by the Environmental Systems Research Institute[1] (Redlands, CA) under contract with the Forest Service through 2002. Stewardship of EMDS was transferred to the Redlands Institute at the University of Redlands in 2005 under a memorandum of understanding between the University and the Forest Service. The EMDS Consortium was organized shortly thereafter as a private, non-profit research and development group to continue system development. Original parties to the Consortium included the Redlands Institute and the Pacific Northwest Research Station, as well as two private companies (Rules of Thumb, Inc. and InfoHarvest, Inc.), that had been instrumental in providing the core decision-support technologies underlying EMDS version 3.0. The most recent addition to the Consortium was Logic Programming Associates (LPA) LLP of London, UK in 2010, which made available a new suite of powerful Prolog programming tools for integration into the EMDS architecture.

From its inception, members of the Consortium have regularly collaborated as an applied research team, using EMDS for knowledge engineering, logic modeling, system assessment, alternative analysis, and prioritization. These decision support projects ranged from local (e.g., forests) to landscape (e.g., watersheds,) to regional and national analysis scales. This applied research serves as a testing and proving ground for EMDS functionality, and as a feedback mechanism for design improvements. The objective of the present chapter is to lay out the Consortium's vision for the next release of EMDS (version 5.0) in terms of major new features and functionality that will take the current, well-established, desktop system for individual users to a powerful, industrial-strength enterprise system for natural resource agencies and large organizations.

2 Background

"An Overview o1.the Ecosystem Management Decision-Support System" (this volume) describes the core functionality of the current EMDS system, including the central roles played by its logic and decision engines in providing spatially enabled decision support for environmental analysis and planning. "NetWeaver" and "Criterium DecisionPlus" provide introductions to the commercial development systems, NetWeaver Developer (NetWeaver) and Criterium DecisionPlus (CDP), which are used to design the respective logic and decision models used in EMDS

---

[1] The use of trade or firm names in this publication is for reader information and does not imply endorsement by the US Department of Agriculture of any product or service.

applications. "An Overview ol.the Ecosystem Management Decision-Support System" also explained that although the EMDS system *per se* is public domain freeware, and is capable of running existing NetWeaver and CDP models, both NetWeaver and CDP are commercial products that must be purchased from their parent companies in order to build new or edit existing logic and decision models used in EMDS.

3 EMDS Version 4.2 Enhancements

The current EMDS software release, version 4.2 (October 2012), represents a comprehensive re-engineering of EMDS. Improved functionality in the 4.x architecture includes:

- an ArcGIS®[2] 10.x add-in component,
- support for printing and exporting data and graphs,
- installers for both 32-bit and 64-bit systems, and
- an increase in the maximum number of features per analysis to 10,000.

New capabilities for the 4.2 product line include:

- support for both ArcGIS® versions 9.3 and 10.x and on Microsoft Windows® 32-bit and 64-bit operating systems,
- an increase in the maximum number of features per analysis to 16,000,
- modularization of the interfaces to the NetWeaver®[3] and Priority Analyst®[4] engines, and
- integration of VisiRule.[5]

Despite all these improvements, EMDS 4.x suffers from four key limitations:

- **It has a fixed analysis *workflow*.**[6] A study area is chosen, a NetWeaver model run, and a CDP model can be executed to generate a prioritization. From the EMDS 4.x user interface, there is no way for an analyst to change that order, or execute one of LPA's VisiRule models.
- **It processes a single feature at a time**. If analysis requires properties of groups of features, such analysis has to be done through geoprocessing of the features datasets before loading EMDS, so no dynamic updates are possible in the course of the analysis.

---

[2] ArcGIS is a product of the Environmental Systems Research Institute, Redlands, CA.

[3] NetWeaver is a product of Rules of Thumb, Inc., North East, PA.

[4] The Priority Analyst is an engine for Criterium DecisionPlus, a product of InfoHarvest, Seattle, WA.

[5] VisiRule is a product of Logic Programming Associates, LLP, London, UK.

[6] A workflow is an organized collection of activities (often, but not necessarily, sequential) that are executed in program code to accomplish some overall task.

- **It is single user only**. There is no way to collaboratively build or share an EMDS project in real time.
- **The number of features it can process is limited**. Due to current architecture in terms of how EMDS is integrated with ESRI's ArcMap, it does not reliably handle feature datasets.

4 EMDS Version 5.0 Design Objectives

The goal of this latest version of EMDS is to transform it into a complete design, analytical, and dynamic *scenario planning*[7] framework (the current sense of the EMDS 4.2 framework is described in "An Overview ol.the Ecosystem Management Decision-Support System"). The purpose of this new version is to:

- Enable EMDS to support one or more analysts and managers to work on one or more projects concurrently across a wide variety of *clients.*[8] The current desktop implementation allows for a single user to work on a project, and there is a limited import/export feature to allow another user to view or alter the current project.
- Allow for more flexible analysis for users of EMDS. Over the years, we have received many requests to add flexibility to the EMDS workflow. For example, users often wanted to create a spatial selection and then run just a CDP analysis. Currently, they would have to create a dummy NetWeaver model and run it before they could do the CDP analysis. In another case, any further geoprocessing done on the results was done outside of EMDS without any tracking. With the new framework, we allow the end user to select from a common set of pre-defined workflows to perform a much wider variety of operations than is possible with EMDS 4.2.

We will be migrating from a traditional desktop architecture into a distributed service-oriented framework. This framework for version 5 will be of *Type II*[9] as defined by the SEI team (i.e., the Carnegie Mellon Software Engineering Institute, SEI). Future versions will move toward a *Type III*[10] architecture to give additional flexibility to the systems.

---

[7] Scenario(-based) planning can inform decision makers by presenting scenarios that combine known or assumed facts with other plausible future conditions to explore the implications of alternative future states of a system.

[8] A client is the part of an application, generally remote, that the user sees and interacts with.

[9] Type II frameworks are "typified by allowing users to customize services in a finite number of commonly understood ways based on shared, community-wide assumptions about what is needed". (Phase 1: Strategic Analysis of Problem; SEI team; http://www.frames.gov/partner-sites/iftdss/phase-i/).

[10] Type III architectures are "typified by supporting the customization of services by users for specific, unique operational situations that may or may not be shared, community-wide ways of solving a particular problem". SEI team; http://www.frames.gov/partner-sites/iftdss/phase-i/.

This framework will support true *provenance tracking,*[11] allowing programs to expose each step of the process and to change conditions dynamically and then view the results. Scientific provenance tracking is defined as having the knowledge of all the steps in producing the result—from design through acquisition of data, manipulation of data, analysis performed, and any additional manipulations. From this information, a user will be able to reproduce a given result consistently, regardless of the complexity of the process. With the new EMDS 5.0 Framework's provenance subsytems, we will be able to track all this information, excluding the model design, which is currently captured in the model building software, and the raw data acquisition.

In addition, the framework will support multiple users and will be *multi-thread safe*,[12] allowing for client applications to be written as standalone applications, ESRI ArcMap add-ins, or as web clients. A defined application programming interface *(API)*[13] is planned that will allow for extending the framework with additional data formats or analytical and modeling engines by the end user. A workflow editor will allow the end user to perform some changes to the execution steps within the EMDS client. This workflow editor will be the same editor the EMDS developers use to create the pre-defined workflows, and will display all the relevant higher level activities the end user selects. This will allow for user-defined workflows for users who have the necessary rights. These workflows can be added to the pre-defined workflow library. All these changes will transform EMDS from being solely an ArcMap add-in into a true suite of products supporting multiple platforms.

The current design enhancement plan for EMDS version 5.0 includes the following objectives:

- **Workflow Foundation**. The EMDS Framework will be powered by workflows, based on the Microsoft Windows Workflow Foundation® (WF). Windows Workflow Foundation is part of the Microsoft .NET Framework and was introduced in version 3 of the framework. WF is a workflow engine, programming model, and set of tools that allows developers and end users to build workflows that coordinate people and software (Chappell 2009). Windows Workflow

[11] Provenance tracking formally documents the ownership(s), origin(s), uses and transformations of computerized data. Provenance tracking is of particular concern with electronic data, because data sets are routinely modified and copied without citation of the originating data set or further documentation of data modifications.

[12] In computer science, a thread of execution is the smallest sequence of programmed instructions that can be managed independently by an operating system scheduler. A piece of code is thread-safe if it only manipulates shared data structures in a manner that guarantees safe execution by multiple threads at the same time. Multiple threads can exist within the same process (i.e. the running application) and share resources such as memory, while different processes do not share these resources.

[13] An application programming interface (API) specifies how software components need to interact with each other. In practice, an API is embodied as a code library that includes specifications for routines, data structures, object classes, and variables.

leverages the concept of activities, which can be simple, with only one activity to execute, or complex, and composed of multiple simple activities. One or more activities can be combined to form a workflow, which is the actual entity that is passed to the workflow engine. In this new workflow architecture, each functional unit of the EMDS platform will be engineered as one or more WF activities, enabling users to organize their analyses as customized process chains. Each engine (NetWeaver®, Priority Analyst®, and VisiRule®) will have a standardized, *pluggable wrapper*[14] that will be exposed within the EMDS Framework as a series of workflow activities. A default set of pre-defined workflows will support traditional EMDS process patterns (e.g., Project > Assessment > Analysis > Scenario > Prioritization).

A library of pre-built workflow activity templates will also be provided to enable users to create their own customized analytical process chains with a minimal amount of coding. A workflow editor will enable users to create, save, and re-use workflows. *Provenance metadata*[15] will be recorded for each workflow activity, enabling users to undo, redo, and pivot from any step and move along an alternate workflow sequence. All workflows will support both synchronous and asynchronous interfaces.[16] In addition, all workflows can be exposed as web services, accessible by both desktop and web clients. The workflow platform will be integrated with Microsoft's Project Trident® workbench (Trident Team 2011). Translators[17] may be built to facilitate integration between Trident and third-party modeling packages such as the IBM Web Process Server.

- **Pure Microsoft .NET® Implementation**. The EMDS core and the integration of the NetWeaver® and Priority Analyst® engines will be re-engineered in .NET® to improve system performance and stability.
- **Multi-core CPU**. To speed up in-memory calculations and operations, EMDS will be updated to support hardware systems with multi-core CPUs.
- **Relational database management system (RDBMS) Support**. Currently EMDS supports SQL Server, Microsoft Access, and SQL Server Compact edition. EMDS 5 will add support for Oracle and Postgres, while supporting the use of Oracle® and Microsoft SQL Server Spatial® RDBMSs.

[14] Pluggable functions (aka "plug-ins") let the user extend the core functionality of an application via software components that add specific features otherwise not present in the application. When an application supports plug-ins, it enables customization and extension.

[15] With provenance tracking, sufficient metadata about current status and configurations are gathered and stored so that operations can be rolled forward and back.

[16] Synchronous operations require the software to wait for the called operation to complete before the current process continues whereas asynchronous operations allow the program to continue doing other things while the asynchronous operation runs. Asynchronous operations generally are more complex to implement, as the program must be notified of their completion so that the program can update accordingly.

[17] Translators convert data and/or programming calls from one convention to another. These are similar to wrappers mentioned below.

- **Graphical User Interface Tools**. The next generation of EMDS will have a new **Project Manager** component for adding, deleting, and updating project metadata, and for importing/exporting multiple projects. A **Report Manager** tool will enable users to create, select, and re-use reports, including support for auto-updating the data behind reports and sending reports on a pre-determined schedule. A new web-based user interface component will enable users to view existing EMDS projects and modeling results in tabular, graphical, and spatial formats.
- **Actions that change the state of the System.**[18] For end users, actions that change the state of the system will be one of the biggest additions to the platform. Supporting actions will allow for running scenarios that are based upon some activity or action that modifies the state of the current system—through altering analytic models or data or both—and then re-running the analysis to see how it affects analysis outcomes. New map comparison tools will be provided to evaluate the change in the systems wrought by such actions.

5 EMDS Version 5.0 Architecture Design

The new architecture of EMDS will transform the platform from a simple ArcMap add-into a complete multi-faceted platform that supports modeling, analysis, actions and scenario-based planning. With the new architecture, instead of a single monolithic application, the work of the EMDS system is now broken into discrete parts set within a systems framework (Fig. 1). There are two low level sections of the framework, which are the Engine Services Tier and the Data Services Tier.

5.1 Engine Services/Wrapping Tier

For the analysis and modeling engines, we now have a layer between the rest of the framework and the individual engines (Fig. 2). For each engine type, we have abstracted out a common set of functions each engine supports and have a query-able interface to call engine-specific functionality via .NET wrappers.[19] The

[18] For example, an EMDS project is created to analyze a set of watershed conditions. After running the models, a possible action is to reforest a stream bank. This would lower the water temperature, which is an effect on the model since it includes temperature. Fish species may be affected due to this change, and the watershed condition may be improved. Another example is for a model for forest fuels management. After running the analysis, one possible action is to remove the low lying brush. If this is done, the particular areas would have a reduced fire danger.

[19] In computer programming a library is a collection of subroutines, usually external to the application. Wrappers are sparse amounts of programming code that translate a library's existing interface into a compatible interface. This is done to allow code or data formats to work together which otherwise cannot, or to enable cross language and/or runtime interoperability.

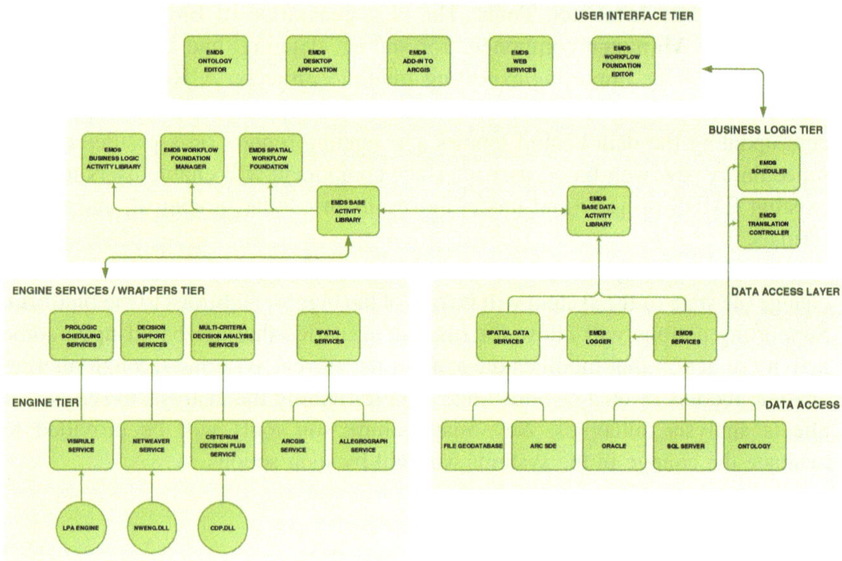

Fig. 1 Overview of the EMDS 5 service-oriented architecture

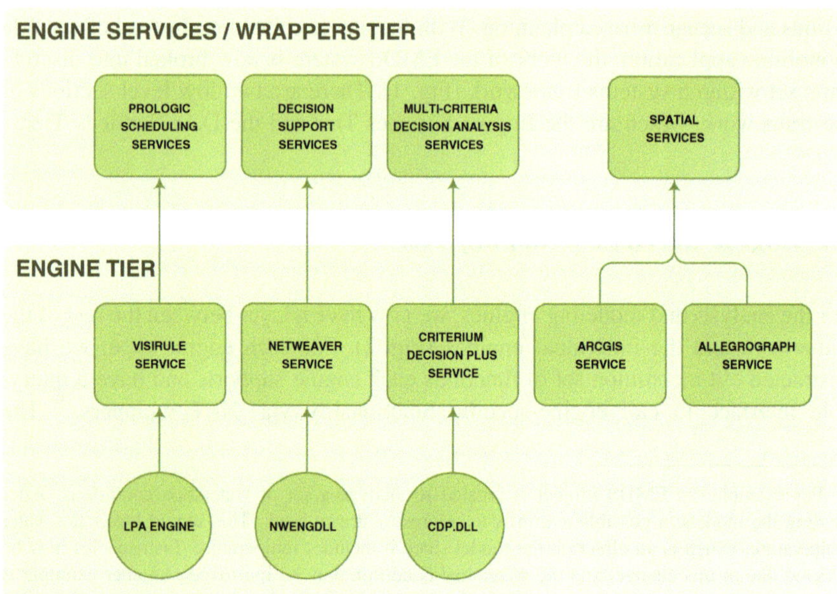

Fig. 2 Engine Services Tier. .NET wrappers provide the interface to each engine in this tier

wrappers provide the generic interface for each engine, as well as a queuing service and work-id management facilities to handle multiple user requests, even for engines that do not support multi-user or are not thread safe. All calls for analysis and processing of models will go through these interfaces. EMDS will support spatial, temporal, ontological,[20] scheduling, decision support, and multi-criteria analysis engines as default components within the new framework.

5.2 Data Services Tier

With the Data Services Tier, all data storage functionality is hidden behind services and objects with which the Business Logic Tier and Presentation Logic will interact (Fig. 3). For EMDS 5, we plan to support spatial, traditional data sources, and ontological data sources. Initial database support will be SQL Server, SQL Server Compact, and Oracle. For spatial storage, we will support file geodatabases, ArcSDE, SQL Server, and AllegroGraph. For ontological sources, initial support will be for Allegrograph, with future support of Oracle already planned. These are mapped to the Data Services and Spatial Data Services.

The key changes to the EMDS database structure will be driven by the need to fully support provenance recording, with assistance from the EMDS Logger service, and the fact that we need to support a more flexible structure than the old Project ⇨ Assessment ⇨ Analysis workflow process when we add the actions capability and the workflow engine. There will be a predefined set of analytical workflows, one of which will match exactly the existing EMDS workflow, along with several other optional workflow process paths to assist in analysis and what-if processing beyond the current limits of the system. Therefore, the database will not only need to keep track of the map and attribute data, but must also include additional metadata to allow the system to handle each of the different workflows within the same schema. The database schema has also been modified to allow for multi-level undo functionality within the system along with ability to view the history of the data changes.

5.3 Business Logic Tier

Both the Engine Services Tier (Fig. 2) and the Data Services Tier (Fig. 3) will interact with the Business Logic Tier (Fig. 4). The Business Logic Tier will expose a series of Windows Communication Foundation (WCF) REST Services

[20] Ontologies allow for the organization of entities, concepts about entities, and relationships between entities. This means we can describe the world or a portion which we wish to deal with in an agreed upon formal vocabulary that allows other people to have the understanding that the original creator of the ontology meant. Once the ontology is created, an ontology engine can be used to infer logical consequences based upon the facts contained within the ontology.

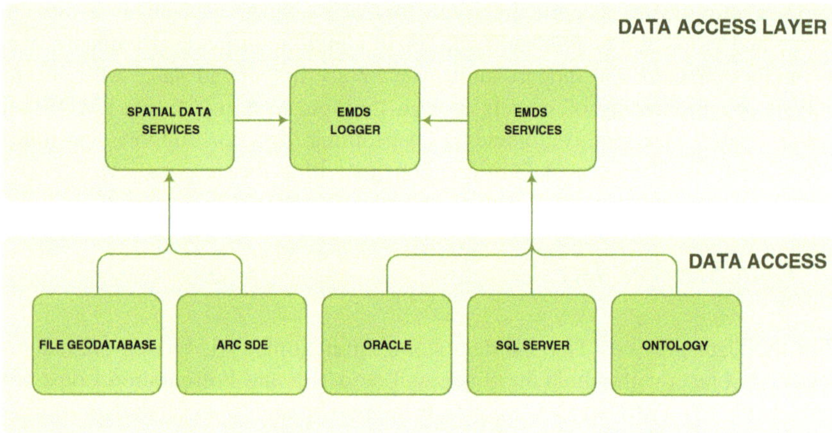

Fig. 3 Data Services Tier. Data storage functionality is hidden behind services and objects that are accessed by the Business Logic Tier (Fig. 4 of Chap. "Measuring Biological Sustainability Via a Decision Support System: Experiences with Oregon Coast Coho Salmon")

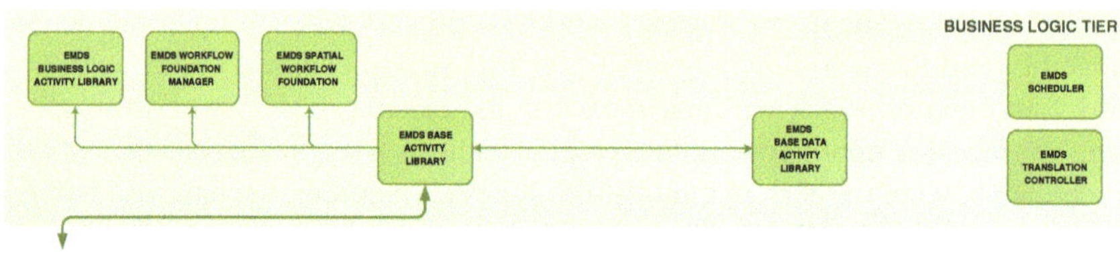

Fig. 4 Business Logic Tier. The Business Logic Tier exposes a series of REST services and workflow activity libraries that allow applications to easily tap into the power of the engines and database

(Flanders 2009) and Workflow Activity libraries to allow end applications to easily tap into the power of the engines and database. The EMDS Base Activity Library will contain the low level workflow activities[21] and WCF REST Services to perform fine grain operations, such as create a new project, do a spatial UNION, or query for a subset of provenance information. This library of activities works along with the EMDS Base Data Activity Library, which handles the low level interactions for data access. These metadata on activities are

[21] Low level workflow activities are the granular operations of the framework. Examples of these include reading and writing from the database, submitting queries to the ontology engines, running a ArcGIS Server service, or routing of messages between tiers.

saved inside the database, which can, in a future update, leverage a reasoning engine such as Allegrograph or LPA,[22] and dynamically create a complex workflow based upon these activities and based on information stored in ontologies.

5.4 EMDS 5 Will Run on Workflows

Activities are chained together using Windows Workflow Foundation to create a complete workflow that is exposed in the EMDS Business Logic Activity Library and EMDS Spatial Workflow Foundation Library. An example would be the Run Priority Analyst Workflow defined in the EMDS Business Logic Library. This multi-step workflow is defined as follows:

- The CDP model is loaded via the EMDS Base Activity Library.
- Another activity is called in the EMDS Base Data Activity Library, which returns the records for the particular dataset.
- A SendandReceive activity in the EMDS Base Activity Library calls the Multi-Criteria Decision Analysis service in the Engine Services Tier. It is called and passes the model and dataset, and waits until the processing is completed and a dataset is returned.
- The EMDS Transaction Controller updates the provenance information.
- The result set is returned to the calling application.

The EMDS Transaction Controller (Fig. 4) is the main sub-system that allows the system and end user to access and manipulate the provenance information. This service handles undo requests, workflow branching due to actions, a true history of work done, user and application state, as well as handling any errors within the Business Logic Activity Libraries. The EMDS Scheduler handles the loading, editing, and processing engine for workflows. This component reads the activity workflows from the other activity libraries and runs the Windows Workflow Engine to perform the actual tasks.

[22] Reasoning engines, which include Ontology Engines and Inference Engines, take a set of facts that is defined as the knowledge model of the system and given a set of conditions. The engine will then infer a logical conclusion or consequence of the action. LPA is a Prolog based Inference Engine system, in which you define several facts and the rules that are applied to it. From this, LPA can evaluate a query that is submitted to the system. LPA has a visual editor called VisiRule to allow for a graphical representation of facts. Allegrograph, an Ontology Engine, takes a standard ontology and allows you to create SPARQL queries to perform the evaluation of facts against.

6 Discussion and Conclusions

From the previous two sections, users of earlier versions of EMDS will hopefully have received a good sense of the major advancement that version 5 represents in terms of the power and flexibility to do environmental analysis and planning in a spatially enabled decision-support framework. Whereas the previous sections provided a more technical explanation of the new features and functionality of EMDS version 5, here we revisit that explanation in somewhat simpler language and in terms of the practical implications for the user experience.

The new workflows architecture moves EMDS from a static, one-size-fits-all, analysis paradigm to one virtually without limit. The new architecture allows the chaining together of any of the supported engines, in any order, to build very complex, maintainable workflows. The user is supported through a powerful workflows editor with the capabilities to flexibly create, maintain, and reuse existing workflows and their constituent workflow activities. EMDS 5 will support users new to the system with a library of workflow activities, lessening the experience necessary to build a solid EMDS application. Now, EMDS 5 will adapt to the user's process model rather than enforcing its traditional process model on the user.

EMDS 5 adds support for the popular relational database management systems Oracle® and SQL Server Spatial®, among others. Through services, data will be able to reside independent of the EMDS deployment. Now EMDS will be able to use corporate data sources directly without the need to import the data into the stricter and harder to maintain EMDS format.

By using web services, EMDS 5 will move from being a strictly desktop application to one with great deployment flexibility. EMDS will still be available as a desktop application, but it will also be available in a server version in which EMDS clients communicate with the EMDS server to provide EMDS access to many remote users concurrently. Data storage and processing reside on the server, while the user interface can be another desktop application or even a web browser. In this context, EMDS modeling software can be developed and maintained centrally where the resources are, and applied where it makes most sense.

Through provenance tracking, EMDS application development and use will be much more interactive and will facilitate developing and using alternative work-flow sequences. The utility is akin to having "go back" and "go forward" buttons on your web browser.

Coding in EMDS makes extensive use of the software technique of abstraction, which combined with the layered systems architecture, will make EMDS 5 easier to maintain and much more "plug and play," in that new features will be much easier to add. The technique of abstraction refers to the hiding of implementation details from the end user, and exposing a representation or interface that is easier to interact with. For example, if a new analysis engine is found that would enhance EMDS usability, it would not require changing the architecture of EMDS. It would, for the most part, require only writing the appropriate wrapper code (a generally small amount of code that translates "engine-speak" to "EMDS-speak") to talk to the new engine. Other external issues such as licensing may exist. Now though, EMDS is easily extendible to add new technologies and data sources.

Finally, readers may be curious to know when EMDS 5 will become available. As of June 2013, approximately 35 % of the version 5 architecture has been implemented. We expect that implementation will be at 65 % by October 2013, given current project support. Beyond that date, the funding picture is less certain, but we hope to have the final production system available through the Redlands Institute (University of Redlands, Redlands, CA) no later than October 2014.

References

Chappell D (2009) The workflow way: understanding windows workflow foundation. http://msdn.microsoft.com/en-us/library/dd851337.aspx. Last accessed on 22 July 2013

Flanders J (2009) An introduction to RESTful services with WCF. MSDN magazine. http://msdn.microsoft.com/en-us/magazine/dd315413.aspx. Last accessed on 22 July 2013

Trident Team (2011) Project trident: a scientific workflow workbench. http://tridentworkflow.codeplex.com/. Last accessed on 22 July 2013

Synthesis and New Directions

Keith M. Reynolds, Paul F. Hessburg and Patrick S. Bourgeron

Abstract Spatial decision support systems (SDSS) are (1) customized software applications that (2) apply analytical constructs to (3) spatial data layers, (4) for the purpose of informing specific decisions and decisions makers. What is special about them is how they enable transparent decision-making processes, effectiveness monitoring, adaptive management, and making better future decisions. Custom SDSS applications can be thought of as snapshots of the logic used to make a decision. As such, they are invaluable to grounding management and its edification through learning. An SDSS clearly reveals the logic and data that decision makers use to derive their best decision to solve a specific set of problems. But at best, it represents the hypothesis—'*this is how we thought to solve these problems, given available information*'. Subsequent decisions can be informed by the portions of a decision that worked/did not work, with little effort to reconstruct the evaluation, only to adapt it. EMDS is a SDSS development tool. It was conceived for application to the decision-making process of ecosystem management because these decisions are typically complex, multi-layered, and difficult to track, once made. EMDS uses spatial data layers, and there is no real limit to the number of dimensions it can consider in decision making. Here, we summarize how EMDS has been used to date and discuss new directions for expanding its utility. We also discuss how users and applications have influenced, and continue to influence,

K. M. Reynolds (✉)
Corvallis Forestry Sciences Laboratory, U.S. Department of Agriculture, Forest Service, Pacific Northwest Research Station, 3200 SW Jefferson Way, Corvallis, OR 97331, USA
e-mail: kreynolds@fs.fed.us

P. F. Hessburg
Wenatchee Forestry Sciences Laboratory, U.S. Department of Agriculture, Forest Service, Pacific Northwest Research Station, 1133 N Western Ave., Wenatchee, WA 98801, USA
e-mail: phessburg@fs.fed.us

P. S. Bourgeron
Institute of Arctic and Alpine Research, University of Colorado at Boulder, 1560 30th Street, Boulder, CO 80309, USA
e-mail: patrick.bourgeron@colorado.edu

K. M. Reynolds et al. (eds.), *Making Transparent Environmental Management Decisions*, Environmental Science and Engineering,
DOI: 10.1007/978-3-642-32000-2_14,

EMDS development, and new versions will no doubt reflect the evolution of decision making and technology.

Keywords Ecosystem management · Adaptive management · Decision support · Environmental analysis · Environmental planning · NetWeaver · Criterium DecisionPlus

1 Introduction

The first three chapters of this volume (Part I) laid out the technological foundations of the Ecosystem Management Decision Support (EMDS) system as currently implemented in v 4.2. The nine chapters that make up Part II presented a broad array of application areas in which EMDS has been applied since its first production release in 1997. In inviting submissions for this volume, we asked authors not only to present a description of their particular application, but to address two questions: (1) in what ways did EMDS work well for their purposes, and (2) in what ways might EMDS be improved to better support their application. Most contributing authors obliged, and we received valuable responses. Thus, in the first part of this final chapter, we synthesize those results, and draw general conclusions. "EMDS 5.0 and Beyond" leads off Part III of this volume, discussing new features and functionality in the upcoming EMDS v 5.0, and we add to this discussion addressing areas where further improvements can be made. The final three sections of this chapter return to broad questions about the future role of EMDS in supporting adaptive management, ecological stewardship, and aiding the advance of decision making in environmental analysis and planning.

2 Versatility of EMDS

Chapters in Part II of this volume cover the most common application areas to which EMDS has been applied since 1997, including:

1. Watershed analysis ("Use of EMDS in Conservation and Management Planning for Watersheds"),
2. Assessment of fire danger in support of forest-fuels management ("Evaluating Wildfire Hazard and Risk for Fire Management Applications"),
3. Integrity and resilience of landscapes ("The Integrated Restoration and Protection Strategy of USDA Forest Service Region 1" and "Landscape Evaluation and Restoration Planning"),
4. Conservation planning ("Forest Conservation Planning"), and

5. Wildlife habitat management ("Wildlife Habitat Management" and "Measuring Biological Sustainability via a Decision Support System: Experiences with Oregon Coast Coho Salmon").

Other, somewhat less common but promising applications presented in this volume include selection of ecological reserves ("Ecological Research Reserve Planning") and planning for urban growth ("Planning for Urban Growth and Sustainable Industrial Development"). An up-to-date list of EMDS publications by application area, and summarized in Table 1, can be found on Wikipedia.

The diversity of spatial scales and analysis topics represented by chapters in Part II, as well as the larger list of applications (Table 1), is indicative of the versatility of EMDS with respect to supporting environmental analysis and planning. As described in "An Overview o1.the Ecosystem Management Decision-Support System", this versatility has its origins in the framework approach, which has underpinned the EMDS system design and implementation from the earliest versions. EMDS provides a constellation of software systems (EMDS per se ["An Overview o1.the Ecosystem Management Decision-Support System"], as well as NetWeaver ["NetWeaver"] and Criterium DecisionPlus ["Criterium DecisionPlus"]) with which users can design spatial decision-support applications for many different kinds of problems, and at whatever spatial scale or scales are necessary. As also described in "An Overview o1.the Ecosystem Management Decision-Support System", though, there is a price to pay for this versatility, because EMDS does not come ready to run "out of the box," and EMDS complexity reflects the complexity of the issues addressed. Users not only have to provide data for an EMDS project (typical of most decision support systems), they also have to design the logic and decision models for their particular problem. From our own experience, we know that this added burden on potential users of EMDS has occasionally been an impediment to its adoption. On the other hand, from the perspective of systems engineering, there are enormous advantages to a general solution framework that can be applied to a broad array of environmental management problems.

Beyond the basic notion of versatility discussed above, sometimes versatility has been manifested in the innovative ways that application developers have used EMDS. For example, the decision-modeling component of EMDS supported by Criterium DecisionPlus[1] ("Criterium DecisionPlus") was not available until v 3.0. However, Ruiz ("Planning for Urban Growth and Sustainable Industrial Development"), using an earlier version of EMDS, combined the use of NetWeaver with the analytic hierarchy process (AHP, Saaty 1992), developing criterion weights in an external AHP system, and then integrating those weights into the NetWeaver models. This was a simple but clever approach to integrating

[1] The use of trade or firm names in this publication is for reader information and does not imply endorsement by the US Department of Agriculture of any product or service.

Table 1 EMDS publications by application area as reported on Wikipedia[a]

Application area	Number of publications[b]
Carbon sequestration	1
Conservation	4
Design and siting of ecological reserves	2
Ecosystem sustainability	4
Land classification	3
Landscape restoration	5
Pollution	2
Social issues in resource management	1
Soil impacts	1
Urban growth and development	2
Watershed analysis	7
Wetlands management	1
Wildlife habitat management	5
Wildland fire danger	4

[a] URL for Wikipedia site is http://en.wikipedia.org/wiki/Ecosystem_Management_Decision_Support
[b] As of 14 May 2013

logic-based reasoning and decision modeling into a single model, and represents an interesting alternative approach to that implemented in EMDS v 3.0 and later.

3 Common Themes

There were several common themes that ran through almost all chapters in Part II:

- The capability of logic-based reasoning to model large, complex, and abstract problems.
- The value of model transparency and documented reasoning in decision support systems.
- The value of fuzzy membership functions as an approach to interpreting ecosystem states and processes.

We briefly explore each of these topics in the following paragraphs.

One of the earliest and most far reaching decisions taken by the original EMDS design team was to implement logic-based reasoning in the EMDS framework as the central technique for interpreting ecosystem states and processes. As discussed at some length in "An Overview o1.the Ecosystem Management Decision-Support System", most contemporary decision support applications in environmental management are routinely confronted by large, complex, and often abstract modeling problems, for which logic-based reasoning was an effective solution. Commentaries in all chapters in Part II seem to bear this out when discussing what worked well. A related comment, often repeated throughout Part II, was the usefulness of an explicit logical formalism for developing a shared understanding of the elements of the problem domain, sometimes across quite disparate disciplines. In other words, the design and implementation of logic models often facilitated common language development, more effective communication, and interdisciplinary reasoning, or more precisely, integrated reasoning.

Two other closely related common themes across Part II were the value of transparency in communicating results to others and the value of logic and decision models as tools for documenting the underlying reasoning of the decision support application. With respect to transparency in particular, there can be significant practical differences between logic and decision models. Multi-criteria decision model systems such as Criterium DecisionPlus ("Criterium DecisionPlus") implement a simple hierarchical structure that is easy to grasp by scientists, planners, and stakeholders and others not directly involved in development of an application. In contrast, the structure of logic models built with NetWeaver ("NetWeaver") can vary from simple to highly complex, depending on the complexity and abstractness of the states and processes modeled. NetWeaver's graphical interface for tracing the derivation of model results certainly supports model transparency, but this is not to say that there may not be challenges associated with explaining a complex logic model to non-modelers, as Gordon aptly describes in "Use of EMDS in Conservation and Management Planning for Watersheds".

"An Overview o1.the Ecosystem Management Decision-Support System" and "NetWeaver" described the potential value of fuzzy membership functions as a tool for interpreting ecosystem states and processes. "Criterium DecisionPlus" similarly discussed the analogous role and usefulness of utility functions in the context of multi-criteria decision models. Whether in the context of logic or decision model development, these functions may be used to define reference conditions that provide a rational basis for interpreting the meaning of model input data. Looking across the array of applications in Part II, these reference conditions can be developed in a variety of ways:

- To produce reference standards or guidelines ("The Integrated Restoration and Protection Strategy of USDA Forest Service Region 1")
- To show the rarity of an outcome or condition relative to a sampled data range ("Use of EMDS in Conservation and Management Planning for Watersheds", "Evaluating Wildfire Hazard and Risk for Fire Management Applications", "Landscape Evaluation and Restoration Planning" and "Forest Conservation Planning")
- To reference biologically relevant values ("Wildlife Habitat Management")
- Or to ground expert judgment or Delphic knowledge extraction processes ("Planning for Urban Growth and Sustainable Industrial Development" and "Measuring Biological Sustainability via a Decision Support System: Experiences with Oregon Coast Coho Salmon").

For purposes of planning and directing management, all federal natural resource management organizations develop standards and guidelines (S&Gs) for (1) prescribing and maintaining suitable terrestrial and aquatic ecosystem states and processes and (2) for environmental protection consistent with trust responsibilities associated with public land management (NEPA 1969). S&Gs are normative because they are generally derived from the collective knowledge and experience of subject-matter experts. In addition, regulatory agencies such as the U.S. Environmental Protection Agency (EPA) and the Council on Environmental Quality (CEQ) commonly circulate regulatory requirements that can be used as a basis for defining reference conditions (Reynolds et al. 2000).

In the absence of well-established S&Gs or regulatory requirements, using the range of input data can be a practical alternative for deriving reference conditions for use in spatial decision support systems (SDSS). This approach works best when the assessment area (e.g., a region or an entire country) and sample sizes are very large (contain hundreds to thousands of observations), so that the reference range for any specific data input is likely to express the full range of system response. Likewise, when reference ranges stem from large databases, the relativized nature of transformed data is of less concern. In fact, for some DSS applications, use of relativized data may not be an issue at all (but see "Wildlife Habitat Management" for a contrary example).

When reference conditions cannot be suitably defined using the first two approaches discussed, their specification can become more challenging for DSS developers. In some cases, biologically relevant values might be gleaned from the scientific literature, but, failing that, there may be no recourse but to resort to reliance on expert judgment. Fortunately, methods for extracting, testing, and using expert knowledge in DSS applications are steadily becoming more rigorous (see Perera et al. 2012). In this context, EMDS can be used when novel ecosystems (*sensu* Hobbs et al. 2006) emerge and *ecosystem stewardship* (Chapin et al. 2009; see Sect. 7) is applied.

Regardless of the method used to define reference conditions, most chapters in Part II agreed with the basic premise, that use of reference conditions to define *degrees* of acceptability or suitability was consistent with the way subject-matter experts provide explicit reasoning about environmental conditions.

4 Enhancements in EMDS v 5.0

The implementation of EMDS v 5.0 adds many new features to the system, including automated report generation and the ability to perform analyses on raster data ("EMDS 5.0 and Beyond"). Some of the most important enhancements focus on improving extensibility and automation of the system, including the abilities to:

1. Easily add new services to the system architecture (e.g., Netica®, VisiRule®),
2. Implement workflows that automate data processing, and
3. Implement workflows that support alternative DSS processes.

Table 2 New EMDS features requested by chapter

Chapter	Issue	Requested feature	EMDS v 5.0 support
4	Automation	Methods to summarize data across scales of assessment	Workflow[a]
4	Automation	Methods to evaluate change in distributions of outcomes across assessments	Workflow
6	Automation	Analysis of multiple weather scenarios for evaluating fire danger	Workflow
6	Uncertainty (automation)	Run multiple alternative models, logic and decision	Workflow
7	Automation	Evaluate multiple spatial realizations under alternative management strategies	Workflow
7	Uncertainty (automation)	Run multiple alternative climate scenarios	Workflow
10	Automation	Evaluation over time series with multiple management strategies	Workflow
10	Spatial analysis	Raster processing	Raster support
10	Reporting	Package and share results with experts, managers and other stakeholders	Report generator
10	Uncertainty (automation)	Assess variability in parameters of fuzzy membership functions	Workflow

[a] Indicates that the requested feature can be implemented with user-designed workflows in the EMDS workflow editor

As we mentioned in the chapter introduction, we asked authors to comment on ways in which EMDS could have better supported their analyses. We summarized the most significant responses by chapter in Table 2, and one can see that most of the comments are requests for various automation upgrades. This was much expected; once end-users become familiar with the advantages associated with implementing a relatively new technology, the need for system upgrades usually exceeds the budget and timeliness of upgrading. Note that even in cases where the primary issue was described in terms of handling uncertainty, the underlying issue was about automation. Table 2 also shows the EMDS v 5.0 features that will provide the requested functionality. Although it is foolhardy to suppose that workflows can solve all automation problems, we believe that all automation requests specifically identified in the table can be satisfactorily addressed through the design of workflows in the EMDS workflow editor. In fact, to give readers some sense of the power of workflows in the EMDS context, the system *itself* has been built from the ground up with workflows. Toward designing automated workflows in the EMDS workflow editor, the EMDS development team anticipates developing a core of generic workflows, similar to those identified in Table 2, as a resource for system developers. They also envision that workflow libraries will be developed by third parties and shared within the user community through an open access workflow library, possibly maintained by the EMDS Consortium. Details are yet to be worked out, but stay tuned. In our experience, there is already ample precedent for sharing of EMDS workflows within the existing user community. Note also the near infectious spread of the R open access statistic package (R Development Core Team 2011) via sharing of non-proprietary, tested (debugged) statistic code. Indeed, much of the rapid growth of R has occurred via the open access facilitated end-user community. Finally, we admit that the EMDS development team cannot begin to conceive of workflows needed to support every environmental analysis and planning effort. There are simply too many possibilities. Thus, just as with the origins of EMDS itself, the practical solution is to provide the tools with which to construct custom workflows. Additionally, more advanced end users will wish to build their own workflows because it will save time and allow them to tailor evaluations to their data and problems.

5 Supporting Adaptive Management

In the final two sections, we consider the role of EMDS in advancing decision support for environmental analysis and planning. As a foundation, we briefly revisit discussion from "An Overview o1.the Ecosystem Management Decision-Support System" concerning the role of EMDS in supporting adaptive management, and we consider new capabilities coming in v 5.0 in this light.

Adaptive management as it has been applied to land and resource planning has been described as a continuous cycle that proceeds from evaluation to planning, to implementation and monitoring, and then back to evaluation (Holling 1978; Walters 1986). Evaluation (or assessment, as it is termed in EMDS parlance) has been a cornerstone of EMDS, because establishing a baseline is the first step to obtaining a new course of action. NetWeaver was adopted for environmental assessment because ecosystems are large, complex, and challenge even the most sophisticated evaluation tools. It is for this reason that many investigators build a new model each time they wish to evaluate a sufficiently new ecological question. The virtue of logic in this context is that if one can reason about the state of a system, or how it works, then it can be modeled with logic.

In most applications described in Part II of this volume, EMDS was used to assess the current state of a system. For example, watershed condition ("Use of EMDS in Conservation and Management Planning for Watersheds"), fire danger ("Evaluating Wildfire Hazard and Risk for Fire Management Applications"), habitat suitability ("Wildlife Habitat Management"), and population viability ("Measuring Biological Sustainability via a Decision Support System: Experiences with Oregon Coast Coho Salmon") were each assessed to establish a current baseline. NetWeaver logic models can be designed to provide a point-in-time assessment, but the time frame may equally well be sometime in the past, present, and/or future. This lack of time dependence of logic-based evaluations provides a strong basis for comparing current with alternative future conditions (e.g., see "Landscape Evaluation and Restoration Planning" and references therein) and for monitoring progress on implementation, by comparing current conditions with earlier points in time (see "Use of EMDS in Conservation and Management Planning for Watersheds"). Regardless of whether the question at hand concerns selection of a strategic alternative for plan implementation or plan performance, the mechanism is essentially the same: comparing distributions of logic-based outcomes, as modeled in NetWeaver. As of EMDS v 4.2, the user performed these comparisons, but as discussed in the previous section, most of the process could be automated in v 5.0 via workflows, thus substantially extending the capabilities of EMDS to support key aspects of monitoring and adaptive management processes, even to include hypothesis testing explicitly in the chain of workflow activities.

Finally, multi-criteria decision models (MCDMs) have been integral to EMDS since v 3.0. Classically, MCDMs are used to rate and select alternatives, and a typical environmental management application would be to select among *aspatial* alternatives (Saaty 1992). However, in the spatial context of EMDS, this technology was adapted to rate landscape or ecosystem elements within Criterium DecisionPlus (CDP) framework ("Criterium DecisionPlus"). In other words, landscape elements and their combinations can be treated as spatial alternatives.

Version 5.0 will provide new possibilities for use of MCDMs within the EMDS framework. First, it will be possible to launch the Priority Analyst component on specific individual landscape elements to guide selection of *tactical* alternatives, based on properties of landscape elements. For example, depending on feature attributes such as forest-floor fuel loadings, tree crown base heights, or crown bulk densities (see "Evaluating Wildfire Hazard and Risk for Fire Management Applications"), a tactical CDP decision model might rate the suitability of various fuel-treatment options such as prescribed fire *versus* thinning in light of the fuel load or forest structural characteristics. This type of tactical analysis could then be combined with spatial prioritization currently available in EMDS to perform both strategic treatment prioritization of all landscape elements and tactical selection of treatment options for individual landscape elements. Moreover, these tactical MCDM models could be automated via workflows to run against a subset of landscape elements meeting some minimum (user specified) requirement for strategic prioritization (such as 20 % of the total forest patches or land area treated).

A more complex use of MCDMs in v 5.0 could involve the design of workflows to compare the *consequences* of implementing strategic and tactical alternatives as discussed above. As a simple example of this more complex case, let us suppose that we have an EMDS application that is designed to assess salmon habitat suitability in subwatersheds, and that recommends strategic selection of subwatersheds as well as tactical actions to take within high priority (selected) subwatersheds. The expert reasoning of fisheries biologists may suggest multiple alternative tactical approaches to salmon habitat restoration, but the reasoning is sufficiently complex that it is not immediately obvious which of the tactical approaches would have the greatest overall effectiveness. In this context, workflows could be used to simulate the effects of the alternative tactical approaches by implementing changes in the input data. Each tactical approach would have its own workflow, generating an alternative assessment set for resubmission to the NetWeaver logic model. Furthermore, a higher order workflow could be designed to (1) run each of the workflows representing the tactical implementations, and (2) perform an automated comparison of the distributions of NetWeaver outcomes across alternatives. This is an example of precisely the kind of automation to which Hessburg et al. ("Landscape Evaluation and Restoration Planning") alluded.

6 Advancing Decision Support for Environmental Analysis and Planning

6.1 Decision Support for Hierarchical NEPA Planning

The National Environmental Policy Act (NEPA 1969) was established on January 1970, after a tumultuous decade of public outcry and demonstrations over oil spills, expanding timber harvests, urban and freeway system development, shrinking wildland area, and increasing numbers of endangered species. NEPA was enacted to ensure that environmental factors were weighted equally in comparison to other factors in the land- or resource-management decision-making processes of all executive branch agencies. The Act, visionary in its depth, established a required multidisciplinary approach to considering environmental effects in decision making, and procedural requirements for preparing environmental assessments of the effects of proposed actions. The NEPA planning process of all federal projects consists of an environmental effects analysis, and includes a pertinent set of alternative actions considered, usually varying in emphasis. There are three levels of analysis that an agency may undertake to comply with the law; the needed level is usually tied to the gravity of environmental concerns surrounding a proposed action. Levels of environmental analysis (in order of increasing documentation and environmental concern) include preparing a Categorical Exclusion (CE), an Environmental Assessment (EA) and Finding of No Significant Impact (FONSI), or an Environmental Impact Statement (EIS).

In "Landscape Evaluation and Restoration Planning" of this volume, Hessburg et al. (and references therein) describe the environmental effects of 20th century management on Pacific Northwest National Forest lands. Likewise, Gordon ("Use of EMDS in Conservation and Management Planning for Watersheds"), Gordon et al. ("Wildlife Habitat Management"), and Keane et al. ("Evaluating Wildfire Hazard and Risk for Fire Management Applications") describe significant changes in Pacific Northwest aquatic and terrestrial species habitat networks and wildfire regimes, respectively, brought about by management actions on public lands. In "Landscape Evaluation and Restoration Planning", Hessburg et al. submitted that 20th century management actions had unintentionally decoupled regional and local landscape functionality by interrupting vitally important patterns and networks of forest and riverine habitats, yielding a long list of terrestrial and aquatic species on the brink, and uncharacteristically severe wildfire, insect, and disease disturbance regimes. These circumstances set the stage for a brand of hierarchal landscape planning for which EMDS is especially well designed and suited.

Suring et al. (2011) developed an 8-step screening process to address the Pacific Northwest native wildlife species for which broad-scale, coarse filter (Hunter et al. 1989) management for ecosystem diversity may be insufficient for providing conditions to sustain viable populations. They applied their screening criteria, identifying >200 species of conservation concern. They aggregated the identified species into families and groups based on habitat associations and risk factors, and selected 36 *focal species* for application in northeastern Washington State, USA. Focal species were broadly emblematic such that if adequate restoration remedies were applied, most other species of concern would sufficiently benefit; essentially acting as an affirmative version of the 'canary in the coal mine' model.

Lee at al. (1997), Thurow et al. (1997) and Rieman et al. (1997) make a similar case for seven native Northwest salmonids, establishing their historical and contemporary subwatershed ranges throughout the Interior Columbia River basin. Gresswell et al. (1999) and Bisson et al. (2003) provide comprehensive syntheses of the historical and contemporary roles of wildfires in fire-prone forests and how anadromous and coldwater fish likely responded, suggesting that fish in general are evolutionarily adapted to fires and their native variability, but that local populations in their current conditions may be exposed to extirpation risks with current wildfires. To protect remaining populations of native fishes and improve the connectivity and quality of their habitats, it is essential to spatially locate existing populations and habitats on the landscape and protect them, and then identify other spatially connected locations that have the inherent capability to become functional habitats, and remove barriers to connectivity. The same is true of terrestrial wildlife, a case well made by both Suring et al. (2011) and Wisdom et al. (2000).

EMDS is well suited to this task of hierarchical planning. Regional NEPA plans (Regional Plans or Guides, formerly in common usage) could be developed whose environmental analyses were grounded in a regional spatial decision support system (SDSS). A Regional SDSS could evaluate maps of existing and potential habitats, existing populations, and impediments to movement for all listed and focal species, throughout their known historical ranges, and their predicted future ranges under climate change. Regional scale analyses could consider and map preferred regional networking solutions to improve habitat connectivity and landscape permeability for each species, and hand these insights down to Forest-level NEPA Plans as regional guidance or standards.

At a second level in the SDSS, National Forests would have a subset of species for which to make specific planning provisions. Forest Plans would develop their own SDSS to jointly consider high-resolution habitat, impediment, and population maps for their subset of species. Each Forest-level SDSS could be later directly linked with the appropriate regional SDSS(s), a relatively straightforward step. Subwatersheds within Forest or logically grouped Forest boundaries could now be prioritized within the Forest SDSS for local restoration activities facilitated by

District-level project planning and landscape restoration projects of the sort described by Hessburg et al. (2013). Forests Plans could enumerate these focal species and subwatersheds and provide particular guidance concerning unique subwatershed conditions and restoration needs. In this way, regional planning guidance for improving terrestrial and aquatic species habitats becomes directly linked in analysis and planning to local District projects whose focus is to restore habitat quality and abundance, and the connectivity and permeability of terrestrial landscapes and aquatic networks.

Given this context, it takes only a little imagination to conceive how regional patterns of vulnerability to forest fuels, wildfires, insect outbreaks, disease pandemics, and major vegetation structure, composition, and fuelbed departures might be co-considered along with the foregoing in Regional, Forest, and District SDSS to provide integrated, multidisciplinary and hierarchical decision support for restoration planning, implementation, and monitoring (see above). Local landscape restoration projects, by definition, would be linked to restoring the connectivity of the regional landscape, while having the added benefit of considering locally high definition data sources and conditions, and experience. Lacking this sort of hierarchical spatial decision support, it is hard to conceive how local restoration projects would have any well-defined capacity to restore broad regional landscapes, their populations, or habitats.

6.2 Expanded Map Exploration Functionality in EMDS

Many chapters in this volume describe planning processes involving the mixed company of knowledge experts, policy analysts, knowledge engineers (people who help build SDSSs), decision makers, and stakeholders. Throughout the planning stages, maps that detail co-occurring conditions are shown and discussed, as are relevant datasets and expert opinion. NetWeaver is often used in meetings to build a rough draft of a logic model (knowledgebase). Through successive meetings, participants review their logic, and incorporate learning from discussions, new map and data evaluations, and presentations of experts. The yield of this process is often a negotiated portrayal of conditions or functionality of a focal system. The act of using NetWeaver 'on the fly' to develop a knowledgebase provides them with a practical understanding of their knowledgebase and, more generally, how to use NetWeaver to codify reasoning. Formally specifying a problem by means of a logic model is the job at hand; tools that speed up schematic visualization of the focal problem are key to an efficient process.

Similarly, once draft NetWeaver and CDP models are developed, these same mixed groups have a deliberate need to check and refine their work, and discover at points along the way whether their reasoning is sensible. They often do this by running the models in EMDS and evaluating their behavior, which is accomplished by visual examination of derived maps and tabular outputs, and conducting sensitivity and decision analyses in CDP.

We have observed that mixed decision making groups readily work out their differences and improve their models by examining maps and discussing particular spatial domains that match (or do not match) expectations. Indeed, most participants are involved because they come equipped with valuable benchmarks from their studies and experiences, and they apply these when fine-tuning models.

This process of model evaluation and fine-tuning could be streamlined if, at all levels in the NetWeaver and CDP models, EMDS provided a simple snapshot utility for rapid, on-screen map visualization by which multiple maps from one or more models could be viewed simultaneously side by side. Users could scrutinize and magnify individual maps, branches, and the complete hierarchy; but most importantly, they would have simultaneous visual representations of model behavior at any and all levels within the model upon which to focus discussions about model performance and revision. To some extent, this multi-map capability is available in the layout view of ArcGIS, but a dedicated custom utility would more readily support fast ad hoc map queries in a workshop environment. Maps could be printed as handouts for closer inspection or used in independent GIS evaluations, which would expand their usefulness as exploratory and performance measurement tools. This utility should be extensible to evaluating and comparing trade-offs among alternatives in CDP as well.

The need for rapid map visualization has grown out of widespread use of EMDS and represents one next step in evolution. The current lack of an expanded utility impedes some exploration of model behavior, and leaves users potentially suspicious that EMDS might exhibit an unquantifiable 'black box' influence, an age old and unnecessary concern. Present-day model builders wish to fully appreciate the consequences of the logic they apply to spatial maps and data, the role and influence of the weightings and logical operators they apply, and how these features influence the partial and final map products, and corresponding numerical evaluations. The need for developing this enhanced utility is now under discussion with the EMDS development team.

6.3 The Need for Automated Scenario Development

At a relatively early stage of SDSS development, NetWeaver model developers become aware of the primary influences on conditions or functionality of a focal ecosystem or landscape. This awareness is due to their perceptive powers and begins when spatial data layers are formatted in a GIS in preparation for NetWeaver parameterization. It continues as NetWeaver parameterization is iteratively checked and successive data layers are assembled into the model. Throughout the process, model developers get a glimpse of the primary factors driving outcomes in their NetWeaver model, and these are later confirmed by a fully operational model, with little added surprise. The same is true for CDP models as variable files are prepared, weights are applied, the Simple Multi-Attribute Rating Technique utility (SMART; Edwards 1977; Edwards and Newman 1982; Kamenetzky 1982) is implemented, and the model is tested to its completion. During model assembly, developers already are forming an idea of driving factors and their relative importance, prior to any weighting, because they are continuously learning.

In Sect. 6.2 above, we present an approach to hierarchical planning, in which key landscape restoration objectives descend from Regional to local landscape planning, enabling local landscape restoration that potentially restores the pattern and functionality of the regional landscape. Under this architecture, primary topics coming from Regional analysis drive geographic differences in Forest- and District-level planning and implementation. Geographic differences in primary topics for Forest SDSSs derive from uniquely combined species habitat connectivity concerns, disturbance regime departures, the history of restorative management, and vegetation departures, which themselves are a legacy of particular management histories and biophysical settings, varying disturbance regimes, home range differences, and the like.

It is intuitive that a core set of spatially explicit Regional, Forest or District planning scenarios[2] would emanate from overarching concerns at each planning level. In fact, the perceptions of model developers, mentioned above, derive from differences they repeatedly observe during model development. These differences could be exploited by an automated analysis routine in EMDS that provided a cross section of "starter scenarios" that build off the key departures as they were originally evaluated in NetWeaver. A range of scenarios would emphasize different key departures. Modelers already tell NetWeaver how to weight primary and secondary topics. From these weights, the key departures may be readily known. For example, if fire regime departure was significant, one scenario would highlight patches or subwatersheds (depending upon the planning level) with high departure, and give them 2:1 or 3:2 weighting preference for restoration priority. In another scenario, vegetation or wildlife habitat departures might be weighted more heavily, and these patches would be highlighted. In this way, planners at each level would have an opportunity to observe an initial range of scenarios that tie directly to key changes or departures noted in their NetWeaver models. From this initial set, planning team members responsible for different resources could further refine scenarios both spatially, in terms of the areas that are selected for emphasis, and by varying the level of emphasis. When all planning team members have accomplished refining their initial scenarios, a large range of planning alternatives will have been considered, hot spots for restoration by emphasis will be clearer, and team members will be better prepared to narrow the range of scenarios (planning alternatives in NEPA parlance) that are considered in detail.

Such a utility would be useful to planning teams and model developers. It would provide a rapid start to scenario development, and it would keep planning

---

[2] Here, we use the term "spatially explicit planning scenario" in a slightly different sense than EMDS scenarios as described in "An Overview o1.the Ecosystem Management Decision-Support System". In the current context, we use scenario (for short) to mean a combination of data inputs and logic outcomes that may suggest a common tactical response for planning purposes. In NEPA parlance, these scenarios might also be described as spatially explicit planning alternatives.

teams honest by reminding them to initially consider a broad set of scenarios with varied emphases, before narrowing the planning scope to a subset of scenarios considered in greater detail. The utility would, no doubt, reveal planning biases and occasionally surprise planners with useful scenarios that would otherwise not have been considered. The need for developing this enhanced utility is also under discussion with the EMDS development team.

7 Final Thoughts

One of the central goals of contemporary environmental assessment is to evaluate interactions among the environmental, social, and economic domains at several essential spatial and temporal scales, with cross-scale interactions among scales and domains, to forecast (*sensu* Clark et al. 2001) the potential effects of these interactions on broad landscape resilience (*sensu* Holling 2001; Holling and Gunderson 2002). In this context, *ecosystem stewardship*[3] (Chapin et al. 2009) has been formulated as an extension of ecosystem management (Jensen et al. 2001) to include all social-ecological systems. *Ecosystem stewardship is* an action-oriented sustainable production framework to provide ecosystem services desired by society, under various scenarios involving uncertainty and change. Current and planned versions of EMDS allow for the explicit evaluation and exposition of realistic strategies that can increase the likelihood of socially beneficial outcomes, while reducing the risk of negative outcomes. More specifically, EMDS can provide a basis to: (1) assess landscape vulnerability[4] to expected changes; (2) define and quantify direct or indirect measures of resilience[5]; (3) foster resilience to sustain desirable conditions in the face of perturbations and uncertainty; (4) identify policies and strategies to navigate undesirable trajectories through transformation[6] when opportunities occur; and (5) assess the capacity[7] to contribute to all three sustainability approaches. By building on prior vulnerability, adaptation,

---

[3] Ecosystem stewardship incorporates four concepts central to framing approaches to sustainability: vulnerability, adaptability, resilience, and transformability (MA 2005; Janssen and Ostrom 2009; IPCC 2007; Jäger et al. 2007; Schneider et al. 2007; Chapin et al. 2009, 2011; Miller et al. 2010).

[4] The degree to which a system is likely to experience harm owing to exposure and sensitivity to a specified hazard or stress and its adaptive capacity to respond to that stress (Chapin et al. 2009).

[5] Capacity of a social–ecological system to absorb a spectrum of shocks or perturbations and to sustain and develop its fundamental function, structure, identity and feedbacks as a result of recovery or reorganization in a new context (Carpenter et al. 2001).

[6] Fundamental change in a social–ecological system resulting in different controls over system properties, new ways of making a living and often changes in scales of crucial feedbacks (Chapin et al. 2009).

[7] Capacity of social–ecological systems, including both their human and ecological components, to respond to, create and shape variability and change in the state of the system (Chapin et al. 2009).

resilience, and transformation research, an EMDS implementation of ecosystem stewardship planning has the potential to provide a clear analytical perspective that better equips society with managing the challenges that it confronts. In this context, the new workflow architecture of EMDS ("EMDS 5.0 and Beyond") includes looping capabilities that provide a way to automate the application of models over time steps, thus allowing for progressive feedback loops, which in turn would support the analysis of emergent system properties and the identification of critical management pathways for maintaining and restoring systems.

As you have seen from foregoing chapters, global land and water ecosystems comprise the human life support system, and ecological stewardship has always been a fundamental civic responsibility. Chapters of this volume have touched on each of the last five points, and we invite our readers to reflect further on the applications presented here, and their ecological and societal underpinnings. It is our hope that most of the best ideas for ecological stewardship and more ecologically and socially attuned management are yet to be designed, and that you will be designing them. Hopefully, it has become clear how EMDS can be useful to management planning and monitoring, for learning, and for improving stewardship of the planet.

References

Bisson PA, Rieman BE, Luce C, Hessburg PF, Lee DC, Kershner J, Reeves GH, Gresswell RE (2003) Fire and aquatic ecosystems of the western USA: current knowledge and key questions. For Ecol Manage 178:213–229

Carpenter SR (2001) Alternate states of ecosystems: evidence and its implications for environmental decisions. In: Press MC, Huntly N, Levin S (eds) Ecology: achievement and challenge. Blackwell, London

Chapin III FS, Carpenter SR, Kofinas GP, Folke C, Abel N, Clark WC, Olsson P, Stafford Smith DM, Walker B, Young OR, Berkes F, Biggs R, Grove JM, Naylor RL, Pinkerton E, SteffenW, Swanson FJ (2009) Ecosystem stewardship: sustainability strategies for a rapidly changing planet. Trends Ecol Evol 25:241–249

Chapin FS, Power ME, Pickett STA, Freitag A, Reynolds JA, Jackson RB, Lodge DM, Duke C, Collins SL, Power AG, Bartuska A (2011) Earth Stewardship: science for action to sustain the human-earth system. Ecosphere 2. doi:http://dx.doi.org/10.1890/ES11-00166.1

Clark JS, Carpenter SR, Barber M, Collins S, Dobson A, Foley JA, Lodge DM, Pascual M, Pielke R Jr, Pizer W, Pringle C, Reid WV, Rose KA, Sala O, Schlesinger WH, Wall DH, Wear D (2001) Ecological forecasts: an emerging imperative. Science 202:657–660

Edwards W (1977) How to use multi-attribute utility measurement for social decision making. IEEE Trans Syst Man Cybern 7(1977):326–340

Edwards W, Newman JR (1982) Multi-attribute evaluation. Sage, Beverly Hills, p 1982

Gresswell RE (1999) Fire and aquatic ecosystems in forested biomes of North America. Trans Am Fish Soc 128:193–221

Hessburg PF, Reynolds KM, Salter RB, Dickinson JD, Gaines WL, Harrod RJ (2013) Landscape evaluation for restoration planning on the Okanogan-Wenatchee National Forest, USA. Sustainability 5:805–840

Hobbs RJ, Arico S, Aronson J, Baron JS, Bridgewater P, Cramer VA, Epstein PR, Ewel JJ, Klink CA, Lugo AE, Norton D, Ojima D, Richardson DM, Sanderson EW, Valladares F, Vilà M,

Zamora R, Zobel M (2006) Novel ecosystems: theoretical and management aspects of the new ecological world order. Glob Ecol Biogeogr 15:1–7
Holling CS (ed) (1978) Adaptive environmental assessment and management. Wiley, Chichester
Holling CS (2001) Understanding the complexity of economic, ecological, and social systems. Ecosystems 4:390–405
Holling CS, Gunderson LH (2002) Resilience and adaptive cycles. In: Gunderson LH, Holling CS (eds) Panarchy: understanding transformations in human and natural systems. Island Press, Washington, DC, pp 25–62
Hunter ML, Jacobson GL, Webb T (1989) Paleoecology and the coarse-filter approach to maintaining biological diversity. Conserv Biol 2:375–385
IPCC (Intergovernmental Panel on Climate Change) (2007) Climate change 2007: impacts, adaptation and vulnerability. In: Parry ML, Canziani OF, Palutikof JP, van der Linden PJ, Hanson CE (eds) Contribution of working group II to the fourth assessment report of the intergovernmental panel on climate change. Cambridge University Press, Cambridge, UK
Jäger J, Kok M, Mohamed-Katerere JC, Karlsson SI, Lüdeke MKB, Dabelko GD, Thomalla F, de Soysa I, Chenje M, Filcak R, Koshy L, Long Martello M, Mathur V, Moreno AR, Narain V, Sietz D (2007) Vulnerability of people and the environment: challenges and opportunities. In: Global environment outlook GEO-4. United Nations Environment Programme, Nairobi, Kenya
Janssen MA, Ostrom E (2006) Resilience, vulnerability, and adaptation: a cross-cutting theme of the international human dimensions programme on global environmental change. Glob Environ Change 16:237–239
Jensen ME, Christensen NL, Bourgeron PS (2001) An overview of ecological assessment principles and applications. In: Jensen MW, Bourgeron PS (eds) A guidebook for integrated ecological assessment. Springer-Verlag, New York, pp 13–28
Kamenetzky R (1982) The relationship between the analytical hierarchy process and the additive value function. Decision Sci 13:702–716
Lee DC, Sedell JR, Rieman BE, Thurow RF, Williams JE (1997) Broad-scale assessment of aquatic species and habitats, Chapter 4. Gen. Technical Report PNW-GTR-405. USDA Forest Service, Pacific Northwest Research Station, Portland, OR
MA (Millennium Ecosystem Assessment) (2005) Ecosystems and human well-being: synthesis. Island Press, Washington DC
Miller F, Osbahr H, Boyd E, Thomalla F, Bharwani S, Ziervogel G, Walker B, Birkmann J, van der Leeuw S, Rockstrôm J (2010) Resilience and vulnerability: complementary or conflicting concepts? Ecol Soc 15. http://www.ecologyandsociety.org/vol15/iss3/art11/. (2010, 11[published online])
NEPA (National Environmental Policy Act) (1969) 42 U.S.C. §§4321-4370 h
Perera AH, Drew CA, Johnson CJ (eds) (2012) Expert knowledge and its application in landscape ecology. Springer, New York
R Development Core Team R (2011) A language and environment for statistical computing. R Development Core Team, Vienna, Austria
Rieman BE, Lee DC, Thurow RF (1997) Distribution, status, and likely future trends of bull trout in the interior Columbia River basin and Klamath River basins. North Am J Fish Manag 17:1111–1125
Reynolds KM, Jensen ME, Andreasen J, Goodman I (2000) Knowledge-based assessment of watershed condition. Comput Electron Agric 27:315–333
Saaty TL (1992) Multicriteria decision making: the analytical hierarchy process. RWS Publications, Pittsburg
Schneider SH, Semenov S, Patwardhan A, Burton I, Magadza CHD, Oppenheimer M, Pittock AB, Rahman A, Smith JB, Suarez A, Yamin F (2007) Assessing key vulnerabilities and the risk from climate change. In: Parry ML, Canziani OF, Palutikof JP, van der Linden PJ, Hanson CE (eds) Climate change 2007: impacts, adaptation and vulnerability. Contribution of working group II to the fourth assessment report of the intergovernmental panel on climate change. Cambridge University Press, Cambridge, UK

Suring LH, Gaines WL, Wales BC, Mellen-Mclean K, Begley JS, Mohoric S (2011) Maintaining populations of terrestrial wildlife through land management planning: a case study. J Wildl Manag 74:945–958

Thurow RF, Lee DC, Rieman BE (1997) Distribution and status of seven native salmonids in the interior Columbia basin and portions of the Klamath River and Great basins. North Am J Fish Manag 17:1094–1110

Walters C (1986) Adaptive management of renewable resources. Macmillan, New York

Wisdom MJ, Holthausen RS, Wales BC, Hargis CD, Saab VA, Lee DC, Hann WJ, Rich TD, Rowland MM, Murphy WJ, Eames MR (2000) Source habitats for terrestrial vertebrates of focus in the interior Columbia basin: broad-scale trends and management implications. USDA Forest Service General Technical Report PNW,-GTR-485, Portland

Index

K. M. Reynolds et al. (eds.), *Making Transparent Environmental Management Decisions*, Environmental Science and Engineering,
DOI: 10.1007/978-3-642-32000-2,

MIX
Papier aus verantwortungsvollen Quellen
Paper from responsible sources
FSC® C105338

If you have any concerns about our products,
you can contact us on
ProductSafety@springernature.com

In case Publisher is established outside the EU,
the EU authorized representative is:
Springer Nature Customer Service Center GmbH
Europaplatz 3, 69115 Heidelberg, Germany

Printed by Libri Plureos GmbH
in Hamburg, Germany